I0056011

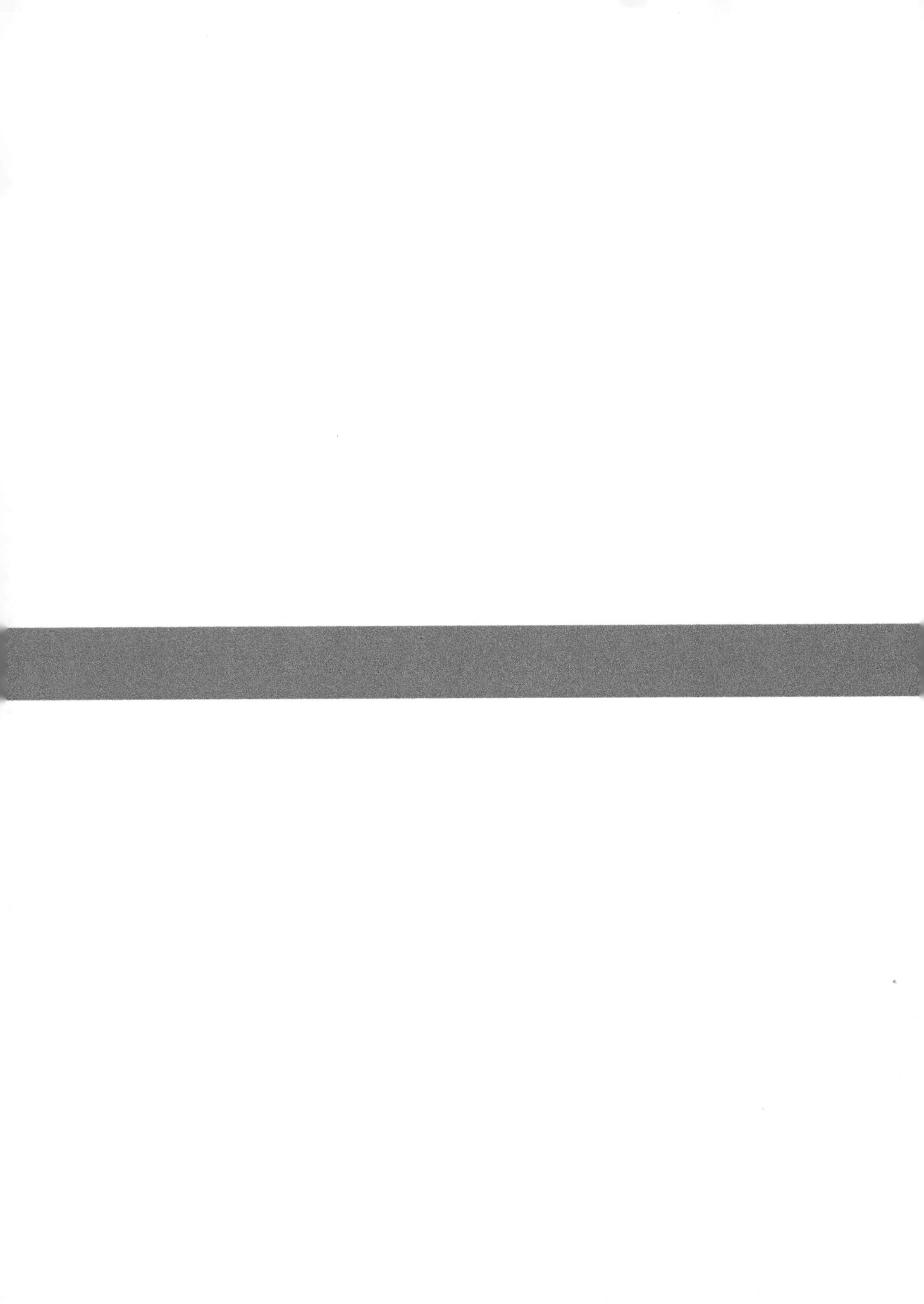

Multiphysics Modeling Using Comsol 4

Multiphysics Modeling Using COMSOL 4

Roger W. Pryor

Mercury Learning and Information
Dulles, Virginia
Boston, Massachusetts

Publisher: David Pallai
Mercury Learning and Information
22841 Quicksilver Drive
Dulles, VA 20166
info@merclearning.com
www.merclearning.com
1-800-758-3756

This book is printed on acid-free paper.

Roger W. Pryor. *Multiphysics Modeling Using COMSOL 4*
ISBN: 9781936420094

Library of Congress Control Number: 2011943234

Printed in Canada

CONTENTS

PREFACE

The purpose of this book is to introduce hands-on model building and solving with COMSOL Multiphysics® software version 4 to scientists, engineers, and others interested in exploring the behavior of physical device structures on a computer (virtual prototype), before going to the workshop or laboratory and trying to build the whatever-it-is (real prototype).

The models presented herein are built within the context of the laws of the physical world (applied physics) and are explored in light of First Principle Analysis techniques. As with any other method of problem solution, the information contained through the solutions to these computer simulations (virtual prototypes) is only as accurate as the materials properties values and the fundamental assumptions employed to build these simulations.

The primary advantage of the combination of computer simulation (virtual prototyping) and First Principles Analysis to explore artifacts (device structures) is that the modeler can try as many different approaches to the solution of the same underlying problem as are needed in order to get it right (or close thereto) before fabrication of the device components and the assembled device (real prototype) in the workshop or laboratory for the first time.

Acknowledgments

I would like to thank David Pallai of Mercury Learning and Information for his ongoing encouragement. I would also like to thank the many staff members of COMSOL, Inc. for their help and encouragement.

I would especially like to thank Beverly E. Pryor, my wife, for the efforts she expended during the reading of the manuscript and verifying of the building instructions for each of the models. Any errors that remain in this work are mine and mine alone.

Roger W. Pryor, PhD

December 2011

INTRODUCTION

COMSOL Multiphysics software is a powerful Finite Element (FEM), Partial Differential Equation (PDE) solution engine. The basic COMSOL Multiphysics 4.x software has fifteen (15) add-on modules that expand the capabilities of the basic software into the following application areas: AC/DC, Acoustics, Batteries and Fuel Cells, CFD, Chemical Reaction Engineering, Electrodeposition, Geomechanics, Heat Transfer, MEMS, Microfluidics, Plasma, RF, Structural Mechanics, and Subsurface Flow. The COMSOL Multiphysics software also has other supporting software, such as: the Optimization Module, the Material Library Module, the CAD Import Module, and LiveLink™ interfaces for several engineering software programs.

In this book, scientists, engineers, and others interested in exploring the behavior of different physical device structures through computer modeling are introduced to the techniques of hands-on building and solving models through the direct application of the basic COMSOL Multiphysics software, along with some samples using the AC/DC, Chemical Reaction Engineering, Heat Transfer, MEMS, and RF modules. The next to the last technical chapter explores the use of Perfectly Matched Layers in the RF Module. The final technical chapter explores the use of the Bioheat Equation in the Heat Transfer and RF Modules.

The models presented herein are built within the context of the physical world (applied physics) and are presented in light of First Principle Analysis techniques. The demonstration models emphasize the fundamental concept that the information derived from the modeling solutions through use of these computer simulations is only as good as the materials coefficients and the fundamental assumptions employed in building the models.

The combination of computer simulation and First Principles Analysis gives the modeler the opportunity to try a variety of approaches to the solution of the same problem as needed in order to get the design right or nearly right in the workshop or laboratory before the first device components are fabricated and tested. The modeler can also use the physical device test results to modify the model parameters and arrive at an improved solution more rapidly than by simply using the cut and try methodology.

Chapter Topics

The eleven (11) technical chapters in this book demonstrate to the reader the hands-on technique of model building and solving. The COMSOL concepts and techniques used in these chapters are shown in Figure Int.1. The COMSOL modules employed in the various models in specific chapters are shown in Figure Int.2, and the Physics concepts and techniques employed in the various models in specific chapters are shown in Figure Int.3.

The information in these three figures link the overall presentation of this book to the underlying modeling, mathematical, and physical concepts. In this book, in contrast to some other books with which the reader may be familiar, key ancillary information is, in most cases, contained in the notes.

NOTE *Please be sure to read, carefully consider, and apply, as needed, each note.*

Chapter 1 Modeling Methodology

Chapter 1 provides an overview of the modeling process by discussing the fundamental considerations involved: the hardware (computer platform), the coordinate systems (physics), the implicit assumptions (lower dimensionality considerations), and First Principles Analysis (physics). Three relatively simple 1D models are presented that build and solve, for comparison, single-, double-, and triple-pane thermal insulation window structures. Comments are also included on common sources of modeling errors.

Chapter 2 Materials Properties

Chapter 2 discusses various sources of materials properties data, including the COMSOL Material Library, basic and expanded module, as well as print and Internet sources. A multi-panel thermal insulation window structure model demonstrates three techniques for entering material

Concept/Technique Chapter:	1	2	3	4	5	6	7	8	9	10	11
0D PDE Modeling			●								
1D PDE Modeling	●	●		●							
2D Axisymmetric Coordinates							●				
2D Axisymmetric Modeling						●					●
2D Modeling					●					●	●
3D Modeling									●		
Animation				●	●	●					
Bioneat Equation											●
Boolean Operations-geometry					●	●		●	●	●	●
Boundary Conditions	●	●		●							
Conductive Media DC					●	●					
Coupled Multiphysics Analysis	●							●	●	●	●
Cylindrical Coordinates						●				●	
Deformed Mesh - Moving Mesh					●			●			
Domain Plot Parameter								●			
Electromagnetics					●						
Electronic Circuit Modeling			●								
Electrostatic Potentials									●		
Filler corners								●			
Floating Contacts					●						
Free Mesh Parameters					●	●	●				
Frequency Domain							●				
Global Equations								●			
Heat Transfer Coefficient	●	●				●					
Lagrange Parameters											●
Laplacian Operator									●		
Lumped Parameters									●		
Magnetostatic Modeling									●		
Materials Library	●	●						●			
Mathematics – Coefficient Form PDE				●							
Mathematics – General Form PDE				●							
Maximum Element Size							●				
Mixed Materials Modeling							●				
Mixed Mode Modeling							●				
Out-of-Plane Thickness							●	●			
Parametric Solutions				●		●				●	
Perfectly Matched Layers										●	
Pointwise Constraints					●						
Quasi-Static Solutions						●					
Reference Frame								●			
Scalar Expressions							●				
Scalar Variables							●				
Spherical Coordinates				●							
Static Solutions						●					
Streamline Plot									●		
Terminal Boundary Condition									●		
Transient Analysis					●		●				
Triangular Mesh						●	●			●	●
Weak Constraints					●						
Work Plane									●		

FIGURE INT.1 COMSOL Concepts and Techniques.

Module Chapter:	1	2	3	4	5	6	7	8	9	10	11
Basic	•	•		•	•	•	•	•	•	•	•
AC/DC			•		•		•		•		
CFD or MEMS								•	•		
Chemical Reaction Engineering								•			
Heat Transfer											•
RF			•							•	

FIGURE INT.2 COMSOL Modules Employed.

Physice Concepts Chapter	1	2	3	4	5	6	7	8	9	10	11
Anisotropic Conductivity					•						
Antennas											•
Bioheat Equation											•
Boltzmann Thermodynamics						•					
Complex AC Theory								•			
Concave Mirror										•	
Coulomb Gauge									•		
Distributed Resistance									•		
Electrochemical Polishing					•						
Electromagnetic induction (Inductance)									•		
Electrostatic Potentials in Different Geometric Configurations									•		
Energy Concentrator										•	
Faraday's Law					•						
Fick's Laws						•					
First Estimate Review	•								•		
Fourier Analysis							•				
Fourier's Law						•	•				
Free-Space Permittivity									•		
Good First Approximation	•						•				
Hall Effect					•						
Heat Conduction	•					•					
Helmholtz Coil									•		
Information Transmission				•							
Insulated Containers						•					
Joule Heating						•	•				
Kirchoff's Laws (Current, Voltage)			•								
Lorentz Force					•						
Magnetic Field					•						
Magnetic Permeability									•		
Magnetic Vector Potential									•		
Magnetostatics									•		
Maxwell-Faraday Equation			•								
Maxwell's Equations						•			•	•	
Microwave Irradiation											•
Nernst-Planck Equations								•			
Newton's Law of Cooling						•					
Ohm's Law			•				•		•		
Optical (Laser) Irradiation											•
Pennes Equation											•
Perfectly Matched Layers: 2D Planar, 3D Cartesian Cylindriacal and Spherical										•	
Perfusion											•
Planck's Constant						•					
Semiconductor Dual Carrier Types					•						
Soliton Waves				•							
Telegraph Equation				•							
Thin-Layer Resistance									•		
Vector Dot Product Current					•						

FIGURE INT.3 Physics Concepts and Techniques.

properties: user-defined direct entry, user-defined parameters, and material definitions. Also included are instructions for building a user defined material library for storage within COMSOL 4.x.

Chapter 3 0D Electrical Circuit Interface

COMSOL 4.x uses zero-dimensional models to provide for the modeling of electrical circuitry. The models in this chapter illustrate techniques for modeling various basic circuits: a resistor-capacitor series circuit, an inductor-resistor series circuit, and a series resistor, parallel inductor-capacitor circuit. Considerations for the proper setup of the circuits are discussed along with the basics of problem formulation and the implicit assumptions built into COMSOL 4.x relative to electrical circuit modeling.

Chapter 4 1D Modeling

The first part of Chapter 4 models the 1D KdV Equation. The KdV Equation is a powerful tool that is used to model soliton wave propagation in diverse media (e.g. physical waves in liquids, electromagnetic waves in transparent media, etc.). It is easily and simply modeled with a 1D PDE mode model.

The second part of Chapter 4 models the 1D Telegraph Equation. The Telegraph Equation is a powerful tool that is used to model wave propagation in diverse transmission lines. The Telegraph Equation can be used to thoroughly characterize the propagation conditions of coaxial lines, twin pair lines, microstrip lines, etc. The Telegraph Equation is easily and simply modeled with a 1D PDE mode model.

The last part of Chapter 4 is a 1D Spherically Symmetric Transport model that illustrates the technique of simplifying models with spherical components from 3D to 1D by assuming that they are essentially symmetrical.

Chapter 5 2D Modeling

The first half of Chapter 5 models the surface smoothing process in 2D Electrochemical Polishing Model. This model is a powerful tool that can be used for diverse surface smoothing projects (e.g. microscope samples, precision metal parts, medical equipment and tools, large and small metal drums, thin analytical samples, vacuum chambers, etc.).

The second half of Chapter 5 models Hall Effect magnetic sensors. The 2D Hall Effect Model is a powerful tool that can be used to model such sensors when used for sensing fluid flow, rotating and/or linear motion, proximity, current, pressure, orientation, etc.

Chapter 6 2D Axisymmetric Modeling

Modeling a 3D device that is symmetrical on one axis by treating it as a 2D Axisymmetric object simplifies the model for quicker first approximation results.

The first half of Chapter 6 discusses a 2D Axisymmetric Heat Conduction in a Cylinder Model, and demonstrates the use of contour plotting of the solver results to show non-linear temperature distribution in the cylinder.

The second half of Chapter 6 models transient heat transfer in a niobium sphere immersed in a medium of constant temperature by using a 2D Axisymmetric model.

Chapter 7 2D Simple Mixed Mode Modeling

In this chapter, simple mixed mode 2D models are presented. Such 2D models are typically more conceptually complex than the models that were presented in earlier chapters of this text. 2D simple mixed mode models have proven to be very valuable to the science and engineering communities, both in the past and currently, as first-cut evaluations of potential systemic physical behavior under the influence of mixed external stimuli. The 2D mixed mode model responses and other such ancillary information can be gathered and screened early in a project for a first-cut evaluation. That initial information can potentially be used later as guidance in building higher-dimensionality (3D) field-based (electrical, magnetic, etc.) models.

The first half of Chapter 7 uses a 2D Electrical Impedance Sensor Model to demonstrate this technique. The concept of electrical impedance, as used in alternating current (AC) theory, is an expansion on the basic concept of resistance as illustrated by Ohm's Law, in direct current (DC) theory.

The second half of Chapter 7 uses a 2D Axisymmetric Metal Layer on a Dielectric Block Model to demonstrate more aspects of the technique. The modeler was introduced to Fick's laws for the diffusion (mass transport) of a first item (e.g. a gas, a liquid, etc.) through a second item (e.g. another gas, liquid, etc.). In the case of this model, the diffusing item is heat.

These models are examples of the Good First Approximation type of models, because they demonstrate the significant power of relatively simple physical principles, such as Ohm's Law, Joule's Laws, and Fick's Laws, when applied in the COMSOL Multiphysics Modeling environment. The equations can, of course, be modified by the addition of new terms, insulating materials, heat loss through convection, etc.

Chapter 8 2D Complex Mixed Mode Modeling

In this chapter, two new primary concepts are introduced to the modeler: mass transport of copper ions through a fluid medium resulting in the electrodeposition (electroplating) of a copper layer and the mass transport and electrocoalescence of water droplets in an oil medium.

The first model in this chapter employs the Nernst-Planck Equation found in the Chemical Species Transport physics interface. The second model in this chapter employs an electrostatic field to induce the electrocoalescence of water droplets in an oil medium under the Laminar Two-Phase Flow, Phase Field physics interface.

Chapter 9 3D Modeling

In this chapter, the modeler is introduced to three new modeling concepts: the Terminal boundary condition, lumped parameters, and coupled thermal, electrical, and structural multiphysics analysis. The Terminal boundary condition and the lumped parameter concepts are employed in the solution of the 3D Spiral Coil Microinductor Model. The fully coupled multiphysics solution is employed in the 3D Linear Microresistor Beam Model.

The lumped parameter (lumped element) modeling approach approximates a spatially distributed collection of diverse physical elements by a collection of topologically (series and/or parallel) connected discrete elements. This technique is commonly employed for first approximation models in electrical, electronic, mechanical, heat transfer, acoustic, and other physical systems.

Chapter 10 Perfectly Matched Layer Models (PML)

One of the fundamental difficulties underlying electromagnetic wave equation calculations is dealing with a propagating wave after the wave interacts with a boundary (reflection). If the boundary of a model domain is terminated in an abrupt fashion, unwanted reflections will typically be incorporated into the solution, potentially creating undesired and possibly erroneous model solution values. Fortunately, for the modeler of today, there is a methodology that works sufficiently well that it essentially eliminates reflection problems at the domain boundary. That methodology is the Perfectly Matched Layer.

Chapter 10 includes two models, the 2D Concave Metallic Mirror PML Model and the 2D Energy Concentrator PML Model, to demonstrate the use of the Perfectly Matched Layer methodology.

Chapter 11 Bioheat Models

The Bioheat Equation plays an important role in the development and analysis of new therapeutic medical techniques (e.g. killing of tumors). If the postulated method raises the local temperature of the tumor cells, without excessively raising the temperature of the normal cells, then the proposed method will probably be successful. The results (estimated time values) from the model calculations will significantly reduce the effort needed to determine an accurate experimental value. The guiding principle needs to be that tumor cells die at elevated temperatures. The literature cites temperatures that range from 42 °C (315.15 K) to 60 °C (333.15).

The first half of Chapter 11 models the Bioheat Equation as applied with a photonic heat source (laser).

The second half of Chapter 11 models the Bioheat Equation as applied with a microwave heat source.

CHAPTER 1

MODELING METHODOLOGY USING COMSOL MULTIPHYSICS 4.X

In This Chapter

- Guidelines for New COMSOL Multiphysics 4.x Modelers
  - Hardware Considerations
  - Simple Model Setup Overview
  - Basic Problem Formulation and Implicit Assumptions
- 1D Window Heat Flow Models
  - 1D 1 Pane Window Heat Flow Model
  - 1D 2 Pane Window Heat Flow Model
  - 1D 3 Pane Window Heat Flow Model
- First Principles as Applied to Model Definition
- Some Common Sources of Modeling Errors
- References
- Suggested Modeling Exercises

GUIDELINES FOR NEW COMSOL MULTIPHYSICS 4.X MODELERS

First, for purposes of clarity in this book, when the term "COMSOL Multiphysics 4.x software" or "4.x" occurs, that term means: COMSOL Multiphysics software version 4.1 or later.

NOTE

Second, the user of this text should be sure to READ ALL NOTES in this book, as the Notes contain information that is not presented elsewhere and should facilitate comprehension and, hopefully, minimize modeling errors for modelers unfamiliar with 4.x.

When building models, new or otherwise, it is VERY IMPORTANT to SAVE EARLY and OFTEN.

Hardware Considerations

There are two basic rules for selecting hardware that will support successful modeling with 4.x. The first rule is that the modeler should be sure to determine the minimum system requirements their version of 4.x requires before borrowing or buying a computer to run his new modeling software. This book introduces the use of 4.x for modeling within the Microsoft Windows® and the Macintosh OS X® operating systems environments.

COMSOL Multiphysics 4.x software supports shared memory parallelism under both the Microsoft Windows and the Macintosh OS X operating systems. Distributed memory parallelism is supported on a Microsoft Windows Cluster, using a COMSOL Floating Network License. Neither distributed memory nor cluster computing will be covered in this book.

The number of platform cores is equal to the number of coprocessors designed into the computer (e.g. 1, 2, 4, 8, 12, etc.).

NOTE

Shared memory parallelism means that multiple cores in the same computer share the same memory array. Distributed memory parallelism means multiple computers with multiple cores are connected in a cluster with the problem divided into multiple parallel computational sub-segments {1.1} distributed over the computational units of the cluster.

The second rule of successful modeling is the modeler should run 4.x on the best platform with the highest processor speed and the most memory obtainable; the bigger, the faster, the better. It is the general rule that the speed of model processing increases in proportion to instruction size (32 bit, 64 bit), the core speed, the number of platform cores, and to the amount of usable, available memory.

The platform this author uses is an Apple Mac Pro®, running Mac OS X version 10.6.x. That platform has 4–3 GHz cores and 16 GB of RAM and is configured for 64-bit processing and runs at the 64-bit rate when using 4.x. If the new modeler desires a different 64-bit operating system than Macintosh OS X, then they will need to choose either a SUN® or a LINUX® platform, using UNIX© or a PC with a 64-bit Microsoft Windows operating system.

NOTE

The 3 GHz specification is the operating speed of each of the cores and the 16 GB is the total shared Random Access Memory (RAM). 64-bit refers to the width of a processor instruction.

Currently available Macintosh Pro hardware can be obtained with up to 12 cores and 64 GB of shared RAM {1.2}.

Once the best available processor is obtained, within the constraints of your budget, install your copy of 4.x, following the installer instructions.

After installation, as a matter of caution, the modeler should restart the processor and, once stable, start the COMSOL Multiphysics 4.x software. 4.x will then present the modeler with a configurable Graphical User Interface (GUI). For computer users not familiar with the GUI concept, information is presented primarily in the form of pictures with supplemental text, not exclusively as text. See Figure 1.1 to observe the details of the 4.x GUI interface.

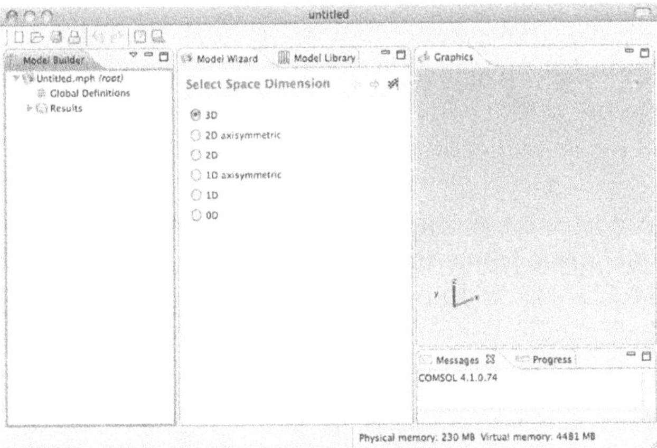

FIGURE 1.1 Default COMSOL Desktop Display Immediately After Startup.

Simple Model Setup Overview

Figure 1.1 shows the default COMSOL Desktop Display immediately after startup. The default coordinate selection for modeling calculations at startup in 4.x is 3D. The 3D coordinate system shown in the Graphics window is based on the Right-Hand-Rule.

NOTE

The Right-Hand-Rule is derived, as it states, from your right hand. Look at your right hand, point the thumb up, point your index (first) finger away from your body, at a right angle (90 degrees) to your thumb and point your middle (second) finger, at a right angle to the thumb, tangent to your body. Your thumb represents the Z-axis, your index (first) finger represents the X-axis, and your middle (second) finger represents the Y-axis.

In a 3D coordinate system that obeys the Right-Hand-Rule, X rotates into Y and generates Z. Z is the direction in which a Right-Handed Screw would advance (move into the material into which it is being screwed).

NOTE

The modeler should note that the default COMSOL Desktop for 4.x displays four active windows: Model Builder, Model Wizard, Graphics, and Messages; and two inactive windows: Model Library and Progress. The Desktop Display can, of course, be modified as needed to show more or fewer (see COMSOL Desktop Environment {1.3}). In this book, the Desktop Display will be used in the default configuration, except when modifications are required.

The most singularly important currently active window for the modeler, at this point, is the Model Wizard window. Before anything else can happen relative to creating a new model, 4.x requires that the modeler either choose a new coordinate system or accept the default 3D Cartesian coordinate system.

The Model Wizard window of the Desktop Display shows six coordinate system (Space Dimension) choices available for selection: 3D, 2D Axisymmetric, 2D, 1D Axisymmetric, 1D, and 0D. See Figure 1.2 to observe the interface details.

Figure 1.2 shows the Model Wizard window for model coordinate system (Space Dimension) selection. The default coordinate selection for modeling calculations at startup in 4.x is 3D.

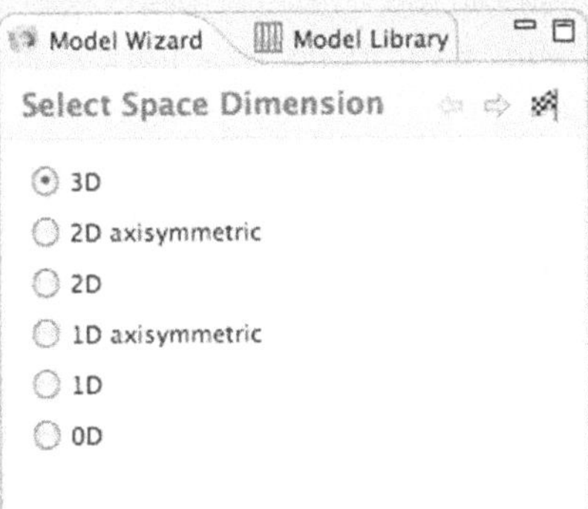

FIGURE 1.2 Default Model Wizard Window for Model Coordinate System (Space Dimension) Selection.

NOTE *Note the Dot in the center of the 3D selection (radio) button.*

Geometric coordinate systems in COMSOL range from the extremely complex to the nominally very simple. The default geometries are the standard Cartesian geometries 3D, 2D, and 1D as shown in Figures 1.3, 1.4, and 1.5.

Figure 1.3 shows a Right-Hand 3D Cartesian Coordinate Geometry Axes configuration.

Figure 1.4 shows a 2D Cartesian Coordinate Geometry Axes configuration.

Figure 1.5 shows a 1D Cartesian Coordinate Geometry Axis configuration.

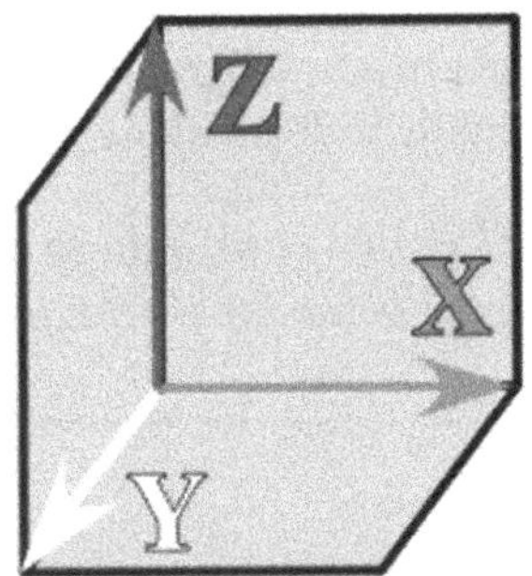

FIGURE 1.3 3D Cartesian Coordinate Geometry.

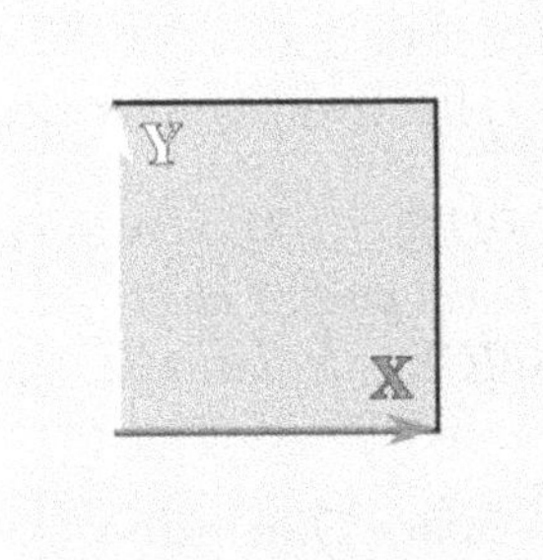

FIGURE 1.4 2D Cartesian Coordinate Geometry.

Comprehension of the 3D, 2D, 1D Cartesian Coordinate Systems is relatively easy for the modeler, assuming that the modeler has some previous engineering or science training. These coordinate systems are simply an orthogonal combination of rectilinearly incremented axes comprising 1, 2, or 3 measurement dimensions.

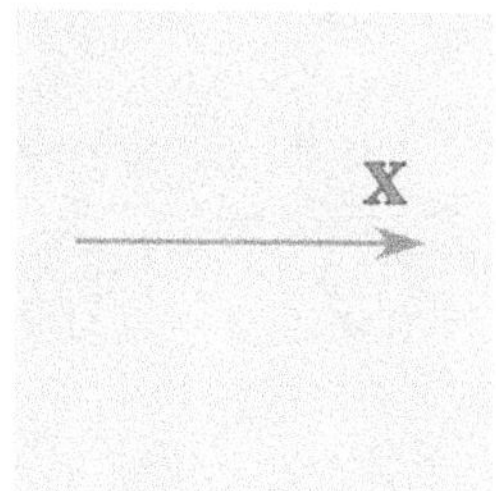

FIGURE 1.5 1D Cartesian Coordinate Geometry.

NOTE

The 2D and 1D Axisymmetric Geometries are somewhat more complex. Both the 2D and 1D Axisymmetric Geometries comprise cylindrical coordinate systems. The primary benefit derived through the use of the axisymmetric coordinate systems is that they allow the modeler to reduce the effective model dimensionality and the calculational complexity of a problem, for example 3D becomes 2D, through the assumption of axial symmetry and rotationally uniform boundary conditions.

Due to the complex nature of the underlying assumptions in the 1D Axisymmetric Coordinate System and its applications, the modeler is referred to the COMSOL literature {1.4} for further details.

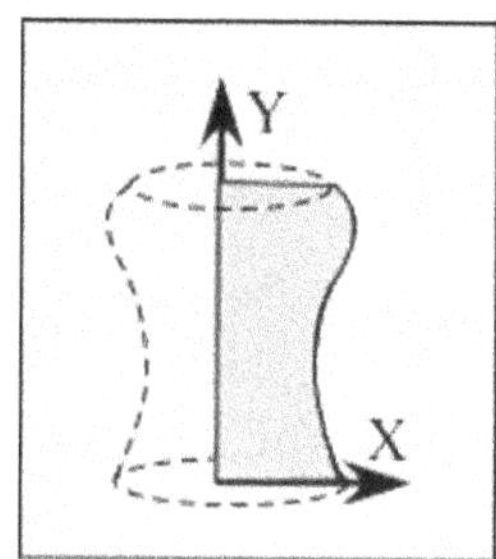

FIGURE 1.6 2D Axisymmetric Coordinate Geometry (Space Dimension).

Figure 1.6 shows a 2D Axisymmetric coordinate geometry axis (Space Dimension) configuration.

The last Space Dimension choice available is 0D. In the case of 0D, the underlying equations in the multiphysics model have either no relational dependence on geometrical factors or react in a homogeneous and isotropic manner (effectively geometrically relationally independent).

NOTE

The 0D Space Dimension will be used in this book only to explore electrical and/or electronic circuit model behavior through the use of SPICE calculations.

Select A Space Dimension

For this first demonstration of the Desktop Display, Select (Click) the 1D selection (radio) button.

NOTE *The term Select (Click) means for the modeler to place the display screen computer cursor in proximity to or superimposed on the item to be selected and then tap (momentarily depress) the physical mouse activation key (button).*

The 1D selection (radio) button will show a Dot in the center once selected. See Figure 1.7.

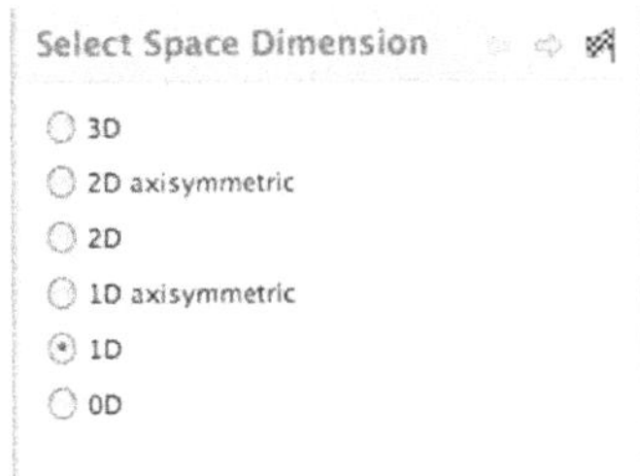

FIGURE 1.7 Selected 1D Radio Button.

Figure 1.7 shows the selected 1D radio button, below the Next arrow (right pointing), the Back arrow (left pointing), and the Finished Flag.

Then, Click the Next arrow (right pointing). See Figure 1.8.

FIGURE 1.8 Next Arrow (Right Pointing).

Figure 1.8 shows the Next arrow (right pointing) located between the Back arrow (left pointing) and the Finished Flag.

Once the Next arrow is clicked, the Desktop Display changes to show the Add Physics window as shown in Figure 1.9.

Figure 1.9 shows the Desktop Display with the Add Physics window ready for the selection of the appropriate Physics equations by selecting the

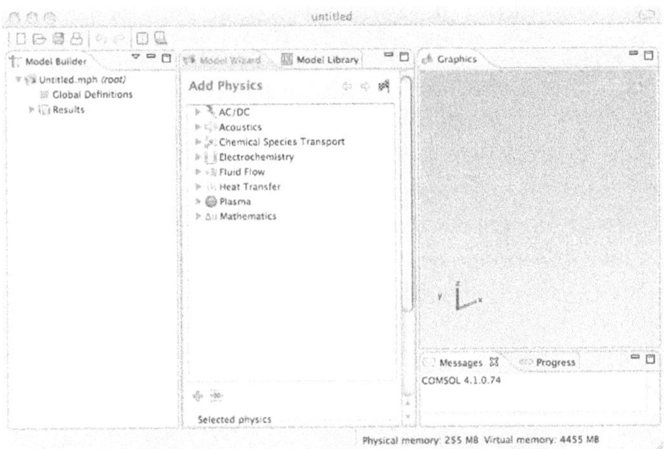

FIGURE 1.9 Desktop Display with the Add Physics Window.

desired Physics Interface. In the case of Figure 1.9, all of the COMSOL Multiphysics Physics Interfaces are displayed, as they would be in a trial license. If the modeler has not licensed all of the available modules, only the licensed modules will be displayed.

A Physics Interface is the set of relevant equations, typical initial conditions, independent variable(s), dependent variable(s), default settings, boundary conditions, etc. that are appropriate for the solution of a particular problem type for a given branch or sub-branch of physics (e.g. acoustics, electromagnetics, heat transfer, etc.).

NOTE *As is indicated by the name Multiphysics, several different branches or sub-branches of physics can be applied, as needed, simultaneously or serially to achieve the solution of a given model. Diverse models that demonstrate the application of the Multiphysics concept through the incorporation of multiple different Physics Interfaces in a given model will be explored in detail as this book progresses.*

The modeler has now been shown fundamentally how to select a set of Space Coordinates and how to access Physics Interfaces.

If the modeler is using a Macintosh, Select > COMSOL Multiphysics menu > Quit COMSOL Multiphysics.

If the modeler is using a PC, Select > File > Exit.

Basic Problem Formulation and Implicit Assumptions

NOTE *A first-cut problem solution is the equivalent of a back-of-the-envelope or on-a-napkin solution. Problem solutions of that type are more easily formulated, more quickly built, and typically provide a first estimate of whether the final solution to the full problem will be deemed to be (should possibly be) within reasonable time constraint or budgetary bounds. Creating first-cut solutions will often allow the 4.x modeler to easily decide whether or not it is worth the additional effort and cost needed to build a fully implemented higher dimensionality model.*

A modeler can generate a first-cut problem solution as a reasonable first estimate, by choosing initially to use a lower dimensionality coordinate space than 3D (e.g. 1D, 2D Axisymmetric, etc.). By choosing a low-dimensionality Space Dimension, a modeler can significantly reduce the ultimate time needed to achieve a detailed final solution for the chosen prototype model. However, both new modelers and experienced modelers must be especially careful to fully understand the underlying (implicit) assumptions, unspecified conditions, and default values that are automatically incorporated into the model by simply selecting a lower dimensionality Space Dimension.

Reality, as we currently understand it, comes in four basic dimensions, three space dimensions (X-Y-Z), and one time dimension (t), (e.g. X-Y-Z-t, r-φ-θ-t, etc.). Relativistic effects can typically be neglected in most cases, except where high velocity or ultra-high accuracy is involved. Models with relativistic effects will not be covered in this book.

NOTE

Relativistic effects typically only become a concern for bodies in motion with a velocity approaching that of the speed of light (~$3.0x10^8$ m/s) or for ultra-high resolution time calculations at somewhat lower velocities.

The types of calculations presented within this book are typically for steady-state models, quasi-static models, and for relatively low-velocity transient model solutions.

NOTE

In a steady-state model, the controlling parameters are defined as numerical constants and the model is allowed to converge to reach the final equilibrium state defined by the specified constants.

In the quasi-static methodology, a model solution to a problem is found by finding an initial steady-state solution, not the final steady-state solution. Then, the modeling constants are incrementally modified. That incremental change in the constants moves the modeling solution toward the desired final steady-state solution.

In a transient solution model, all of the appropriate variables in the model are a function of time. The model solution builds from a set of suitable initial conditions, through a set of incremental intermediate solutions, to a final solution.

1D WINDOW HEAT FLOW MODELS

Consider, for example, a brief comparison between a relatively simple 1D heat flow model and the identical problem as a 3D model. The following models are those of a single pane, a dual pane, or a triple pane window mounted in the wall of a building on a typical winter's day. The basic question to be answered is: Why use a dual pane window rather than a single pane window or a triple pane window?

1D 1 Pane Window Heat Flow Model

Run 4.x > Select 1D in the Model Wizard.

Click > Next (right pointing arrow).

Click the Heat Transfer twistie.

Select > Heat Transfer in Solids *(ht)* in the Add Physics window.

Click the Add Selected (plus sign) (see Figure 1.10).

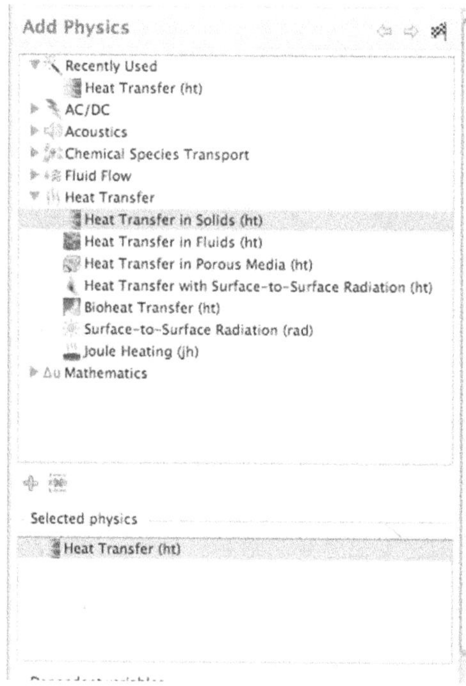

FIGURE 1.10 Model Wizard – Add Physics Window.

Figure 1.10 shows the Model Wizard – Add Physics window.

NOTE *The modeler should note that Clicking on the twistie for any Physics Interface expands the menu associated with that Physics Interface. That expansion allows the modeler to then Select the appropriate Physics Interface necessary to perform the needed model calculations. The Selected Physics Interface is added to the Selected Physics window by Clicking the Add Selected Physics (plus sign) icon. If necessary, the added Selected Physics Interface can be removed by Clicking the Remove Selected icon.*

Click > Next (right pointing arrow).

Select > Preset Studies > Stationary.

NOTE *The modeler should note that Stationary and Steady-State are equivalent terms.*

Click > Finish Flag (see Figure 1.11).

Figure 1.11 shows the Finish Flag in the Select Study Type window for the completion of Model Wizard process, resulting in the Initial Model Build. See Figure 1.12.

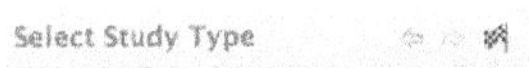

FIGURE 1.11 Finish Flag in the Select Study Type Window.

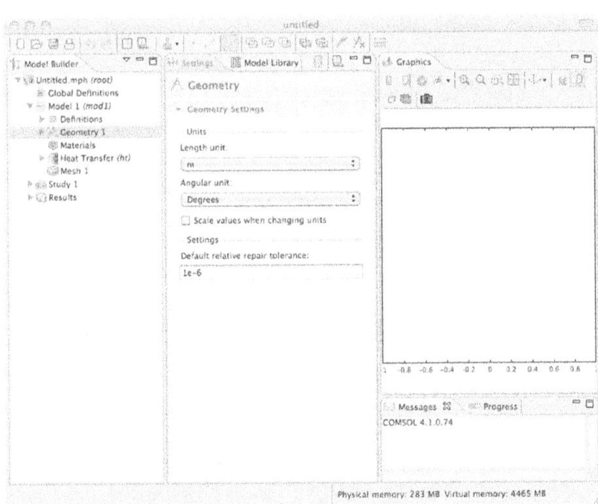

FIGURE 1.12 Initial Model Build for the 1D Window Heat Transfer.

Figure 1.12 shows Initial Model Build for the 1D Window Heat Transfer Desktop Display.

Save the Model as > MMUC4_1D_W1Pane.mph. See Figure 1.13.

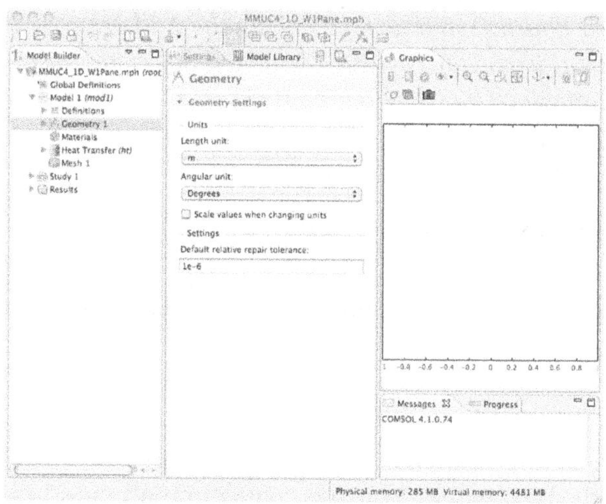

FIGURE 1.13 Saved Initial Model Build MMUC4_1D_W1Pane.mph.

Figure 1.13 shows the saved Initial Model Build MMUC4_1D_W1Pane.mph.

NOTE *The modeler should note that both the Desktop Display and the Model Builder windows show the saved file name.*

Now that the Initial Model Build has been saved, the modeler needs to enter the constant values needed for this model.

In the Model Builder window of the Desktop Display, Right-Click > Global Definitions > Select Parameters. See Figure 1.14.

NOTE *The term Right-Click means for the modeler to place the display screen computer cursor in proximity to or superimposed on the item to be selected and then tap (momentarily depress) the right physical mouse activation key (button).*

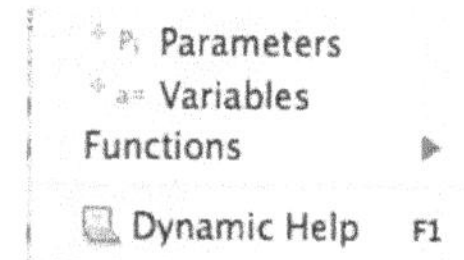

FIGURE 1.14 Global Definitions Selection Menu.

Figure 1.14 shows the Global Definitions Selection menu.

Once the Parameters menu item has been selected, the Desktop Display will be modified to incorporate a Settings window. The Settings window contains a Parameters Entry Table, listing the following attributes for each parameter: Name, Expression, Value, and Description (see Figure 1.15).

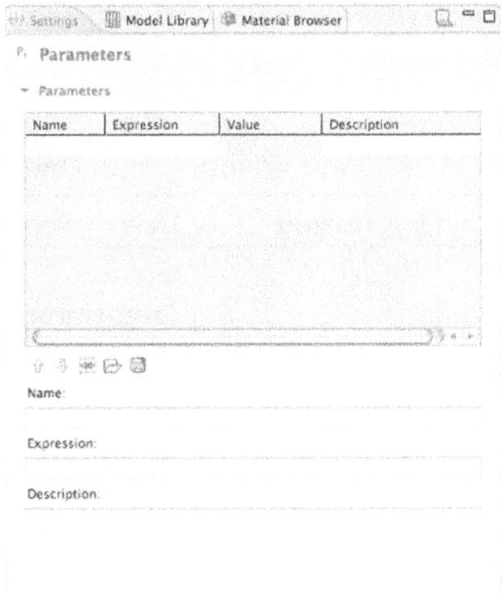

FIGURE 1.15 The Desktop Display – Settings Window with Parameters Entry Table.

Figure 1.15 shows the Desktop Display – Settings window with a Parameters Entry Table.

Parameter information can be entered directly into the fields in the Parameters Entry Table or into the parameter entry fields below the Parameters Entry Table. Control buttons are located between the Parameters Entry Table window and the parameter entry fields. See Figure 1.16.

Figure 1.16 shows the control button array for the Parameters Entry Table.

FIGURE 1.16 Control Button Array for the Parameters Entry Table.

These are the control button array functions, from left to right: Up Arrow (Move Up), Down Arrow (Move Down), Red X (Delete), Open Folder Icon (Load from File), and Disk Icon (Save to File).

To enter the parameters for the model, Click on the first entry window in the Name column and enter (type) each piece of the information, as shown in Table 1.1.

When entering information into a Parameters field, first Click on the field desired and then observe the presence of a blinking cursor. The blinking cursor verifies that 4.x is ready to accept the information. Then type the desired piece of information into the activated field.

The modeler can transition to the next information entry point in the tables by clicking on the next field.

TABLE 1.1 1D 1 Pane Window Parameters

Name	Expression	Description
T_in	70[degF]	Interior Temperature
T_out	0[degF]	Exterior Temperature
p	1[atm]	Air Pressure

These entries in the Parameters Entry Table define the interior temperature, the exterior temperature, and the air pressure for use in this model. See Figure 1.17.

Figure 1.17 shows a close-up of the Settings window with a filled Parameters Entry Table.

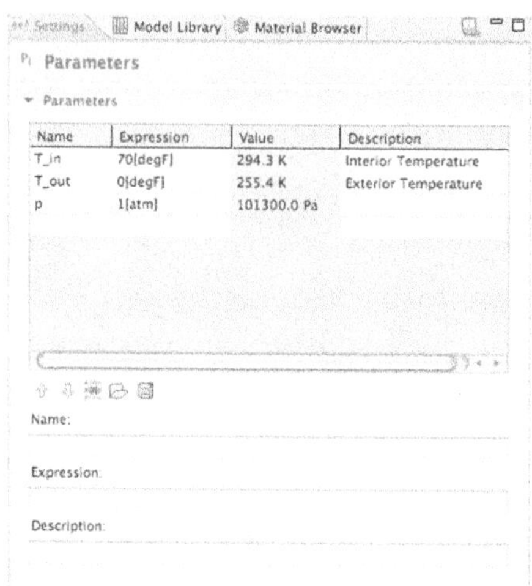

FIGURE 1.17 The Settings Window with a Filled Parameters Entry Table.

Click on the Save to File icon (disk), the rightmost symbol in the control button array, and enter MMUC4_1D_W1Pane_Pram.txt to save these parameters for future use.

The next step in building a model of a single pane window is to define the geometry through which the heat will flow from the heated interior to the less heated exterior.

NOTE *The modeler should note at this point, that in the next few steps a 1D path for heat flow will be created that is represented by a line progressing from left to right. Shortly, later, boundary conditions will be added.*

Right-Click in the Model Builder window on Geometry 1 under Model 1 *(mod1)* to display the selection list. See Figures 1.18 and 1.19.

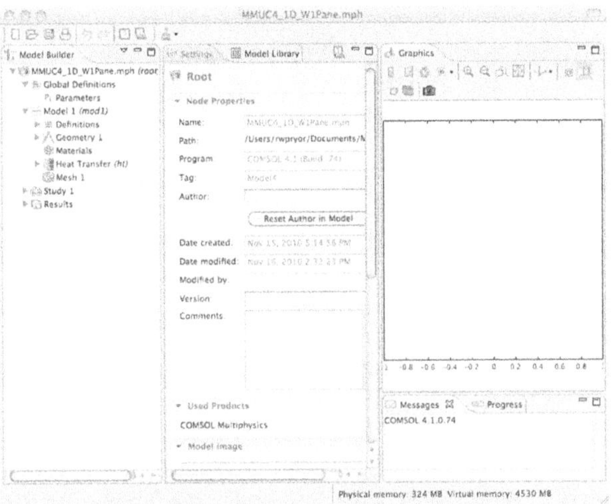

FIGURE 1.18 Model Builder Showing Geometry 1.

Figure 1.18 shows the Geometry 1 entry in the Model Builder window.

Figure 1.19 shows the Geometry 1 selection list for selecting the Line Interval option.

Select > Interval. See Figure 1.20.

FIGURE 1.19 Geometry 1 Selection List.

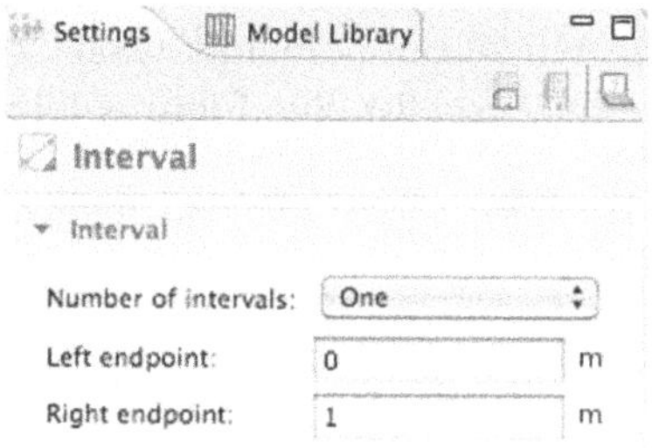

FIGURE 1.20 Settings - Interval Window.

Figure 1.20 shows the Settings – Interval window.

Enter 0[m] as the Left endpoint.

Enter 5e-3[m] as the Right endpoint, as shown in Figure 1.21.

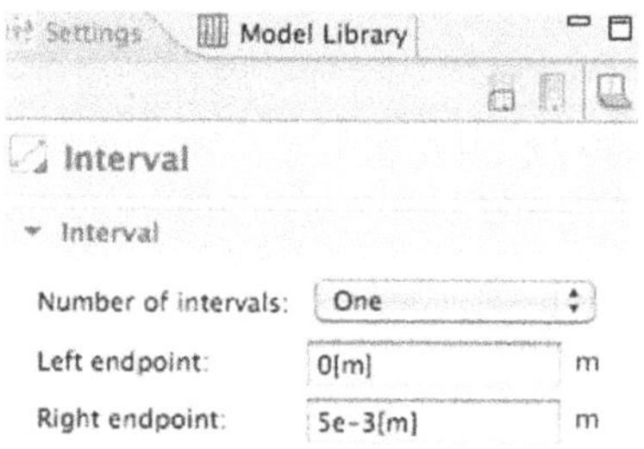

FIGURE 1.21 Settings - Interval Window, Filled.

Figure 1.21 shows the Settings – Interval window with the values filled in.

Click > Build All.

The geometry for the interval will be built and displayed in the Graphics window.

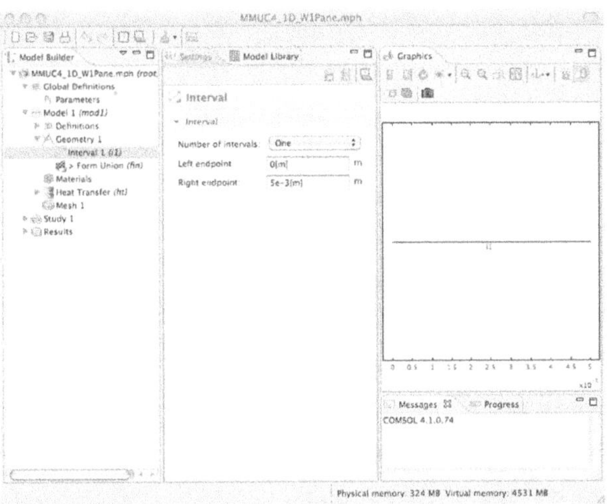

FIGURE 1.22 Desktop Display Window with Built Geometry.

Figure 1.22 shows the Desktop Display window with the Built Geometry.

Now that the Geometry has been built, the next step is to define the material(s) which comprise the model. In this case, it is relatively easy because the model only needs to use one (1) material. Also, the material properties of that material are resident in the Built-In Materials Library.

Right-Click in the Model Builder window on Materials under Model 1 *(mod1)* to display the Materials Selection List. See Figures 1.23 and 1.24.

Figure 1.23 shows the Materials entry in the Model Builder window.

Figure 1.24 shows the Materials Selection List for selecting the Material Browser.

Select > Open Material Browser. See Figure 1.25.

Figure 1.25 shows the Open Material Browser.

Click the twistie {1.5} pointing at Built-In in the Material Browser window. See Figures 1.26 and 1.27.

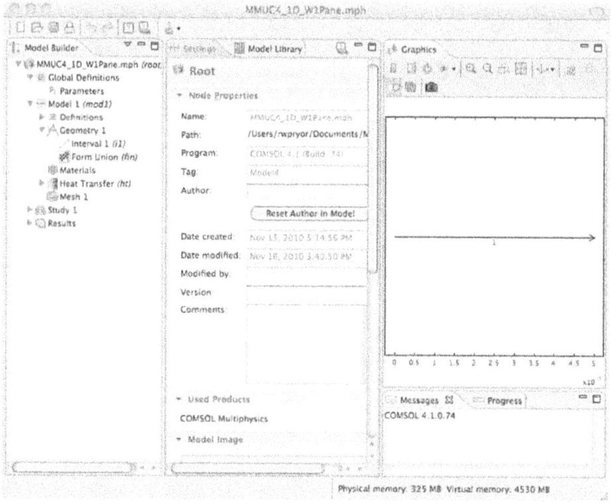

FIGURE 1.23 Model Builder Showing Materials.

Figure 1.26 shows the twistie before the entry for the Built-In material library.

Figure 1.27 shows the beginning of the expanded list of materials in the Built-In material library.

Material
Open Material Browser
Dynamic Help F1

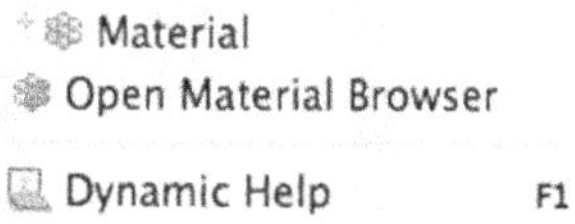

FIGURE 1.24 Materials Selection List.

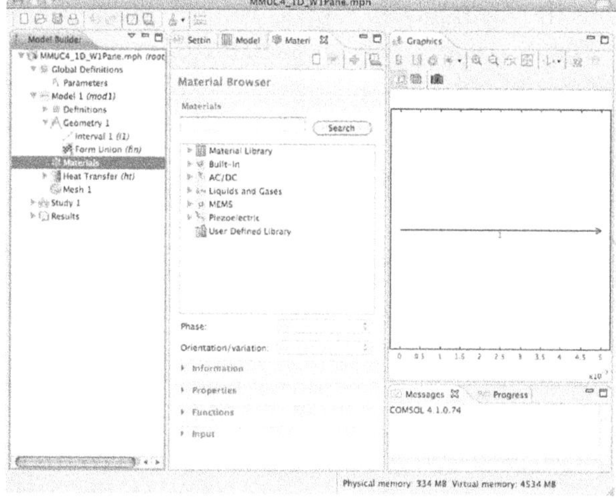

FIGURE 1.25 Open Material Browser.

FIGURE 1.26 Twistie Before Built-In.

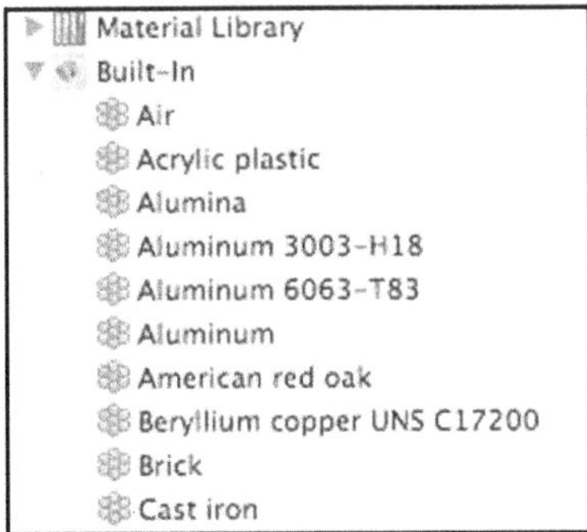

FIGURE 1.27 Expanded Built-In Material List.

Scroll the material list until Silica glass appears. See Figure 1.28.

Figure 1.28 shows the Scrolled List with the Silica glass entry visible.

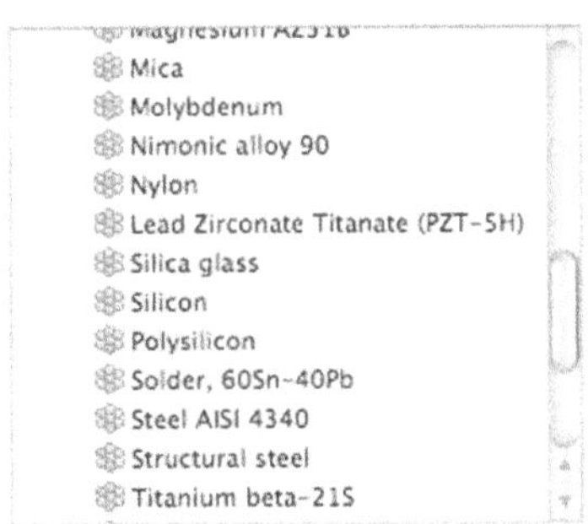

FIGURE 1.28 Scrolled List with Silica Glass Visible.

Right-Click > Silica glass. See Figure 1.29.

Figure 1.29 shows the Add Material to Model selection in the pop-up menu.

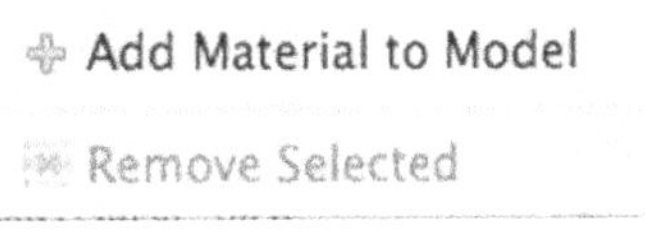

FIGURE 1.29 Add Material to Model.

Select > Add Material to Model. See Figure 1.30.
Figure 1.30 shows the Silica glass material added to the model.

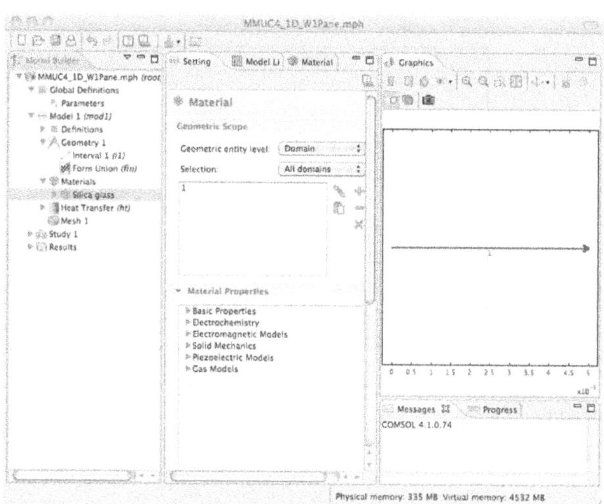

FIGURE 1.30 Silica Glass Material Added to Model.

Models in 4.x may comprise multiple geometric entities. Each geometric entity comprises connected manifolds. The geometric entities of the maximum dimension are called Domains. The order of precedence for 3D is: Domain, Boundary, Edge, and Point. As the dimensionality is reduced (e.g. 3D → 2D), the list is shortened from the right {1.6}.

NOTE

Since this model is 1D and there are only Domain and Boundary in this model, 4.x automatically assigns the Silica Glass material's properties to the available domain. If there were more than 1 domain, 4.x would assign this first material selection to all of the available domains.

As the modeler builds more complex and difficult models, he will need to pay close attention to the assignment of materials properties and to verify that the correct materials are added to the model and that they are assigned to the proper domains.

Also, the modeler should verify that the material selected has the necessary physical properties to allow for the solution of the problem in question. In this case, 4.x has determined which physical properties are needed for the current model and displays a green check mark next to the needed properties in the Material Contents window of the Settings – Material window. See Figure 1.31.

NOTE *The modeler should scroll-down in the Settings – Material window to see the green check marks in the Material Contents window.*

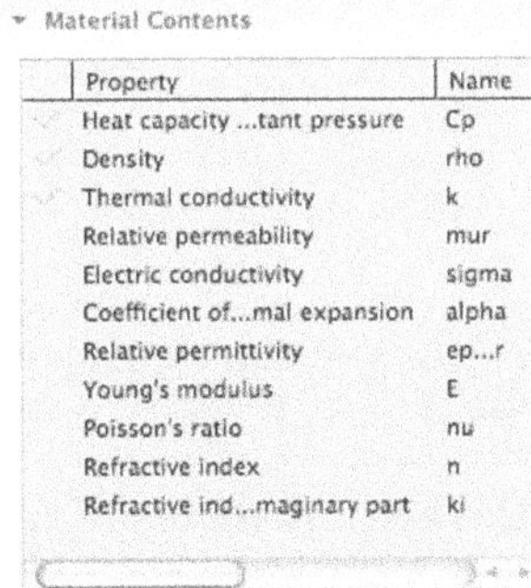

FIGURE 1.31 Checked Silica Glass Material Properties in Material Contents Window.

Figure 1.31 shows the checked Silica glass material properties in the Material Contents window.

Since the model geometry has been built and the appropriate material properties have been assigned to Domain 1, the next step in the modeling process is to specify both the domain and the boundary conditions for this model.

Click on the twistie next to the Heat Transfer *(ht)* Physics Interface to expand that section of the Model Builder. See Figure 1.32.

NOTE *The modeler should notice that when the twistie is Clicked, the menu expands and the implicit assumptions are made available for the Heat Transfer (ht) Module. The modeler will need to explore this region for each model created and each Physics Interface used to verify that the implicit assumptions made by 4.x either satisfy his needs or need to be modified.*

For the inquisitive experienced modeler, further information about the underlying equations, the implicit assumptions, and the initial values can be obtained by Clicking each associated twistie and perusing the available information. However, the new modeler should save some of his inquisitiveness for later explorations.

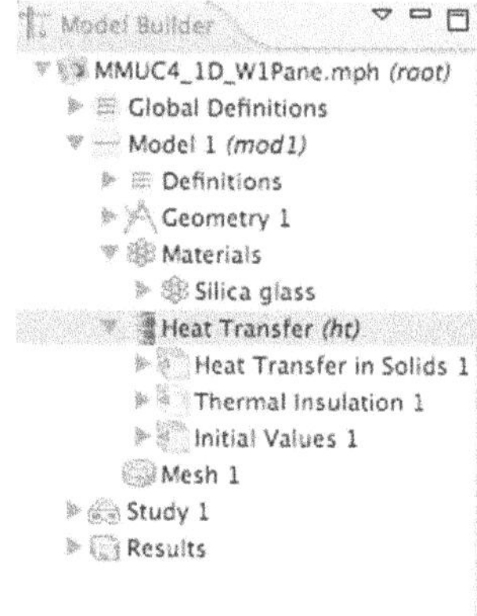

FIGURE 1.32 Expanded Heat Transfer *(ht)* in Model Builder Window.

Figure 1.32 shows the expanded Heat Transfer *(ht)* entry in the Model Builder window.

At this point in the development of this model, the modeler needs to verify the assignment of values to the domain. In the Model Builder window under Heat Transfer *(ht)*,

Click > Heat Transfer in Solids 1. See Figure 1.33.

Figure 1.33 shows the Settings for the Heat Transfer in Solids window.

The modeler can now verify that the selected material has been assigned to Domain 1 and that the material's properties are available for use in the equations to calculate the model solution.

If the modeler were now to Click on Thermal Insulation 1, he would find that both end points (1, 2) are insulated.

Also, if the modeler were to Click on Initial Values 1, he would find that the Temperature is set to Room (20°C = 293.15) and the Radiative Intensities are set to 0.

Those values are the result of the implicit assumptions remarked on earlier.

At this point, the modeler needs to set the desired modeling conditions for each endpoint (Boundary) and utilize the previously entered parameters that are needed to calculate the transfer of heat through this single pane window.

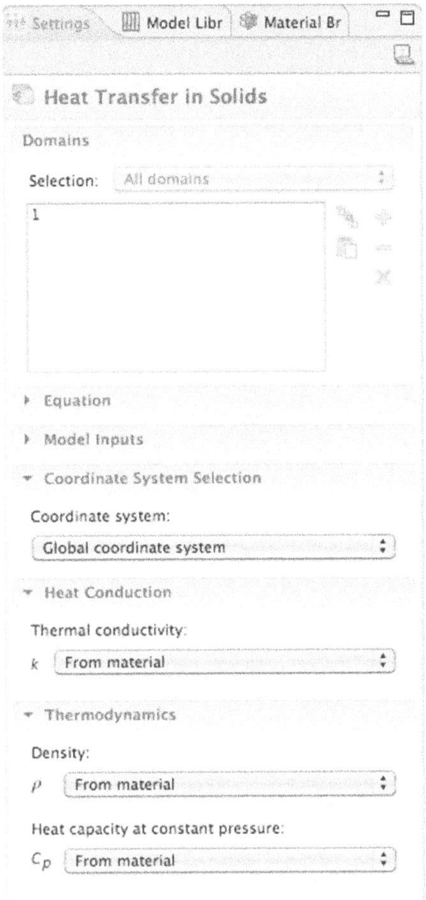

FIGURE 1.33 Settings for the Heat Transfer in Solids Window.

Right-Click > Heat Transfer *(ht)* in the Model Builder. See Figure 1.34.

Figure 1.34 shows the Heat Transfer Selection window, which lists the conditions that may be defined for this Physics Interface.

Select > Heat Flux. See Figure 1.35.

Figure 1.35 shows the Initial Boundaries Heat Flux window.

In the Graphics window, Click > Point 1 (leftmost boundary).

In the Settings – Heat Flux window, Click > Plus (Add to Selection).

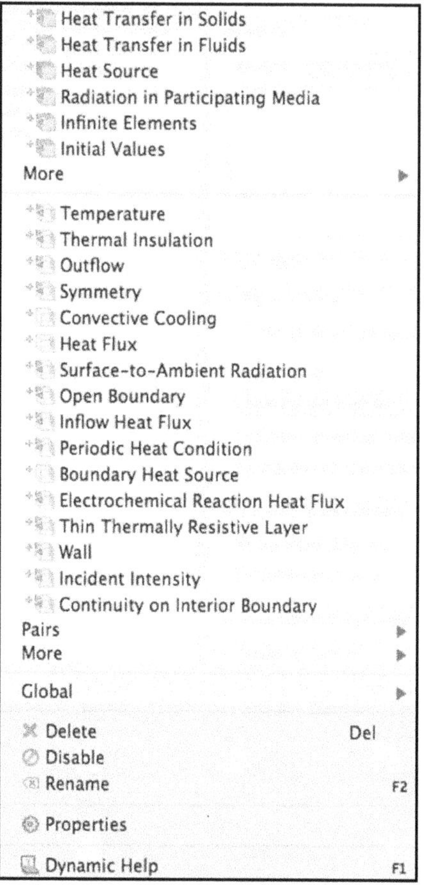

FIGURE 1.34 Heat Transfer Selection Window.

In the Settings – Heat Flux window, Click > Inward heat flux.

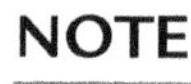

A typical value for the convection heat transfer coefficient (h) in the presence of a gas ranges from 2 to 25 W/(m^2 · K){1.7}. In this case, a typical value of 15 W/(m^2 · K) is chosen.

Type 15 in the Heat transfer coefficient entry window.

Type T_in in the External temperature entry window.

Figure 1.36 shows the Settings – Heat Flux window as finally configured for Boundary Point 1.

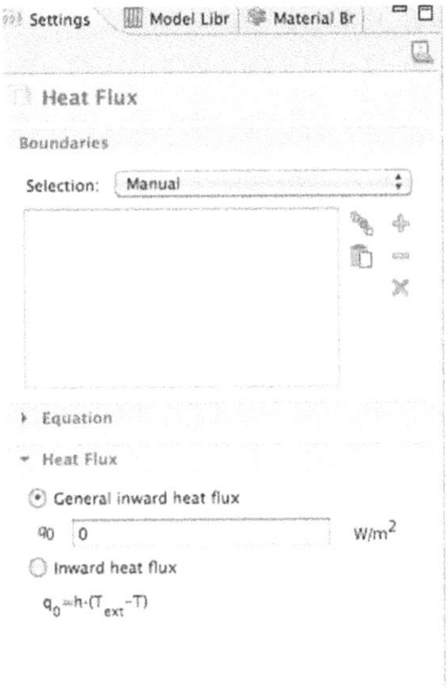

FIGURE 1.35 Initial Boundaries Heat Flux Window.

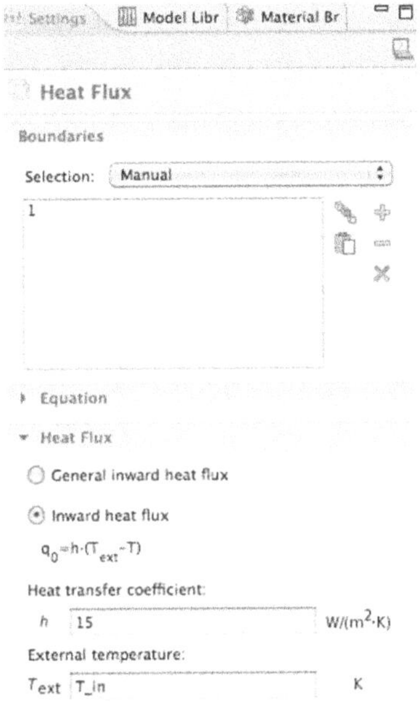

FIGURE 1.36 Filled Boundary Point 1 Heat Flux Settings.

Right-Click > Heat Transfer *(ht)*.

Select > Heat Flux.

In the Graphics window, Click > Point 2 (rightmost boundary).

In the Settings – Heat Flux window, Click > Plus (Add to Selection).

In the Settings – Heat Flux window, Click > Inward Heat Flux.

Type 15 in the Heat transfer coefficient entry window.

Type T_out in the External temperature entry window.

Figure 1.37 shows the Settings – Heat Flux window as finally configured for Boundary Point 2.

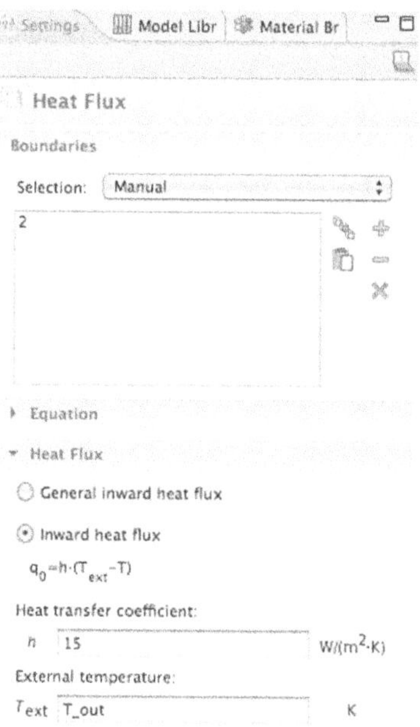

FIGURE 1.37 Filled Boundary Point 2 Heat Flux Settings.

Since this first model is relatively simple, the easiest path is to allow 4.x to automatically mesh the model.

Right-Click > Mesh 1.

Select > Build All.

The meshing results are shown in Figure 1.38.

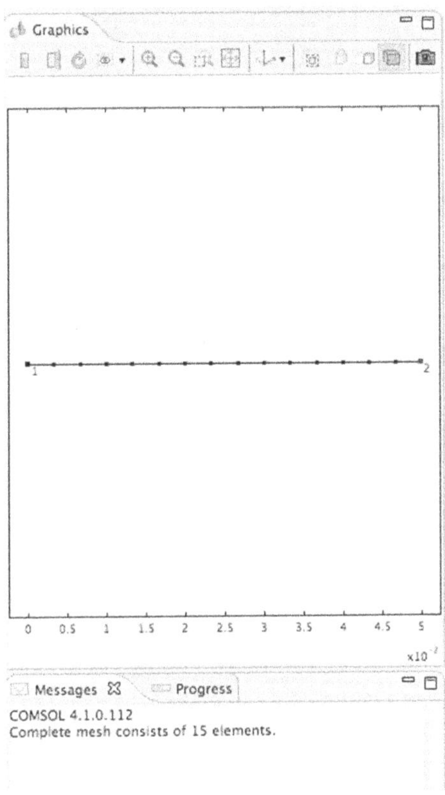

FIGURE 1.38 Meshed 1D Domain.

Figure 1.38 shows the meshed 1D domain.

NOTE

Meshing is the process by which the active geometrical space (domain) of a model is sub-divided into a collection of sufficiently smaller spaces so that linear or higher order polynomial approximations can be used as a reasonable analog of the functional physical behavior being modeled {1.8, 1.9}.

In this case, based on a First Principles Analysis, the heat flow is linear within the domain.

Now that the model has been meshed, the modeler can proceed to compute the solution.

Right-Click > Study 1 in the Model Builder window.

Select > Compute.

NOTE *4.x automatically selects the appropriate Solver for this problem, computes the solution, and displays the results in the Graphics window.*

The results of this modeling computation are shown in Figure 1.39.

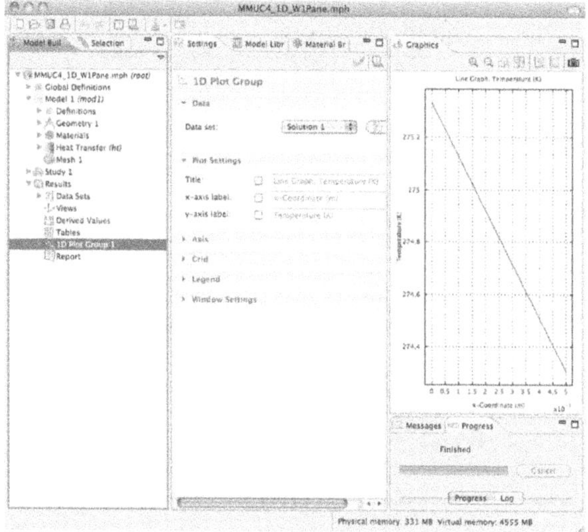

FIGURE 1.39 Initial Calculated Results for a 1 Pane Window.

Figure 1.39 shows the Initial Desktop Display window of the calculated results for a 1 Pane Window.

The 1D Plot displayed in the Desktop Display Graphics window shows the temperature in degrees Kelvin as a function of the distance from the inside surface of the Silica glass of the 1 Pane Window to the outside surface.

Since the modeler originally specified the initial input parameters in degrees Fahrenheit, the modeler now needs to adjust the instructions of 4.x so that the Graphics window will display the results of this modeling calculation in the desired units.

Under Results, Click > 1D Plot Group 1 twistie.

Click > Line Graph 1.

In the Settings – Line Graph window,
Under Y-Axis Data, Click > Unit Pull-down menu > Select > degF >
Click > Plot. See Figure 1.40.

NOTE *The Plot button is one of the symbols located at the upper right corner of the Settings window. The Plot symbol appears next to the Help Button (Computer Screen) symbol.*

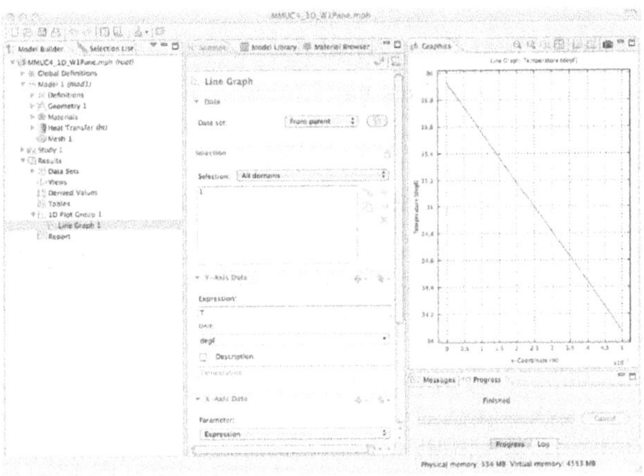

FIGURE 1.40 Calculated Results for a 1 Pane Window in Degrees Fahrenheit.

Figure 1.40 shows the Desktop Display Graphics window with the calculated results for a 1 Pane Window in degrees Fahrenheit.

1D 1 Pane Window Model Analysis and Conclusions

Table 1.2 shows a summary of the calculated model results at critical analytical points. The modeler should notice that this calculation shows that for a 1 Pane Window, the temperature drop across the Pane is approximately 2 degrees Fahrenheit. It also shows that the Interior Surface will be perceptibly cold to the touch. The 1 Pane Window will, due to the temperature difference between the Interior Window Pane surface and the room air, cause condensation of the water vapor in the room on the surface.

TABLE 1.2 1 Pane Window Calculation Results

Name	Coordinate Location (x)	Temperature
Interior Surface	0.0e-3[m]	≈35.926 °F
Midpoint	2.5e-3[m]	≈35.000 °F
Exterior Surface	5.0e-3[m]	≈34.074 °F
Interior ΔT	0.0e-3[m]	≈34.074 °F
Exterior ΔT	5.0e-3[m]	≈34.074 °F
Window ΔT	0.0–5.0e-3[m]	≈1.852 °F
Room ΔT	0.0 to -∞ [m]	≈34.074 °F
Exterior ΔT	5.0e-3 to ∞ [m]	≈34.074 °F

1D 2 Pane Window Heat Flow Model

Now that the modeler has had an initial experience in model development, it is time to consider, for example, how the addition of a second Pane in series with the first Pane will alter the heat flow through the window, given the same environmental conditions. That comparison is relatively simple to implement in a 1D heat flow model. The following model is similar to the first model, but sufficiently different that the modeler should build it with care.

Run 4.x > Select 1D in the Model Wizard.

Click > Next (right pointing arrow).

Select > Heat Transfer in Solids *(ht)* in the Add Physics window.

Click > Add Selected (plus sign).

Click > Next (right pointing arrow).

Select Preset Studies > Stationary.

Click > Finish Flag.

Figure 1.41 shows the initial 4.x model build for the 1D 2 Pane Window Heat Transfer Desktop Display.

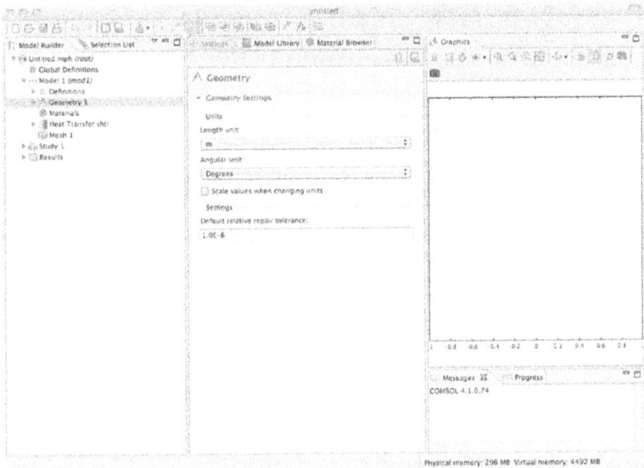

FIGURE 1.41 Initial 4.x Model Build for the 1D 2 Pane Window Heat Transfer.

Save the Model as > MMUC4_1D_W2Pane.mph. See Figure 1.42.

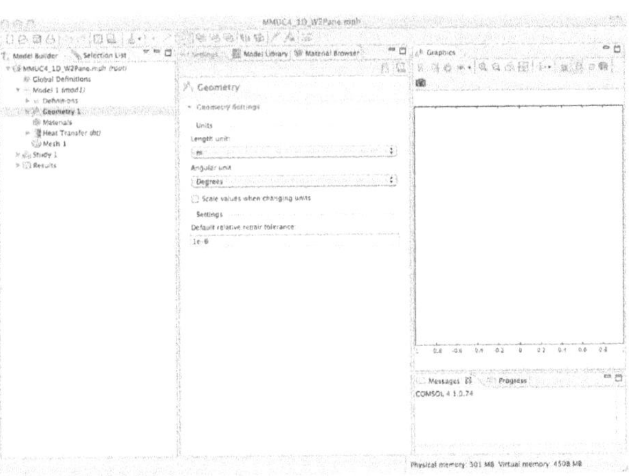

FIGURE 1.42 Saved Initial Model Build MMUC4_1D_W2Pane.mph.

Figure 1.42 shows the saved initial model build for MMUC4_1D_W2Pane.mph.

Now that the model has been saved, the modeler needs to enter the constant values needed for this model.

In the Model Builder window of the Desktop Display,

Right-Click > Global Definitions > Select Parameters.

Once Parameters has been selected, the Desktop Display will be modified to incorporate a Settings window. The Settings window contains a Parameters Entry Table where the following attributes can be entered for each Parameter: Name, Expression, Value, and Description.

Parameter information can be entered directly into the fields in the Parameters Entry Table or into the parameter entry fields below the table. Control buttons are located between the Parameters Entry Table window and the parameter entry fields.

To enter the parameters, Click on the first entry window in the Name column and enter (type) each piece of the information, as shown in Table 1.3.

TABLE 1.3 1D 2 Pane Window Parameters

Name	**Expression**	**Description**
T_in	70[degF]	Interior Temperature
T_out	0[degF]	Exterior Temperature
h_sg	15[W/(m^2*K)]	Heat Transfer Coefficient
p	1[atm]	Air Pressure

These entries in the Parameters Table define the interior temperature, the exterior temperature, the heat transfer coefficient, and the air pressure for use in this model.

Click on the Save to File icon (Disk), the rightmost symbol in the control button array, and enter MMUC4_1D_W2Pane_Pram.txt to save these parameters for future use.

The next step in building a model of a 2 Pane Window is to define the geometry through which the heat will flow from the heated interior to the less heated exterior.

Right-Click in the Model Builder window on Geometry 1 under Model 1 *(mod1)*.

Select > Interval.

Enter 0.0 as the Left endpoint.

Enter 5e-3[m] as the Right endpoint.

Click > Build Selected.

Right-Click in the Model Builder window on Geometry 1 under Model 1 *(mod1)*.

Select > Interval.

Enter 5e-3[m] as the Left endpoint.

Enter 20e-3[m] as the Right endpoint.

Click > Build Selected.

Right-Click in the Model Builder window on Geometry 1 under Model 1 *(mod1)*.

Select > Interval.

Enter 20e-3[m] as the Left endpoint.

Enter 25e-3[m] as the Right endpoint.

Click > Build All.

Click > Zoom Extents (the square icon with green arrows in the Graphics window).

See Figure 1.43.

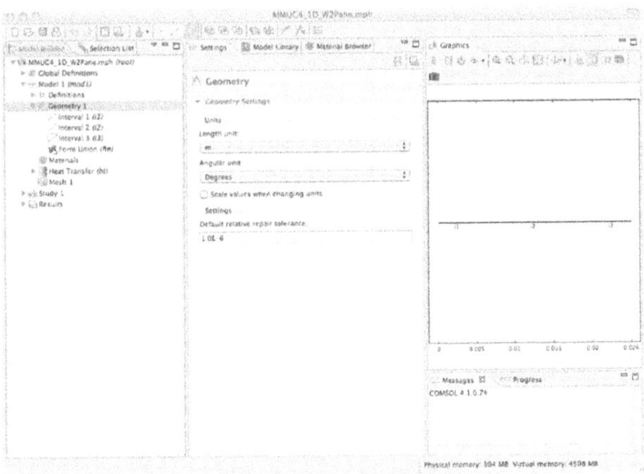

FIGURE 1.43 Desktop Display Window with the 2 Pane Built Geometry.

Figure 1.43 shows the Desktop Display window with the 2 Pane Built Geometry.

Now that the 2 Pane Geometry has been built, the next step is to define the materials comprising the model. In this case, it is relatively easy because the model only needs to use two (2) materials. Also, the material properties of those materials are resident in the Built-In Materials Library.

NOTE

The 2 Pane Window Model physically comprises 2 Panes of Silica glass separated by a "Dead Air Space." The dimensions of the "Dead Air Space" are such that convection currents are suppressed. Thus, thermal conduction plays the major role in the conduction of heat from the interior surface of the Window to the exterior surface of the Window.

Right-Click in the Model Builder window on Materials under Model 1 *(mod1)*.

Select > Open Materials Browser.

Click the twistie pointing at Built-In in the Material Browser window.

Scroll the Materials List until Silica glass appears.

Right-Click > Silica glass.

Select > Add Material to Model. See Figure 1.44.

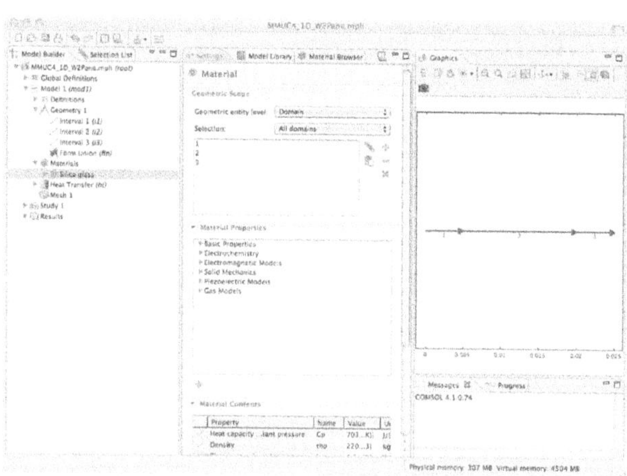

FIGURE 1.44 Silica Glass Material Added to Model.

Figure 1.44 shows the Silica glass material has been added to the model.

NOTE

Because this model is 1D and there are only Domain and Boundary in this model, 4.x automatically assigns the Silica glass material's properties to all the available domains (1, 2, 3).

In this case, the materials assignment of Silica glass to Domain 2 is not correct and must be adjusted.

Click > 2 in the Settings – Material – Geometric Scope Selection window.

Click > Minus (–) button in the rightmost vertical control button array. See Figure 1.45.

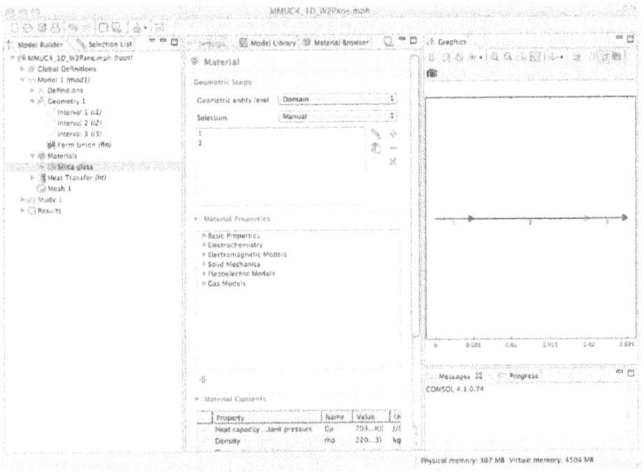

FIGURE 1.45 Silica Glass Material Removed from Domain 2.

Figure 1.45 shows that the Silica glass material has been removed from Domain 2.

The next step is to add the material Air to the model.

Right-Click in the Model Builder window on Materials under Model 1 *(mod1)*.

Select > Open Materials Browser.

Click the twistie pointing at Built-In in the Material Browser window.

Scroll the materials list until Air appears.

Right-Click > Air.

Select > Add Material to Model. See Figure 1.46.

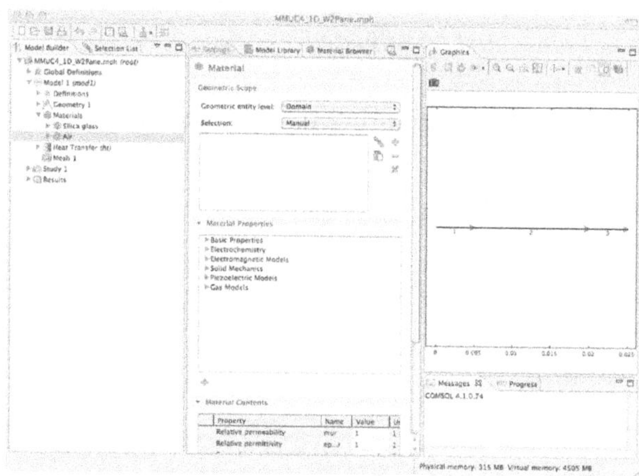

FIGURE 1.46 Material Air Added to Model.

Figure 1.46 shows the material Air has been added to the model.

NOTE

In this case, since there are multiple Domains, 4.x does not know where to assign the material Air. Thus the material Air is left unassigned.

The materials assignment of Air to Domain 2 must be performed manually.

In the Graphics window, Click > Domain 2.

In the Settings – Material – Geometric Scope Selection window.

Click > Plus (+) button in the rightmost vertical control button array. See Figure 1.47.

Figure 1.47 shows the material Air has been added to Domain 2.

NOTE

The modeler should verify that the material selected has the necessary physical properties to allow for the solution of the problem in question. In this case, 4.x has determined which physical properties are needed for the current model and displays a green check mark next to the needed properties in the Material Contents window of the Settings - Material window. See Figure 1.48.

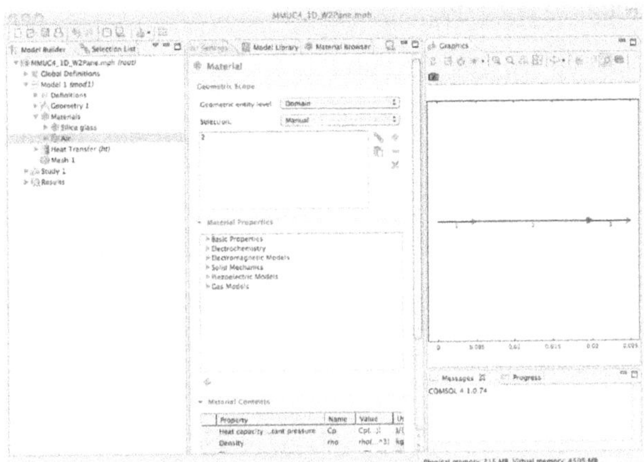

FIGURE 1.47 Material Air Added to Domain 2.

▾ Material Contents

Property	Name	Value	Ur
Heat capacity ...tant pressure	Cp	Cp(...)]	J/(
Density	rho	rho(...^3]	kg
Thermal conductivity	k	k(T[...K)]	W,
Relative permeability	mur	1	1
Relative permittivity	ep...r	1	1
Dynamic viscosity	mu	eta...*s]	Pa
Ratio of specific heats	gamma	1.4	1
Electric conductivity	sigma	0[S/m]	S/
Speed of sound	c	cs(.../s]	m

FIGURE 1.48 Material Properties of Air in Domain 2 Verified.

Figure 1.48 shows the material properties of Air in Domain 2 have been verified.

Since the model geometry has been built and the appropriate material properties have been assigned to the Domains 1, 2, and 3, the next step in the modeling process is to specify both the Domain and the Boundary conditions for this model.

Thus, the modeler needs to utilize the previously entered parameters that are needed to calculate the transfer of heat through this 2 Pane Window.

Right-Click > Heat Transfer *(ht)*.

Select > Heat Flux.

In the Graphics window, Click > Point 1 (leftmost boundary).

In the Settings – Heat Flux window, Click > Plus (Add to Selection).

In the Settings – Heat Flux window, Click > Inward Heat Flux.

Type h_sg in the Heat transfer coefficient entry window.

Type T_in in the External temperature entry window.

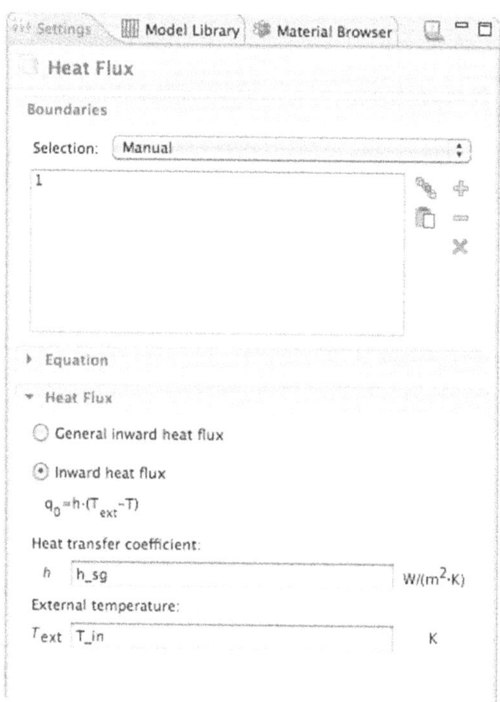

FIGURE 1.49 Filled Boundary Point 1 Heat Flux Window.

Figure 1.49 shows the Settings – Heat Flux window as finally configured for Boundary Point 1.

Right-Click > Heat Transfer *(ht)*.

Select > Heat Flux.

In the Graphics window, Click > Point 4 (rightmost boundary).

In the Settings – Heat Flux window, Click > Plus (Add to Selection).

In the Settings – Heat Flux window, Click > Inward Heat Flux.

Type h_sg in the Heat transfer coefficient entry window.

Type T_out in the External temperature entry window.

Figure 1.50 shows the Settings – Heat Flux window as finally configured for Boundary Point 4.

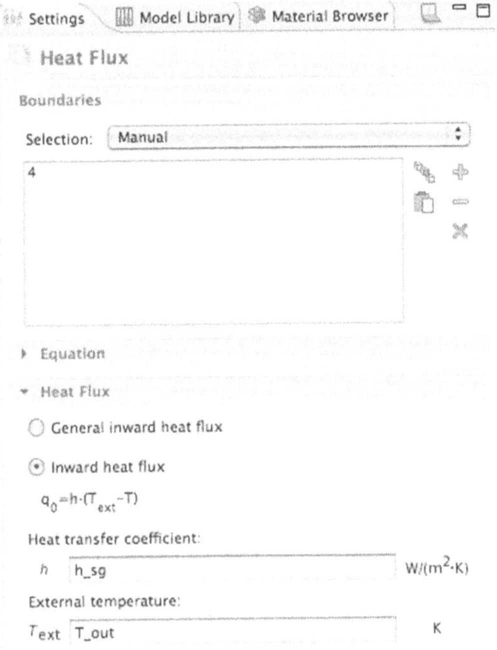

FIGURE 1.50 Filled Boundary Point 4 Heat Flux Window.

Since this first model is relatively simple, the easiest path is to allow 4.x to automatically mesh the model.

Right-Click > Mesh 1.

Select > Build All.

The meshing results are shown in Figure 1.51.

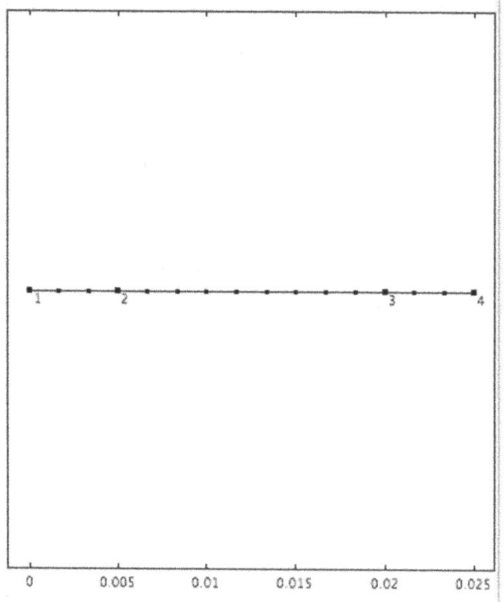

FIGURE 1.51 Meshed 1D Domain.

Figure 1.51 shows the meshed 1D domain.

Now that the model has been meshed, the modeler can proceed to compute the solution.

Right-Click > Study 1.

Select > Compute.

4.x automatically selects the appropriate Solver for this problem, computes the solution and displays the results in the Graphics window.

The results of this modeling computation are shown in Figure 1.52.

Figure 1.52 shows the initial Desktop Display window of the calculated results for a 2 Pane Window.

The 1D Plot displayed in the Desktop Display Graphics window shows the temperature in degrees Kelvin as a function of the distance from the inside surface of Silica glass of the First Pane of the 2 Pane Window to the outside surface of Silica glass of the second Pane of the 2 Pane Window.

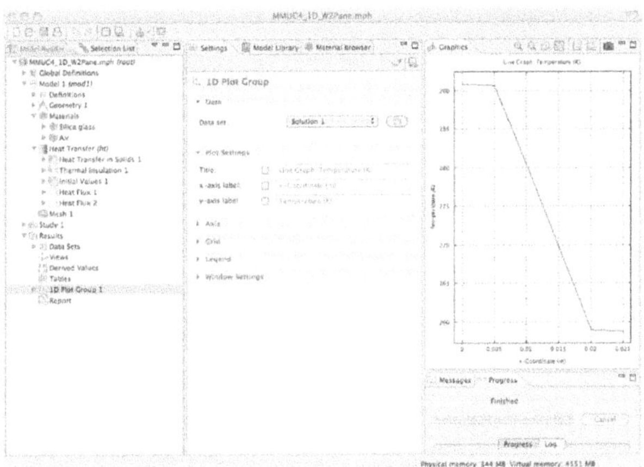

FIGURE 1.52 Initial Calculated Results for a 2 Pane Window.

Since the modeler originally specified the initial input parameters in degrees Fahrenheit, the modeler now needs to adjust the instructions of 4.x so that the Graphics window will display the results of this modeling calculation in the desired units.

Under Results, Click > 1D Plot Group 1 twistie.

Click > Line Graph 1.

In the Settings – Line Graph window,

Under Y-Axis Data, Click > Unit Pull-down menu > Select > degF.

Click > Plot.

Figure 1.53 shows the Desktop Display window of the calculated results for a 2 Pane Window in degrees Fahrenheit.

1D 2 Pane Window Model Analysis and Conclusions

Table 1.4 shows a summary of the calculated model results at critical analytical points. The modeler should notice this calculation shows that for a 2 Pane Window, the temperature drop across the 2 Panes in this configuration is approximately 58 degrees Fahrenheit. It also shows that the interior surface will be perceptibly cool but not cold to the touch. The 2 Pane Window will, due to the small temperature difference between the interior Window Pane surface and the room air, not cause condensation of the water vapor in the room on the surface, except under highly saturated (high relative humidity) conditions.

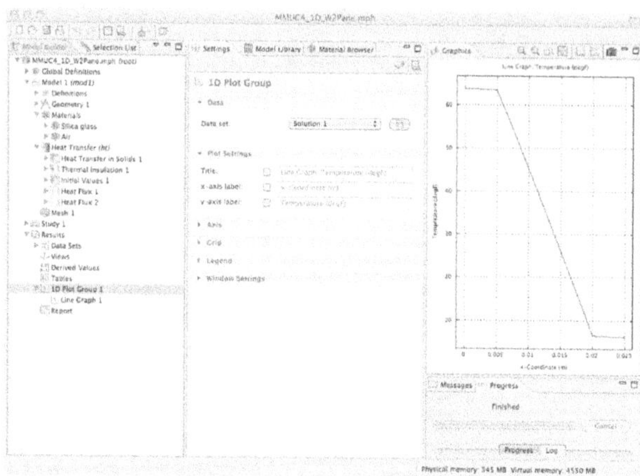

FIGURE 1.53 Calculated Results for a 2 Pane Window in Degrees Fahrenheit.

TABLE 1.4 2 Pane Window Calculation Results

Name	Coordinate Location (x)	Temperature
Interior Surface	0.0e-3[m]	≈63.84 °F
Midsurface 1	5.0e-3[m]	≈63.50 °F
Midsurface 2	20.0e-3[m]	≈6.50 °F
Exterior Surface	25.0e-3[m]	≈6.16 °F
Midpoint	12.5e-3[m]	≈35.000 °F
Interior Pane DT	0.0e-3[m]	≈0.34 °F
Exterior Pane DT	5.0e-3[m]	≈0.34 °F
Window DT	0.0–5.0e-3[m]	≈57.68 °F
Room DT	0.0 to -∞ [m]	≈6.16 °F
Exterior DT	5.0e-3 to ∞ [m]	≈6.16 °F

1D 3 Pane Window Heat Flow Model

Now that the modeler has had some experience in model development, it is time to consider, for example, how the addition of a third Pane in series with the first two Panes will alter the heat flow through the window, given the same environmental conditions. That comparison is relatively simple to implement in a 1D heat flow model. The following model is similar to the first and second models, but sufficiently different that the modeler should build it with care.

Run 4.x > Select 1D in the Model Wizard.

Click > Next (right pointing arrow).

Select > Heat Transfer in Solids *(ht)* in the Add Physics window.

Click > Add Selected (plus sign).

Click > Next (right pointing arrow).

Select > Preset Studies > Stationary.

Click > Finish Flag.

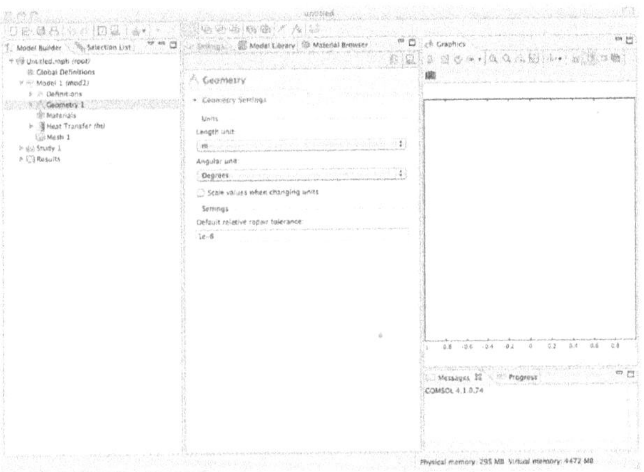

FIGURE 1.54 Initial 4.x Model Build for the 1D 3 Pane Window Heat Transfer.

Figure 1.54 shows the initial 4.x model build for the 1D 3 Pane Window Heat Transfer Desktop Display.

Save the Model as > MMUC4_1D_W3Pane.mph. See Figure 1.55.

Figure 1.55 shows the Saved Initial Model Build MMUC4_1D_W3Pane.mph.

Now that the model has been saved, the modeler needs to enter the constant values needed for this model.

In the Model Builder window of the Desktop Display,

Right-Click > Global Definitions > Select Parameters.

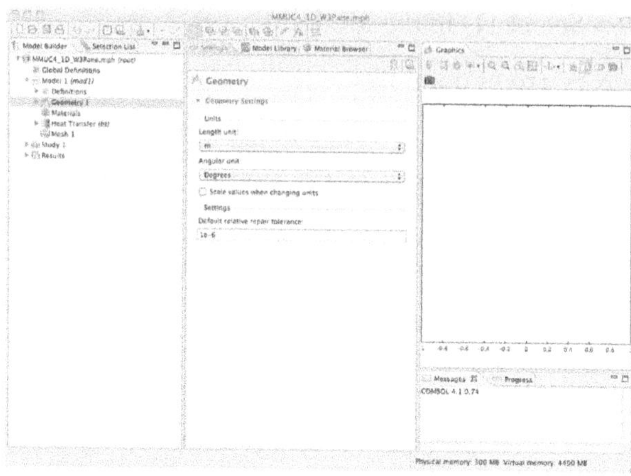

FIGURE 1.55 Saved Initial Model Build MMUC4_1D_W3Pane.mph.

To enter the parameters, Click on the first entry window in the Name column and enter (type) each piece of the information, as shown in Table 1.5.

TABLE 1.5 1D 2 Pane Window Parameters

Name	Expression	Description
T_in	70[degF]	Interior Temperature
T_out	0[degF]	Exterior Temperature
h_sg	15[W/(m^2*K)]	Heat Transfer Coefficient
p	1[atm]	Air Pressure

These entries in the Parameters Entry Table define the interior temperature, the exterior temperature, the heat transfer coefficient, and the air pressure for use in this model.

Click on the Save to File (Disk) icon, the rightmost symbol in the control button array, and enter MMUC4_1D_W3Pane_Pram.txt to save these parameters for future use.

The next step in building a model of a 3 Pane Window is to define the geometry through which the heat will flow from the heated interior to the less heated exterior.

Right-Click in the Model Builder window on Geometry 1 under Model 1 *(mod1)*.

Select > Interval.

Enter 0.0 as the Left endpoint.

Enter 5e-3[m] as the Right endpoint.

Click > Build Selected.

Right-Click in the Model Builder window on Geometry 1 under Model 1 *(mod1)*.

Select > Interval.

Enter 5e-3[m] as the Left endpoint.

Enter 20e-3[m] as the Right endpoint.

Click > Build Selected.

FIGURE 1.56 Zoom Extents Button.

Click > Zoom Extents. See Figure 1.56.

Figure 1.56 shows the Zoom Extents button, which changes the dimensions of the axes in the Graphics window to reflect the value range of the geometry.

Right-Click in the Model Builder window on Geometry 1 under Model 1 *(mod1)*.

Select > Interval.

Enter 20e-3[m] as the Left endpoint.

Enter 25e-3[m] as the Right endpoint.

Click > Build Selected.

Click > Zoom Extents.

Right-Click in the Model Builder window on Geometry 1 under Model 1 *(mod1)*.

Select > Interval.

Enter 25e-3[m] as the Left endpoint.

Enter 40e-3[m] as the Right endpoint.

Click > Build Selected.

Click > Zoom Extents.

Right-Click in the Model Builder window on Geometry 1 under Model 1 *(mod1)*.

Select > Interval.

Enter 40e-3[m] as the Left endpoint.

Enter 45e-3[m] as the Right endpoint.

Click > Build All.

Click > Zoom Extents.

Now that the 3 Pane Geometry has been built, the next step is to define the materials comprising the Model. In this case, it is relatively easy because the model only needs to use two (2) materials. Also, the material properties of those materials are resident in the Built-In Materials Library.

NOTE

The 3 Pane Window Model physically comprises 3 Panes of Silica Glass separated by 2 "Dead Air Spaces." The dimensions of the "Dead Air Spaces" are such that convection currents are suppressed. Thus, thermal conduction plays the major role in the conduction of heat from the interior surface of the Window to the exterior surface of the Window.

Right-Click in the Model Builder window on Materials under Model 1 *(mod1)*.

Select > Open Materials Browser.

Click the twistie pointing at Built-In in the Material Browser window.

Scroll the materials list until Silica glass appears.

Right-Click > Silica glass.

Select > Add Material to Model. See Figure 1.57.

Figure 1.57 shows the Silica glass material has been added to the 3 Pane Model.

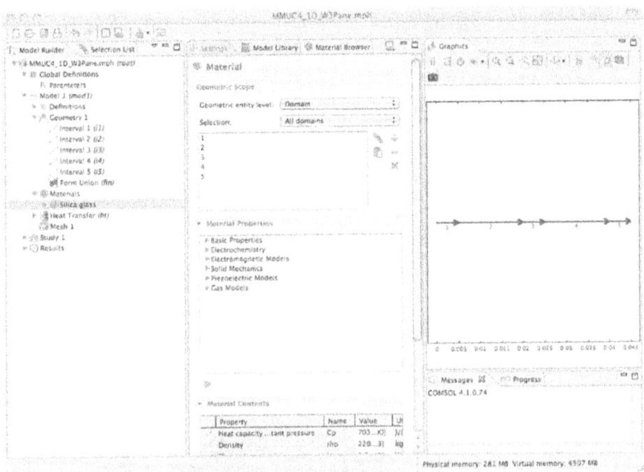

FIGURE 1.57 Silica Glass Material Added to 3 Pane Model.

NOTE *Since this model is 1D, there are 5 Domains and this is the first material added, 4.x automatically assigns the Silica glass material's properties to all the available domains (1, 2, 3, 4, 5).*

In this case, the materials assignment of Silica glass to Domains 2 & 4 is not correct and must be adjusted.

Click > 2 in the Settings – Material – Geometric Scope Selection window.

Click > Minus (–) button in the rightmost vertical control button array.

Click > 4 in the Settings – Material – Geometric Scope Selection window.

Click > Minus (–) button in the rightmost vertical control button array.

The next step is to add the material Air to the model.

Right-Click in the Model Builder window on Materials under Model 1 (*mod1*).

Select > Open Materials Browser.

Click the twistie pointing at Built-In in the Material Browser window.

Scroll the materials list until Air appears.

Right-Click > Air.

Select > Add Material to Model.

NOTE

In this case, since there are multiple Domains, 4.x does not know where to assign the material Air. Thus, the material Air is left unassigned.

In this case, the materials assignment of Air to Domains 2 & 4 must be performed manually.

In the Graphics window, Click > Domain 2.

In the Settings – Material – Geometric Scope Selection window.

Click > Plus (+) button in the rightmost vertical control button array.

In the Graphics window, Click > Domain 4.

In the Settings – Material – Geometric Scope Selection window.

Click > Plus (+) button in the rightmost vertical control button array.

Since the model geometry has been built and the appropriate material properties have been assigned to the Domains 1, 2, 3, 4, & 5, the next step in the modeling process is to specify both the domain and the boundary conditions for this model.

Thus, the modeler needs to utilize the previously entered parameters that are needed to calculate the transfer of heat through this 2 Pane Window.

Right-Click > Heat Transfer *(ht)*.

Select > Heat Flux.

In the Graphics window, Click > Point 1 (leftmost boundary).

In the Settings – Heat Flux window, Click > Plus (Add to Selection).

In the Settings – Heat Flux window, Click > Inward Heat Flux.

Type h_sg in the Heat transfer coefficient entry window.

Type T_in in the External temperature entry window.

Right-Click > Heat Transfer *(ht)*.

Select > Heat Flux.

In the Graphics window, Click > Point 6 (rightmost boundary).

In the Settings – Heat Flux window, Click > Plus (Add to Selection).

In the Settings – Heat Flux window, Click > Inward Heat Flux.

Type h_sg in the Heat transfer coefficient entry window.

Type T_out in the External temperature entry window.

Since this first model is relatively simple, the easiest path is to allow 4.x to automatically mesh the model.

Right-Click > Mesh 1.

Select > Build All.

The meshing results are shown in Figure 1.58.

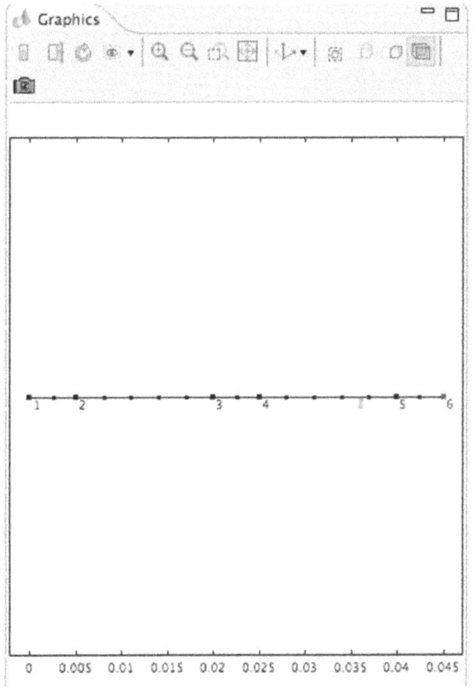

FIGURE 1.58 Meshed 1D 3 Pane, 5 Domain Window Model.

Figure 1.58 shows the meshed 1D 3 Pane, 5 Domain Window model.

Now that the model has been meshed, the modeler can proceed to compute the solution.

Right-Click > Study 1.

Select > Compute.

NOTE *4.x automatically selects the appropriate Solver for this problem, computes the solution, and displays the results in the Graphics window.*

The results of this modeling computation are shown in Figure 1.59.

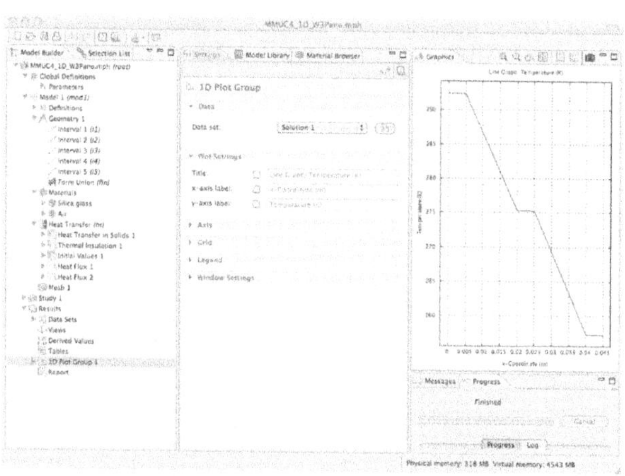

FIGURE 1.59 Initial Calculated Results for a 3 Pane Window.

Figure 1.59 shows the Desktop Display window of the initial calculated results for a 3 Pane Window.

The 1D Plot displayed in the Desktop Display Graphics window shows the temperature in degrees Kelvin as a function of the distance from the inside surface of Silica glass of the first Pane of the 2 Pane Window to the outside surface of Silica glass of the second Pane of the 2 Pane Window.

Since the modeler originally specified the initial input parameters in degrees Fahrenheit, the modeler now needs to adjust the instructions of 4.x so that the Graphics window will display the results of this modeling calculation in the desired units.

Under Results, Click > 1D Plot Group 1 twistie.

Click > Line Graph 1.

In the Settings – Line Graph window.

Under Y-Axis Data, Click > Unit Pull-down menu > Select > degF.

Click > Plot.

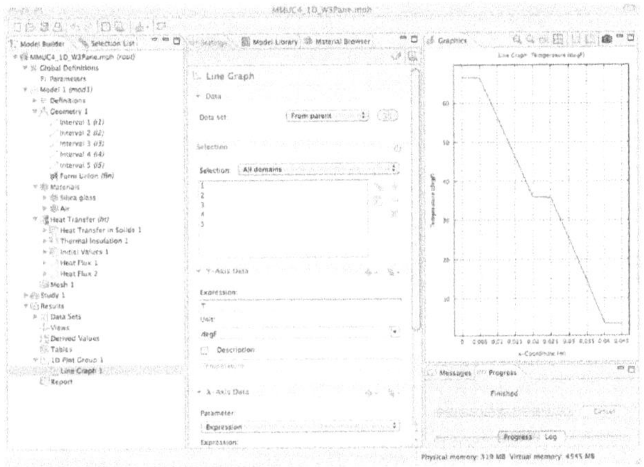

FIGURE 1.60 Calculated Results for a 3 Pane Window in Degrees Fahrenheit.

Figure 1.60 shows the Desktop Display window of the Calculated Results for a 3 Pane Window in degrees Fahrenheit.

1D 3 Pane Window Model Analysis and Conclusions

Table 1.6 shows a summary of the calculated model results at critical analytical points. The modeler should notice that this calculation shows that for a 3 Pane Window, the temperature drop across the 3 Panes in this configuration is approximately 63 degrees Fahrenheit. It also shows that the interior surface will be perceptibly cool but not cold to the touch. The 3 Pane Window will, due to the small temperature difference between the interior Window Pane surface and the room air, not cause condensation of the water vapor in the room on the surface, except under highly saturated (high relative humidity) conditions.

TABLE 1.6 3 Pane Window Calculation Results

Name	**Coordinate Location (x)**	**Temperature**
Interior Surface	0.0e-3[m]	≈66.62 °F
Midsurface 1	5.0e-3[m]	≈66.43 °F
Midsurface 2	20.0e-3[m]	≈35.96 °F
Midsurface 3	25.0e-3[m]	≈35.77 °F
Midsurface 4	40.0e-3[m]	≈3.57 °F
Exterior Surface	45.0e-3[m]	≈3.38 °F
Midpoint	22.5e-3[m]	≈35.87 °F
Interior Pane ΔT	0.0e-3[m]	≈0.19 °F
Middle Pane ΔT	20.0e-3[m]	≈0.19 °F
Exterior Pane ΔT	40.0e-3[m]	≈0.19 °F
Window ΔT	0.0-45.0e-3[m]	≈63.24 °F
Room ΔT	0.0 to -∞ [m]	≈3.38 °F
Exterior ΔT	45.0e-3 to ∞ [m]	≈3.38 °F

1D Window Model 1 Pane, 2 Pane, and 3 Pane Analysis and Conclusions

Table 1.7 shows a comparison of the 1, 2, and 3 pane differential temperatures. It can be readily seen that the 2 Pane Window adds significant advantage in heat loss reduction and reduces condensation (fogging) on the window. The incremental advantage compared to the added cost of adding the third Pane is probably not a good relative investment. Other factors would need to be considered to derive significant perceived benefit.

TABLE 1.7 All Window Calculation Results

Name	**Window Δ Temperature**
1 Pane	≈1.85 °F
2 Pane	≈57.68 °F
3 Pane	≈63.24 °F

FIRST PRINCIPLES AS APPLIED TO MODEL DEFINITION

First Principles Analysis derives from the fundamental laws of nature. In the case of models in this book or from any other source, the modeler should be able to demonstrate that the calculated results derived from the models are consistent with the laws of physics and the basic observed properties of materials. In the case of this Classical Physics Analysis, the laws of conservation in physics require that what goes in (as mass, energy, charge, etc.) must come out (as mass, energy, charge, etc.) or must accumulate within the boundaries of the model. To do otherwise violates fundamental principles.

NOTE *In the COMSOL Multiphysics software, the default interior boundary conditions are set to apply the conditions of continuity in the absence of sources (e.g. heat generation, charge generation, molecule generation, etc.) or sinks (e.g. heat loss, charge recombination, molecule loss, etc.).*

The careful modeler must be able to determine by inspection of the model that the appropriate factors have been considered in the development of the specifications for the various geometries, for the material properties of each domain and for the boundary conditions. He must also be knowledgeable of the implicit assumptions and default specifications that are normally incorporated into the COMSOL Multiphysics software model, when a model is built using the default settings.

Consider, for example, the three Window models developed earlier in this chapter. By choosing to develop those models in the simplest 1D geometrical mode, the implicit assumption is the heat flow occurs in only one direction. That direction is basically normal to the surface of the window and from the high temperature (interior temperature) to the low temperature (exterior temperature). That assumption essentially eliminates the consideration of heat flow along other paths. It also assumes the materials are homogeneous and isotropic and there are no thin thermal barriers at the surfaces of the panes. None of these assumptions are typically true in the general case. However, by making such assumptions, it is possible to easily build a 1D First Approximation Model.

NOTE

A First Approximation Model is one that captures all the essential features of the problem that needs to be solved, without dwelling excessively on small details. A good First Approximation Model will yield an answer that enables the modeler to determine if he needs to invest the time and the resources required to build a more highly detailed model.

SOME COMMON SOURCES OF MODELING ERRORS

There are four primary sources of modeling errors: insufficient preparation, insufficient attention to detail, insufficient understanding of the basic principles required for the creation of an adequate model, and the lack of a comprehensive understanding of what is required to define an adequate model to fulfill the needs.

Primarily, the most frequent modeling errors are those that result from the modeler exercising insufficient attention to either the development of the details of the model or the incorporation into the model of conceptual errors and/or the generation of keying errors during data/parameter/formula entry.

NOTE

One major source of errors occurs during the process of naming variables. The modeler should be careful to NEVER GIVE THE SAME NAME TO HIS VARIABLES AS COMSOL GIVES TO THE DEFAULT VARIABLES. COMSOL Multiphysics software seeks a value for the designated variable everywhere within its operating domain.

If two or more variables have the same name, an error is created. Also, it is best to avoid human errors by using uniquely distinguishable characters in variable names. For example, avoid using the lower case L, the number 1, and the upper case I, which in some fonts are relatively indistinguishable. Similarly, avoid the upper case O and the number 0. Give your variables meaningful names (T_in, T_out, T_hot, etc.). Also, variable names are case-sensitive, i.e. T_in is not the same as T_IN.

The first rule in model development is to clearly understand the exact problem to be solved and to specify in detail what aspects of the problem the model will address. The specification of the nature of the problem should include a list of the magnitude of the relative contributions from the physical properties vital to the functioning of the anticipated model and their relative degree of interaction. Always build the model to solve for the minimum desired information. It will be faster and the model can always be expanded later.

NOTE *Examples of typical physical properties that are potentially/probably coupled in any developed model are heat and geometrical expansion/contraction (liquid, gas, solid), current flow and heat generation/reduction, phase change and geometrical expansion/contraction (liquid, gas, solid) and/or heat generation/reduction, chemical reactions, etc. Be sure to investigate your problem and understand what you are building first, then build your model carefully.*

Having done the primary analysis and written a hierarchical list, the modeler should then estimate the best physical, least coupled, lowest dimensionality modeling approach to achieve the most meaningful First Approximation Model.

REFERENCES

1.1 Blaise Barney, Lawrence Livermore National Laboratory, Introduction to Parallel Computing, https://computing.llnl.gov/tutorials/parallel_comp/

1.2 http://www.apple.com/macpro/

1.3 COMSOL Desktop Environment, COMSOL Multiphysics Users Guide, pp. 52

1.4 COMSOL Multiphysics Users Guide, pp. 151

1.5 http://www-03.ibm.com/support/techdocs/atsmastr.nsf/Web/TwHelp

1.6 COMSOL Multiphysics Users Guide, pp. 171

1.7 F. P. Incropera and D. P. Dewitt, *Fundamentals of Heat and Mass Transfer*, Fifth Edition, John Wiley & Sons, Hoboken, NJ, 2002, ISBN 0-471-38650-2, p. 8

1.8 D. W. Pepper and J. C. Heinrich, *The Finite Element Method*, Second Edition, Taylor & Francis, New York, 2006, ISBN 1-59169-027-7, p. 1

1.9 COMSOL Multiphysics Users Guide, pp. 608

Suggested Modeling Exercises

1. Build, mesh, and solve the 1D 1 Pane Window problem presented earlier in this chapter.
2. Build, mesh, and solve the 1D 2 Pane Window problem presented earlier in this chapter.
3. Build, mesh, and solve the 1D 3 Pane Window problem presented earlier in this chapter.
4. Change the material comprising the Panes, then build, mesh, and solve the problems. Analyze, compare and contrast the results with the new material to the results of Problems 1, 2, and 3.
5. Change the material (Gas) between the Panes, then build, mesh, and solve the problems. Analyze, compare and contrast the results with the new material to the results of Problems 1, 2, and 3.

CHAPTER 2

MATERIALS PROPERTIES USING COMSOL MULTI-PHYSICS 4.X

In This Chapter

- Materials Properties Guidelines and Considerations
- COMSOL Materials Properties Sources
- Other Materials Properties Sources
- Material Property Entry Techniques
- Multi-Pane Window Model
- References

MATERIALS PROPERTIES GUIDELINES AND CONSIDERATIONS

The selection of materials for a given device or process is crucial to the success of the model for that device or process. When that model reflects the content and behavior of the actual device or process, then the conclusions drawn from the model will be most reliable. Once the first approximation of the device or process has been defined, begin choosing the materials.

NOTE *The modeler does not need to describe all of the properties of the material. Once the essential physical functions have been determined, only those properties that are used by the model's physics interfaces need to be supplied.*

A large number of materials and their properties are available, as general searches of the Internet will demonstrate. The modeler will need to exercise some selection criteria based on practical considerations to narrow the material choices to a manageable number.

NOTE *The search for reliable and accurate material property values may encounter one or more of the following issues: 1) a source may not have all of the properties needed; 2) a source may supply the value in an unconventional, or at least not the desired, unit of measure, requiring conversion; 3) once in the desired units, the property value may prove to be incorrect based on a knowledge of values for other almost identical materials; and 4) errors may occur during transcription from the source into the model. Care must be taken to verify all property values before running the model.*

This chapter will discuss the two COMSOL material property libraries available in 4.x. This chapter also covers other sources of property values, how to introduce those values into a model, and how to create your own User Defined Library of materials properties with 4.x.

COMSOL MATERIALS PROPERTIES SOURCES

COMSOL provides two sources of materials properties: a basic library that is included in the COMSOL Multiphysics software, and a Material Library module that provides many more materials and referenced property functions, related to some variable, such as temperature {2.1}. Access to both of these sources is through the Material Browser that was used in the models in Chapter 1. Use the search engine or browse to determine whether the materials needed for the model are already available in the Materials Library.

Run 4.x > Select the Material Browser tab to display the contents of the Materials Library.

In the Materials entry window, Enter > steel in the Material Browser.

NOTE *The search term may be the name of a material, or in the case of the Material Library module, a UNS number {2.2} or a DIN number {2.3}.*

Click > Search.

The results of a successful search will be displayed in the Materials section of the Material Browser window. See Figure 2.1.

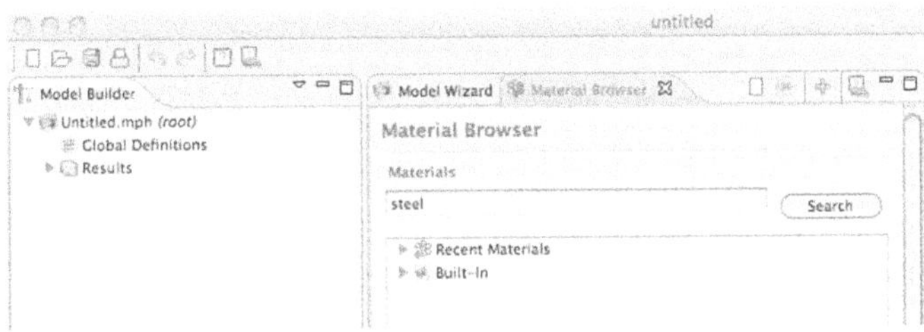

FIGURE 2.1 Material Browser Search Results.

Figure 2.1 shows the results of a Material Browser search.

To view the properties available for one of the selected materials,

Click > Built-In twistie.

Select > Structural steel.

In the Material Browser – Materials, Click > Properties twistie.

A list of properties and their values are displayed. See Figure 2.2.

Figure 2.2 shows the list of available properties for the selected material, Structural steel.

OTHER MATERIALS PROPERTIES SOURCES

Materials and their properties are available from many sources, either in printed form or on the Internet.

Manufacturers provide such information for the products in their catalogs.

Some professional societies, such as ASM International {2.4}, provide subscriptions to handbooks of materials information, including material properties.

Online encyclopedias, such as Wikipedia {2.5}, may contain data for some of the more common materials.

Specialized sites, such as MatWeb {2.6}, provide searchable online databases of materials properties, which may be downloadable in a format compatible with particular versions of modeling software.

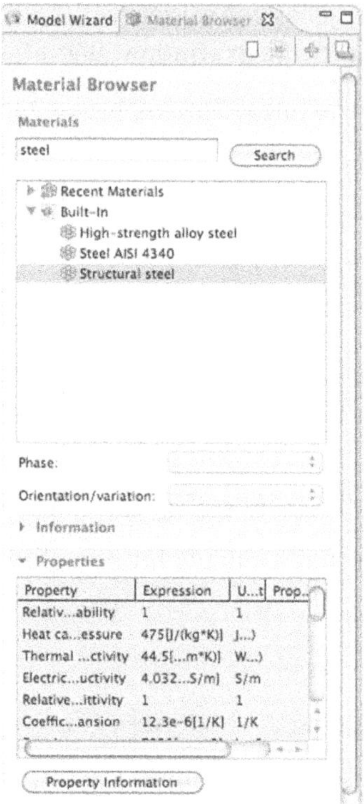

FIGURE 2.2 Material Properties List.

Some search engines, both general purpose Internet tools and specialized tools designed to be used with a particular Website, will allow the modeler to search by a property value range or material type, thus assisting the modeler in finding materials that fit the design criteria.

MATERIAL PROPERTY ENTRY TECHNIQUES

The modeler enters the materials properties data into the model using one of the three techniques: building a material property entry, setting user defined parameters, or entering user defined property values directly.

This chapter illustrates the three techniques by means of a model similar to the 1D 3 Pane model in Chapter 1, using the properties of materials not found in the basic COMSOL Multiphysics materials library.

Multi-Pane Window Model

NOTE

The Multipane Window Model physically comprises 3 panes of Vycor silica glass separated by 2 "Dead Air Spaces," one containing argon and the other krypton. The dimensions of the "Dead Air Spaces" are such that convection currents are suppressed. Thus, thermal conduction plays the major role in the conduction of heat from the interior surface of the Window to the exterior surface of the Window.

Run 4.x > Select 1D in the Model Wizard – Select Space Dimension window.

To display the list of available physics interfaces, Click > Next (right pointing arrow).

In the Add Physics window, Click > Heat Transfer twistie.

Select > Heat Transfer in Solids (ht).

Click > Add Selected button (plus sign) below the physics interface list.

The Heat Transfer (ht) selection will be displayed in the Selected physics section of the Add Physics window (see Figure 2.3).

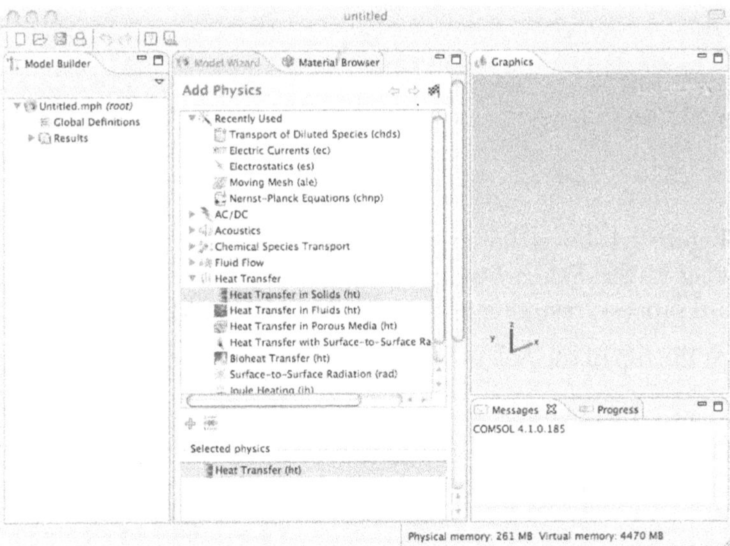

FIGURE 2.3 Model Wizard – Add Physics Window.

Figure 2.3 shows the Model Wizard – Add Physics window.

In the Add Physics window, Click > Next (right pointing arrow).

Select > Preset Studies > Stationary (see Figure 2.4).

NOTE *The modeler should note that "Stationary" and "Steady State" are equivalent terms.*

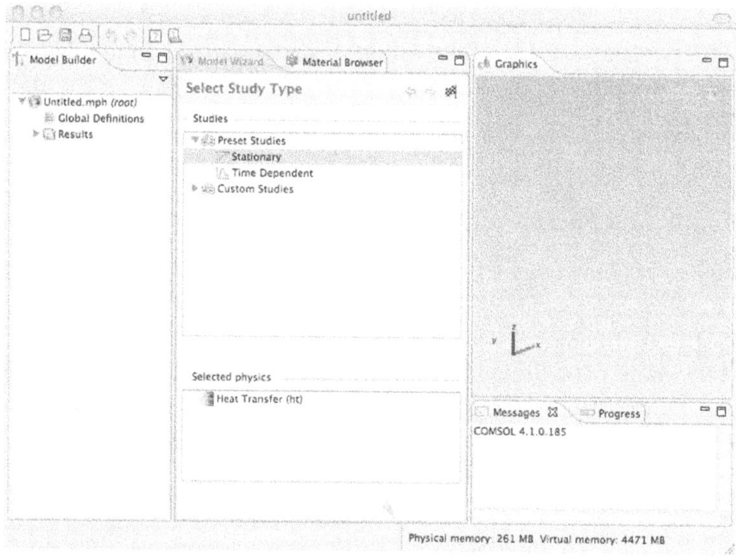

FIGURE 2.4 The Select Study Type Window.

Figure 2.4 shows the selection of the Stationary study type and the Finish Flag in the Select Study Type window for the completion of Model Wizard process, resulting in the Initial Model Build.

At the top of the Select Study Type window, Click > Finish Flag.

Click > File menu > Save as… to save the Model as > MMUC4_1D_Multipane.mph. See Figure 2.5.

Figure 2.5 shows the saved Initial Model Build of the 1D_Multipane model.

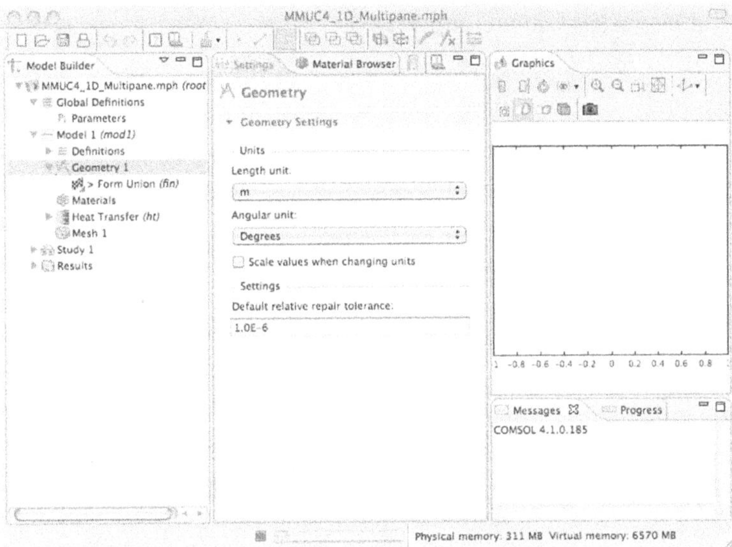

FIGURE 2.5 Saved Initial Model Build of the 1D_Multipane Model.

Parameter Setting

One way to enter a material property into a model is to provide it as a parameter, along with the other constant values needed for a model. For example, in this model the properties for the Vycor silica glass to be used in the panes will be entered using this technique. This approach allows the modeler to save the property along with any other constants needed for the model in one file, but it does mean that the property parameter names will need to be entered manually wherever they will be used in the model.

NOTE *Care must be taken to ensure that the name assigned to the material property parameter is unique to the model and is not one of the parameter names already in use by 4.x. An error message will warn the modeler if the name is already defined when entry is attempted.*

Now that the Initial Model Build has been saved, the modeler needs to enter the constant values needed for this model.

In the Model Builder window of the Desktop Display,

Right-Click > Global Definitions.

Select > Parameters. This action will display the Settings – Parameters window.

To enter the parameters for the model,

Click on the first entry window in the Name column and enter (type) each piece of the information, as shown in Table 2.1.

NOTE

When entering information into a Parameters field, first Click on the field desired and then observe the presence of a blinking cursor. The blinking cursor verifies that 4.x is ready to accept the information. Then type the desired piece of information into the activated field.

The modeler can transition to the next information entry point in the tables by clicking on the next field.

TABLE 2.1 Multi-Pane Window Parameters

Name	Expression	Description
T_in	70[degF]	Interior Temperature
T_out	0[degF]	Exterior Temperature
h_sg	15[W/(m^2*K)]	Heat Transfer Coefficient
p	1[atm]	Air Pressure
kVycor	1.38[W/(m*K)]	Thermal Conductivity of Vycor
CVycor	753.624[J/(kg*K)]	Heat Capacity of Vycor
rhoVycor	2180[kg/m^3]	Density of Vycor

These entries in the Parameters Entry Table define the interior temperature, the exterior temperature, the heat transfer coefficient, the air pressure, and the thermal conductivity, heat capacity, and density of the Vycor glass for use in this model. See Figure 2.6.

Figure 2.6 shows a close-up of the Settings window with a filled Parameters Entry Table.

Click on the Save to File icon (disk), the rightmost symbol in the control button array below the parameter list, and enter MMUC4_1D_Multipane_Pram.txt to save these parameters for future use in another model.

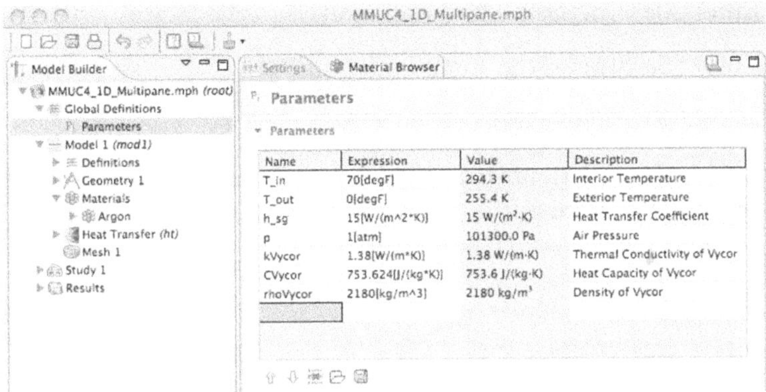

FIGURE 2.6 The Settings Window with a Filled Parameters Entry Table.

Geometry Building

The next step in building a model of a Multi-Pane Window is to define the geometry through which the heat will flow from the heated interior to the less heated exterior.

Right-Click in the Model Builder window on Geometry 1 under Model 1 *(mod1)*.

Select > Interval.

Enter 0.0 as the Left endpoint.

Enter 5e-3[m] as the Right endpoint.

Click > Build Selected.

Right-Click in the Model Builder window on Geometry 1 under Model 1 *(mod1)*.

Select > Interval.

Enter 5e-3[m] as the Left endpoint.

Enter 20e-3[m] as the Right endpoint.

Click > Build Selected.

Right-Click in the Model Builder window on Geometry 1 under Model 1 *(mod1)*.

Select > Interval.

Enter 20e-3[m] as the Left endpoint.

Enter 25e-3[m] as the Right endpoint.

Click > Build Selected.

Click > Zoom Extents to see the first three intervals in the Graphics window. See Figure 2.7.

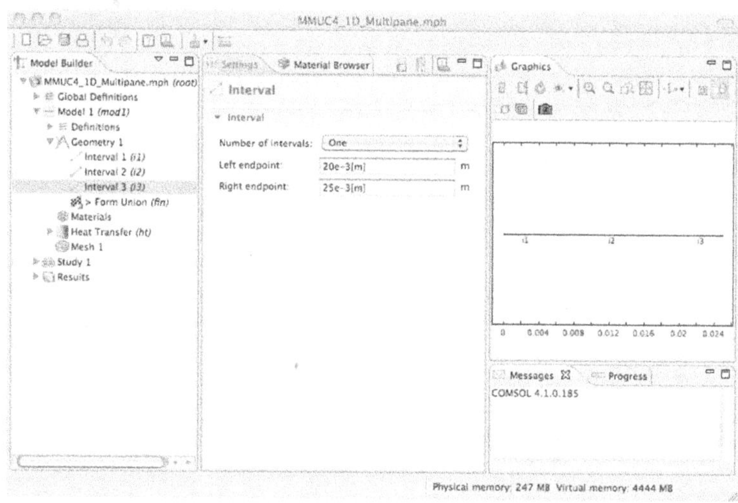

FIGURE 2.7 Zoom Extents Results.

Figure 2.7 shows the results of clicking the Zoom Extents button, which changes the dimensions of the axes in the Graphics window to reflect the value range of the geometry.

Right-Click in the Model Builder window on Geometry 1 under Model 1 *(mod1)*.

Select > Interval.

Enter 25e-3[m] as the Left endpoint.

Enter 40e-3[m] as the Right endpoint.

Click > Build Selected.

Click > Zoom Extents.

Right-Click in the Model Builder window on Geometry 1 under Model 1 *(mod1)*.

Select > Interval.

Enter 40e-3[m] as the Left endpoint.

Enter 45e-3[m] as the Right endpoint.

Click > Build All.

Click > Zoom Extents.

Material Definition

Now that the Multi-Pane Geometry has been built, the next step is to define the materials comprising the Model. Because we are assuming the three materials being used are not already stored in the COMSOL Material Library, each material will be defined by a different approach to illustrate the three techniques: material building, user defined parameters, and user defined direct entry.

Materials Properties from a Newly Built Material

The required material properties for the Argon gas will be defined using the Materials Settings Window.

Right-Click in the Model Builder window on Materials under Model 1 *(mod1)*.

Select > Material.

A Material Settings Window will appear on the desktop.

Right-Click > Material 1.

Select > Rename.

Enter Argon in the Rename Material window. See Figure 2.8.

Figure 2.8 shows the Rename Material window related to the Material Settings for Material 1.

Click > OK.

To close the property selection list, Click > the Material Properties twistie.

The Material Contents section of the Material Settings window shows the properties required for the Heat Transfer physics interface used in this model. All entries are marked with red stop sign symbols to indicate that the properties have not yet been assigned a value, as shown in Figure 2.9.

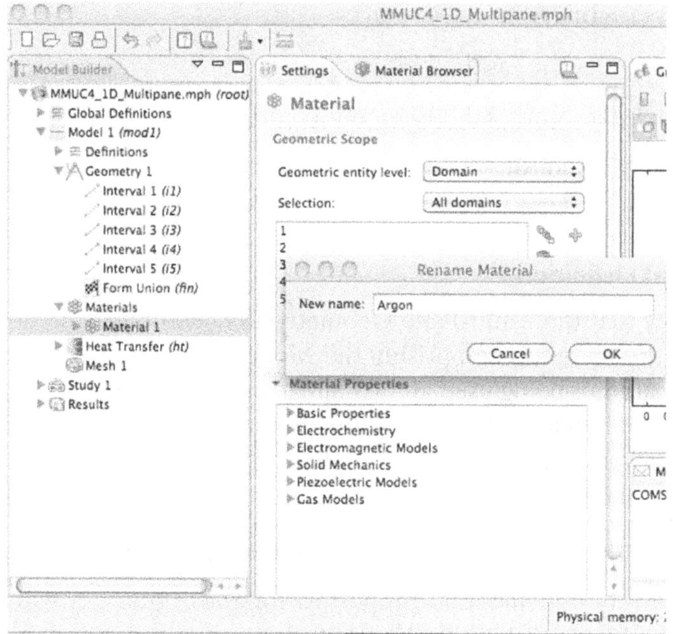

FIGURE 2.8 Material Settings - Rename Material Window.

Figure 2.9 shows the Material Contents section of the Material Settings for Argon.

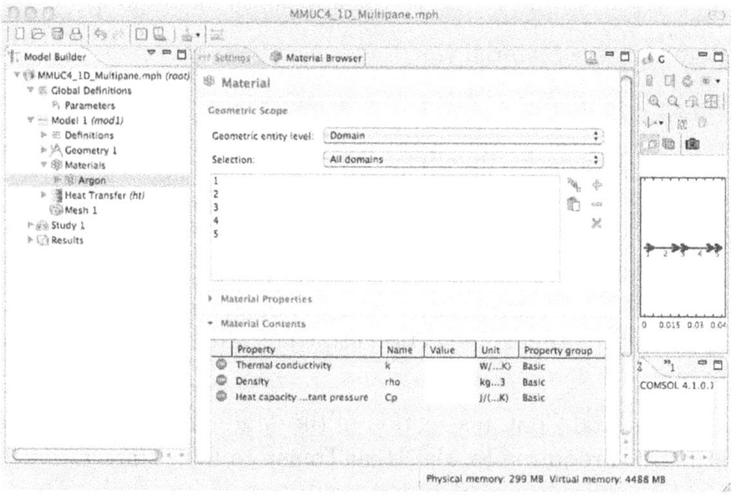

FIGURE 2.9 Material Contents Section for Argon.

Enter the property values for Argon into the entry windows in the Value column of the Material Contents section: thermal conductivity is 1.7e-2[W/(m*K)], density is 1.784[kg/m^3], and heat capacity is 523.0[J/(kg*K)].

The red stop signs change to green check marks when a value has been entered for the property, as shown in Figure 2.10.

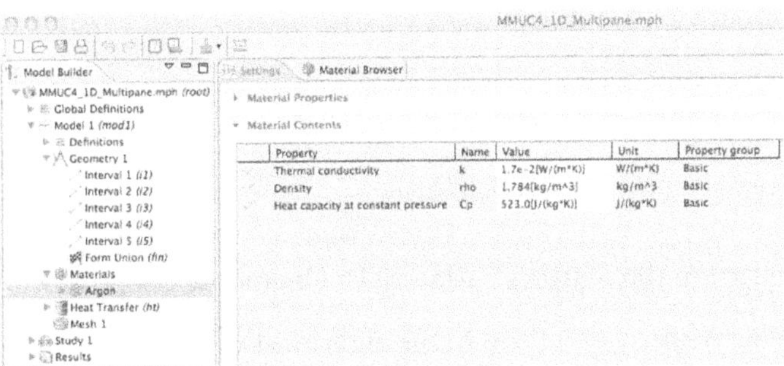

FIGURE 2.10 Material Settings - Material Contents Section Completed.

Figure 2.10 shows the property values entered in the Material Contents section of the Material Settings for Argon.

A material definition can be saved for use in future models in the User Defined Library in the Materials Browser.

To save a definition,

Right-Click > Model 1 – Materials – Argon.

Select > Add Material to User Defined Library from the pop-up menu.

To verify the entry,

Click > Material Browser Tab.

In the Materials window, Click > User Defined Library twistie to display the material list (see Figure 2.11).

Figure 2.11 shows the material list in the User Defined Library.

NOTE

Since this model is 1D, there are 5 Domains and this is the first material added, 4.x automatically assigns the Argon material's properties to all the available domains (1, 2, 3, 4, 5).

In this case, the materials assignment of Argon to Domains 1, 3, 4, & 5 is not correct and must be adjusted.

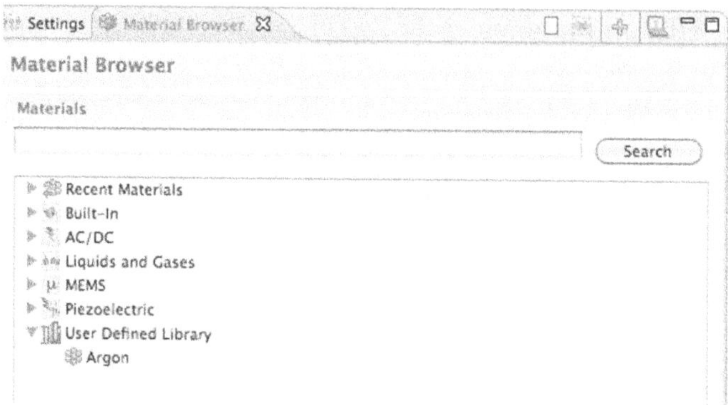

FIGURE 2.11 Material Browser - User Defined Library.

Click > 1 in the Settings – Material – Geometric Scope Selection window.

Click > Minus (–) button in the rightmost vertical control button array.

Click > 3 in the Settings – Material – Geometric Scope Selection window.

Click > Minus (–) button in the rightmost vertical control button array.

Click > 4 in the Settings – Material – Geometric Scope Selection window.

Click > Minus (–) button in the rightmost vertical control button array.

Click > 5 in the Settings – Material – Geometric Scope Selection window.

Click > Minus (–) button in the rightmost vertical control button array.

The Argon material has now been defined, saved, and assigned to the appropriate domain.

Materials Properties from User Defined Parameters

The default source for the properties needed by a model is defined as coming from the material.

To verify this setting,

Click > Model 1 – Heat Transfer *(ht)* twistie in the Model Builder window to display the Heat Transfer entries.

Click > Heat Transfer in Solids 1.

Note that all five domains are included by default in the Domains – Selections list at the top of the Settings – Heat Transfer in Solids window.

Also note that the properties listed at the bottom of the Settings – Heat Transfer in Solids window are all set to the "From material" option from the pop-up menu next to each property. See Figure 2.12.

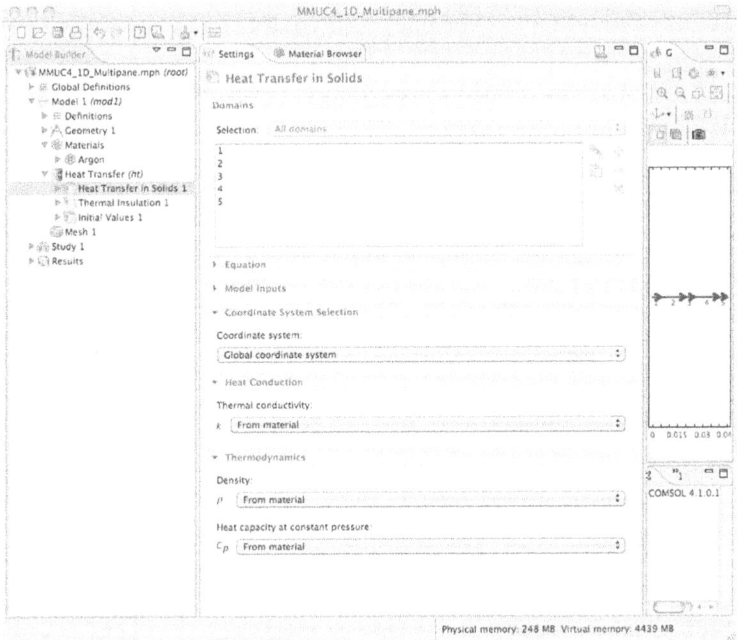

FIGURE 2.12 Settings Window – Heat Transfer in Solids.

Figure 2.12 shows the default property settings for the Heat Transfer in Solids physics interface.

That default setting is appropriate for the Argon gas since its material property values do come from the material definition. However, since the properties for the Vycor glass were entered as parameters in a previous section and therefore will not be coming from a material definition, the model needs another set of Heat Transfer in Solids settings to override the "From materials" default.

Add the new settings.

Right-Click > Model 1 > Heat Transfer *(ht)* in the Model Builder window.

Select > Heat Transfer in Solids from the pop-up menu.

Click > Heat Transfer in Solids 2.

Enter the material property values.

In the Heat Transfer in Solids window, Click > the pop-up menu in the Heat Conduction > Thermal conductivity.

Select > User defined.

Type kVycor in the Thermal conductivity entry window.

Click > the pop-up menu in the Thermodynamics > Density.

Select > User defined.

Type rhoVycor in the Density entry window.

Click > the pop-up menu in the Thermodynamics > Heat capacity at constant pressure section.

Select > User defined.

Type CVycor in the Heat Capacity entry window.

To assign the domains that will use these property values,

In the Graphics window, Click > Interval 1.

Shift-Click > Interval 3.

Shift-Click > Interval 5.

In the Settings – Heat Transfer in Solids – Domains window, Click > Add to Selection button (blue plus sign).

See Figure 2.13.

Figure 2.13 shows the property value sources for the Vycor glass domains.

Materials Properties by User Defined Direct Entry

The properties for the Krypton gas were neither entered as parameters nor as a material definition. Therefore, the values will be entered directly into another set of Heat Transfer in Solids settings.

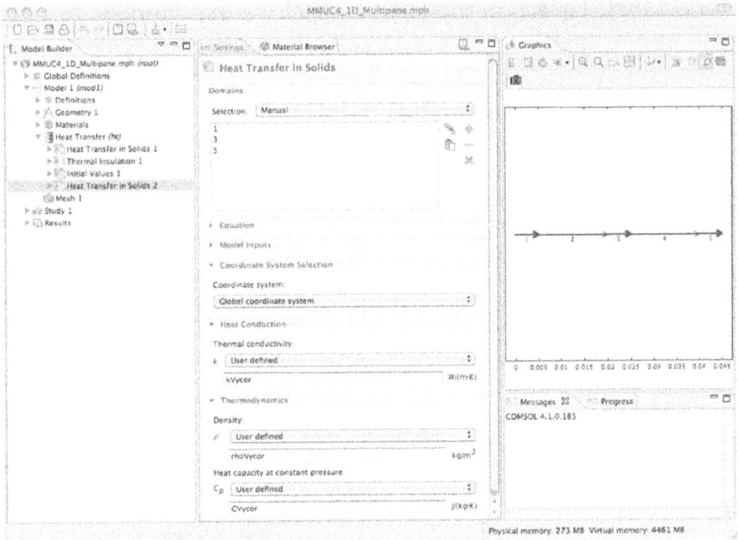

FIGURE 2.13 Property Value Sources for Vycor Glass.

Right-Click > Model 1 > Heat Transfer *(ht)* in the Model Builder window.

Select > Heat Transfer in Solids from the pop-up menu.

Click > Heat Transfer in Solids 3.

Click > the pop-up menu in the Heat Conduction > Thermal conductivity section.

Select > User defined.

Type 8.8e-3[W/(m*K)] in the Thermal conductivity entry window.

Click > the pop-up menu in the Thermodynamics > Density section.

Select > User defined.

Type 3.743[kg/m^3] in the Density entry window.

Click > the pop-up menu in the Thermodynamics – Heat capacity at constant pressure section.

Select > User defined.

Type 248[J/(kg*K)] in the Heat Capacity entry window.

To assign the domain that will use these property values,

In the Graphics window, Click > Interval 4.

In the Domains window, click the Add to Selection button (blue plus sign).

See Figure 2.14.

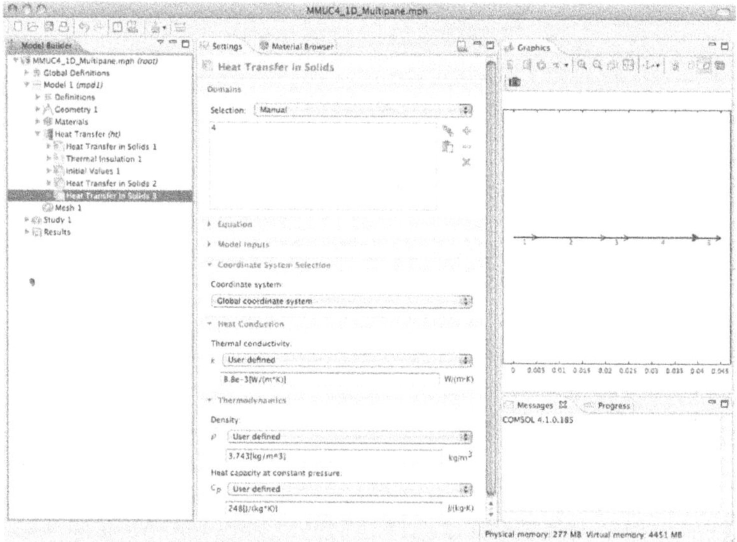

FIGURE 2.14 Property Values for Krypton Gas.

Figure 2.14 shows the material property values for the Krypton gas domain.

In the Model Builder window, Click > Heat Transfer *(ht)* > Heat Transfer in Solids 1.

Now the settings in the Domains window show that the material property value sources for all of the domains except Domain 2, the Argon gas, have been overridden by the subsequent entries for the Vycor glass and Krypton gas, as shown in Figure 2.15.

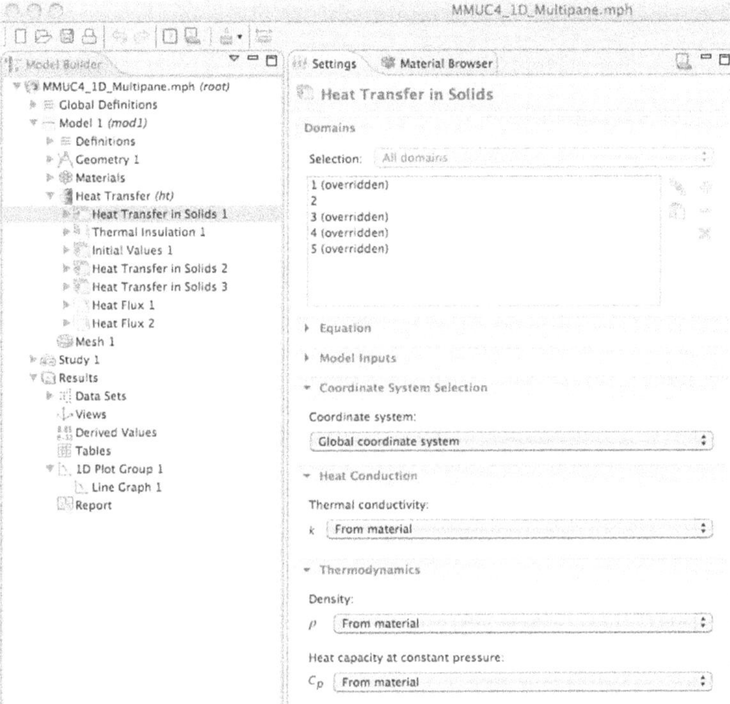

FIGURE 2.15 Default Property Value Source Settings – Overridden.

Set Boundary Conditions

Since the model geometry has been built and the appropriate material properties have been assigned to the Domains 1 through 5, the next step in the modeling process is to specify the boundary conditions for the Heat Transfer physics in this model.

Right-Click > Heat Transfer *(ht)* in the Model Builder window.

Select > Heat Flux.

Click > Heat Flux 1.

Click > Point 1 (leftmost boundary) in the Graphics window.

Click > Plus (Add to Selection) in the Boundaries window.

Click > Inward Heat Flux in the Heat Flux window.

Type h_sg in the Heat transfer coefficient entry window.

Type T_in in the External temperature entry window.

Right-Click > Heat Transfer *(ht)*.

Select > Heat Flux.

Click > Heat Transfer *(ht)* – Heat Flux 2 in the Model Builder window.

Click > Point 6 (rightmost boundary) in the Graphics window.

Click > Plus (Add to Selection) in the Boundaries window.

Click > Inward Heat Flux in the Heat Flux window.

Type h_sg in the Heat transfer coefficient entry window.

Type T_out in the External temperature entry window.

Meshing and Solution Computations

Since this model is relatively simple, allow 4.x to automatically mesh the model.

Right-Click > Mesh 1 in the Model Builder window.

Select > Build All.

The meshing computation results in a complete mesh of 16 elements.

Now that the model has been meshed, the modeler can proceed to compute the solution.

Right-Click > Study 1 in the Model Builder window.

Select > Compute.

NOTE *4.x automatically selects the appropriate Solver for this problem, computes the solution, and displays the results in the Graphics window.*

The 1D Plot displayed in the Graphics window shows the temperature in degrees Kelvin as a function of the distance from the inside surface of Vycor glass of the first pane of the Multi-Pane window to the outside surface of Vycor glass of the third pane of the Multi-Pane window.

Since the modeler originally specified the input parameters in degrees Fahrenheit, the modeler now needs to adjust the instructions of 4.x so that the Graphics window will display the results of this modeling calculation in the desired units.

Click > the Results -1D Plot Group 1 twistie in the Model Builder window.

Click > Line Graph 1.

In the Settings – Line Graph window,

Under Y-Axis Data, Click > Unit Pull-down menu > Select > degF.

Click > the Plot button at the top of the Settings – Line Graph window.

Figure 2.16 shows the Graphics window of the Calculated Results for a Multi-Pane Window in degrees Fahrenheit.

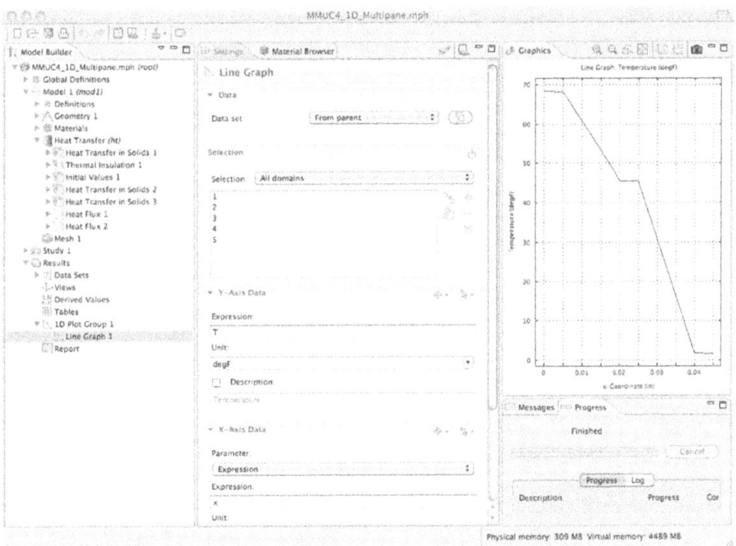

FIGURE 2.16 Calculated Results for a Multi-Pane Window in Degrees Fahrenheit.

The results of this model that demonstrates material property entry techniques are very similar to the 1D 3 Pane model demonstrated in Chapter 1 using equivalent materials.

REFERENCES

2.1 http://www.comsol.com/products/material-library/

2.2 http://en.wikipedia.org/wiki/Unified_numbering_system/

2.3 http://en.wikipedia.org/wiki/DIN/

2.4 http://www.asminternational.org/portal/site/www/

2.5 http://www.wikipedia.org/

2.6 http://www.matweb.com/

CHAPTER 3

0D ELECTRICAL CIRCUIT INTERFACE MODELING USING COMSOL MULTIPHYSICS 4.X

In This Chapter

GUIDELINES FOR ELECTRICAL CIRCUIT INTERFACE MODELING IN 4.X

In this chapter, the modeler is introduced to the development and analysis of electrical and electronic circuit models. Such "0D" models have proven very valuable to the science and engineering communities, both in the past and currently, as first-cut evaluations of possible power supplies, signal-drivers, and control circuit models. Those and other such ancillary circuits are developed and screened early in a project for potential later use in higher-dimensionality (2D, 3D, etc.) field-based (electrical, magnetic, etc.) geometrically dependent models.

NOTE *Once the 0D button has been selected and the modeler has clicked Next and moved to the Add Physics Interface page, the Physics Interfaces that may be selected have no geometric dependence [no (x, y, z) etc.]. That lack of geometric dependence results from the incorporated underlying assumption in 4.x that a 0D model calculation reflects either a homogeneous, isotropic, universal (throughout all model space) reaction or comprises a model formed of a collection of connected lumped-constant devices, such as an electrical or electronic circuit.*

For information on the other Physics Interfaces that may be used to implement non-circuit based analyses in 0D, the modeler should consult the literature {3.1, 3.2, 3.3} that accompanies 4.x.

Electrical/Electronic Circuit Considerations

Electrical circuits {3.4} are typically those comprising passive (lumped-constant) devices, such as simple (non-active) resistors, capacitors, and inductors. Electronic circuits {3.5} comprise both passive and active (lumped-constant) devices (e.g. transistors, integrated circuits, thermistors, and many other device types). In this chapter, the 4.x Electrical Circuit Interface will be used to present an introduction to the modeling of the most important basic electrical circuits. Further circuit development by the curious, innovative modeler is recommended and encouraged. That task is left to the modeler to pursue through exploration of the published literature.

A basic understanding of electrical circuits began with the work of Georg Simon Ohm {3.6}. Ohm discovered and published the relationship between current, voltage, and resistance in 1827 {3.7}, approximately four (4) years before the birth of James Clerk Maxwell {3.8}. Ohm's Law, as it is commonly used throughout the science and engineering communities and as a 4.x modeler would typically employ it, is stated as follows:

$$V = IR \tag{3.1}$$

Where: V is the electromotive force in Volts (V).

I is the current in Amperes (A).

And R is the resistance in Ohms (Ω).

NOTE *The modeler may find in other sources that either of the symbols E and V have been employed to represent the Electromotive Force {3.9}. In this text, E will be used for Electric Field (Volts per meter). Ohm's Law, as presented in the previous equation, is formulated such that the resistance (R) has a constant nominal value. That nominal value is, to first-order, independent of temperature (T), frequency (f), and all other potential functional dependence variables.*

Figure 3.1 shows an example of a typical simple battery powered circuit with a resistive-load (R).

NOTE *The modeler should note that, based on a First Principles Analysis, the principle of charge conservation is employed in the analysis that follows. The Law of Charge Conservation {3.10} states that charge is neither created nor destroyed in the absence of sinks or sources, where a sink decreases the magnitude of charge and a source increases the magnitude of charge.*

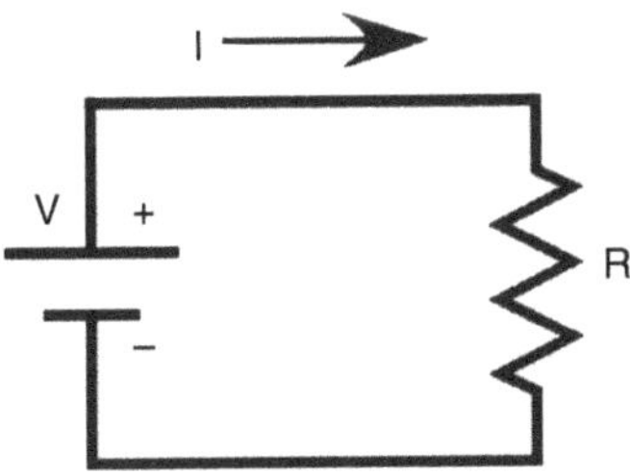

FIGURE 3.1 A Typical Simple Battery Powered Circuit with a Resistive-Load.

In configuring simple circuits, there are two basic connection sequences that are most commonly employed. In the following examples, resistors are employed for simplicity and clarity. However, other basic electrical components are also later discussed and are also so configured in the basic arrangements.

The first connection sequence discussed herein is that of the series resistive circuit. In the series connection case, two or more resistors are connected in a chain. In this case, the two serially connected resistors form a voltage divider. See Figure 3.2.

Figure 3.2 shows an example of a simple battery powered series resistive circuit (a voltage divider).

In order to calculate the ohmic resistance value of the series equivalent resistor (R_{SE}), consider the following argument. Since the two resistors are connected in series and there are no sinks, sources, or branch paths, the same magnitude of current (I) must flow through both resistors.

Thus:

$$V = V_1 + V_2 = I(R1 + R2) \tag{3.2}$$

Where: I is the common circuit current in Amperes (A).

V is the electromotive force in Volts (V).

V1 is the voltage dropped across R1 by the current I.

V2 is the voltage dropped across R2 by the current I.

R1 is the resistance in Ohms (Ω) of the first resistor in the series.

And R2 is the resistance in Ohms (Ω) of the second resistor in the series.

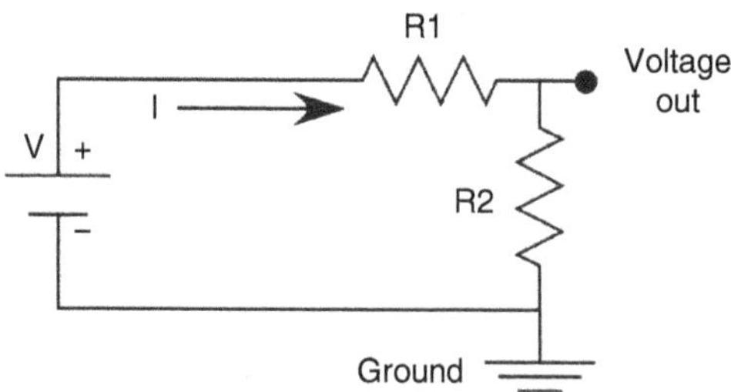

FIGURE 3.2 An Example of a Simple Battery Powered Series Resistive Circuit (a Voltage Divider).

And therefore:

$$R_{SE} = R1 + R2 \tag{3.3}$$

Where: R_{SE} is the series equivalent resistance of resistors R1 and R2.

R1 is the resistance in Ohms (Ω) of the first resistor in the series.

And R2 is the resistance in Ohms (Ω) of the second resistor in the series.

The voltage at the output of this resistive divider is:

$$V_{out} = \frac{V_2}{V}.V = \frac{I.R2}{I(R1+R2)}.V = \frac{R2}{R1+R2}.V = \frac{R2}{R_{SE}}.V \tag{3.4}$$

Where: V_{out} is the output voltage of the resistive divider in Volts (V).

V_2 is voltage dropped across resistor R2 in Volts (V).

And I is the common circuit current in Amperes (A).

The output voltage of the resistive divider is the product of the source voltage (battery voltage) and the ratio of the ohmic value of the grounded resistor to the ohmic value of the circuit series equivalent resistance.

The next basic connection sequence discussed herein is that of the parallel resistive circuit. In the parallel connection sequence, two or more resistors are connected in a parallel configuration. In this case, the resistors have a common (the same) voltage and the circuit acts as a current divider. See Figure 3.3.

Figure 3.3 shows an example of a simple battery powered parallel resistive circuit (a current divider).

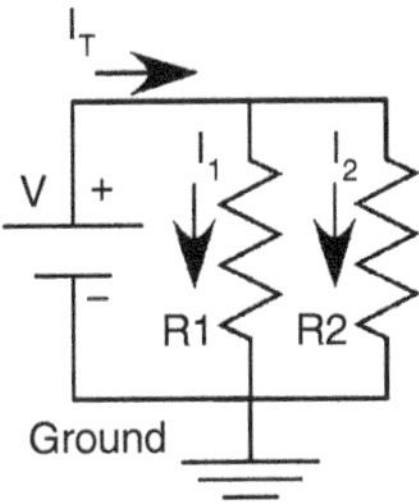

FIGURE 3.3 An Example of a Simple Battery Powered Parallel Resistive Circuit (a Current Divider).

The value of the equivalent series resistance as viewed from the battery is calculated as follows:

$$I = I_1 + I_2 = \frac{V}{R_1} + \frac{V}{R_2} = V\left(\frac{1}{R_1} + \frac{1}{R_2}\right) = V\left(\frac{R_1 + R_2}{R_1 R_2}\right) = \frac{V}{R_{SE}} \tag{3.5}$$

Thus:

$$\frac{1}{R_{SE}} = \left(\frac{R_1 + R_2}{R_1 R_2}\right) \rightarrow R_{SE} = \frac{R_1 R_2}{R_1 + R_2} \tag{3.6}$$

Where: R_{SE} is the series equivalent resistance of resistors R1 and R2 in parallel.

R1 is the resistance in Ohms (Ω) of the first resistor connected in parallel.

And R2 is the resistance in Ohms (Ω) of the second resistor connected in parallel.

NOTE *The modeler can easily see that the calculations for the equivalent series resistance at any point in the circuit can rapidly become lengthy and tedious for even a somewhat complicated circuit.*

In Figure 3.4 resistors are shown connected into a series-parallel configuration.

Figure 3.4 shows an example of a simple battery powered series-parallel resistive circuit.

The next major expansion in the understanding and analysis of electrical circuits, building on the work of Ohm, occurred through the work of

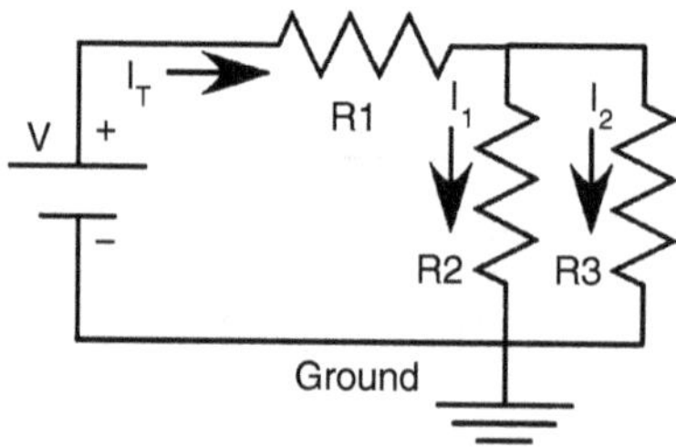

FIGURE 3.4 An Example of a Simple Battery Powered Series-Parallel Resistive Circuit.

Gustav Robert Kirchhoff {3.11}. Kirchhoff developed two circuit laws, both of which are fundamentally employed in 4.x, one for voltage and one for current. He first described his circuit laws in 1845, when he was still a student {3.12}.

Kirchhoff's first law, the Current Law:

$$\sum_{n=1}^{n} A_{mn} I_{mn} = 0 \tag{3.7}$$

Where: I_{mn} is the current flowing into or out of the particular junction (node).

A_{mn} has a value of +1 for each in-flowing nodal current and a value of −1 for each out-flowing nodal current.

m is the index number of the node under analysis.

And n is the index number of a particular branch on the mth node.

Figure 3.5 shows an example of a five-branch node.

Kirchhoff's second law, the Voltage Law:

$$\sum_{n=1}^{n} B_{mn} V_{mn} = 0 \tag{3.8}$$

Where: V_{mn} is the voltage dropped at the nth element of the mth loop as determined by the direction of the elemental current flow.

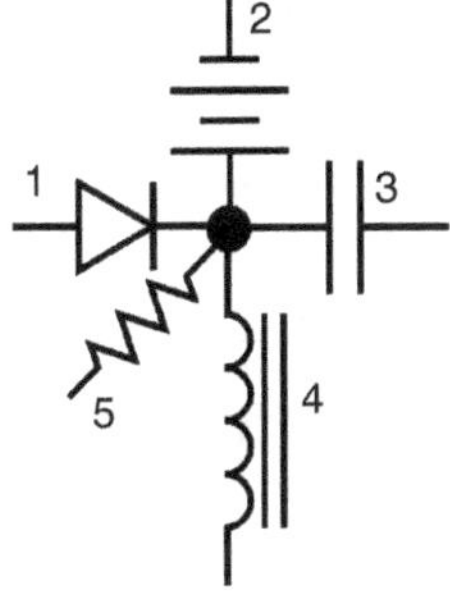

FIGURE 3.5 An Example of a Five-Branch Node.

B_{mn} has a value of +1 for each non-source element and a value of -1 for each source element.

m is the index number of the loop under analysis.

And n is the index number of a particular element in the mth loop.

Loop 1 of Figure 3.6 shows an example of a four-element loop.

The modeler can easily see that Figure 3.6 Loop 1 shows:

$$V_{ab} + V_{be} + V_{ef} - V_{af} = 0 \tag{3.9}$$

and that:

$$V_{ab} + V_{bc} + V_{cd} + V_{de} + V_{ef} - V_{af} = 0 \tag{3.10}$$

and that Loop 2 shows:

$$V_{bc} + V_{cd} + V_{de} - V_{be} = 0 \tag{3.11}$$

The basics of Ohm's Law and Kirchhoff's Laws have been introduced using resistive circuits. To expand the modeler's analytical capability, he needs to consider more than simply circuit resistance.

There are two other very important lumped-constant electrical components employed in most circuits. They are the capacitor and the inductor. Ewald Georg von Kleist {3.13} first reported the capacitive charge storage effect in 1745. Pieter van Musschenbroek invented the first capacitor {3.14}, the Leiden Jar, in 1746 {3.15}.

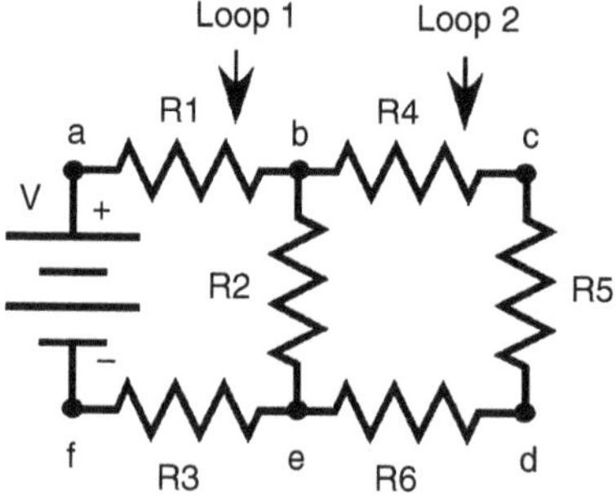

FIGURE 3.6 An Example of a Four-Element Loop.

Capacitors comprise two conductors, separated by a dielectric (insulating) {3.16} medium (material). The charge stored is as follows:

$$Q = CV \tag{3.12}$$

Where: Q is the charge stored on the capacitor.

C is the capacitance in Farads.

And V is the electromotive potential measured between the two conductors in Volts.

NOTE *Q = CV is correct for an IDEAL capacitor. For the equivalent circuit of real capacitors, there will generally be some additional terms that include a small series or parallel resistance and a small series inductance, depending upon the operating frequency of the circuit or model. Each of those terms depends upon the physical specification and geometrical configuration of the actual component. (See the manufacturer's specification data sheet.)*

Michael Faraday {3.17} and Joseph Henry {3.18} discovered electromagnetic induction in 1831 {3.19}. The effect of electromagnetic induction is to induce an electromagnetic force into a conductor that is proportional to the rate of change of the magnetic flux. The Maxwell-Faraday formulation is as follows {3.20}:

$$\nabla \times E = -\frac{\partial B}{\partial t} \tag{3.13}$$

Where: E is electromagnetic force (electric field) in Volts (V) per meter (m).

B is the magnetic flux in webers per square meter.

And t is the time in seconds.

These two components are inherently time dependent in their behavior {3.21, 3.22, 3.23}.

The self-inductance of a circuit is defined as follows:

$$V_L = L\frac{dI}{dt} \tag{3.14}$$

Where: V_L is the induced voltage in Volts.

L is the inductance in webers per ampere (henries).

I is the current flow in amperes (A).

And t is the time in seconds.

NOTE *Self-inductance occurs when a voltage is induced into the same circuit by a current flowing within that circuit.*

Mutual-inductance occurs when a voltage is induced into a second circuit by a current flowing in a first circuit (transformers, etc.).

The capacitance of a circuit is defined as follows:

$$V_C = \frac{1}{C}\int i dt = \frac{Q}{C} \tag{3.15}$$

Where: V_C is the voltage between the terminals of the capacitor in Volts.

C is the capacitance in Farads (1 Coulomb per Volt).

i is the current flow in Amperes.

And t is the time in seconds.

The current flow through a capacitor is defined as:

$$I = C\frac{dV_C}{dt} \tag{3.16}$$

Where: V_C is the voltage between the terminals of the capacitor in Volts.

C is the capacitance in Farads (1 Coulomb per Volt).

I is the current flow in Amperes.

And t is the time in seconds.

Utilizing Kirchhoff's Voltage Law for a simple series resistor-capacitor-inductor circuit yields the following equation:

$$V_R + V_C + V_L - V(t) = 0 \tag{3.17}$$

Where: V_R is the voltage-drop across the resistor in Volts.

V_C is the voltage-drop across the capacitor in Volts.

V_L is the voltage-drop across the inductor in Volts.

And V(t) is the time-varying source voltage driving the circuit.

Now that the modeler has been introduced to the underlying physical concepts involved in the functioning of electrical circuits, the modeler will now be shown examples of how to solve for the physical behavior of some of the basic circuits using 4.x. These examples can, of course, be used as guidance to the development of more complex circuits by the modeler at some later time when the need arises.

Simple Electrical Circuit Interface Model Setup Overview

Figure 3.7 shows the default COMSOL Desktop Display immediately after startup. The default coordinate selection for modeling calculations at startup in 4.x is 3D. The bottom button in the Space Dimension window is 0D. In the case of 0D, the underlying equations in the multiphysics model have either no relational dependence on geometrical factors or react in a homogeneous and isotropic manner (effectively geometrically relationally independent).

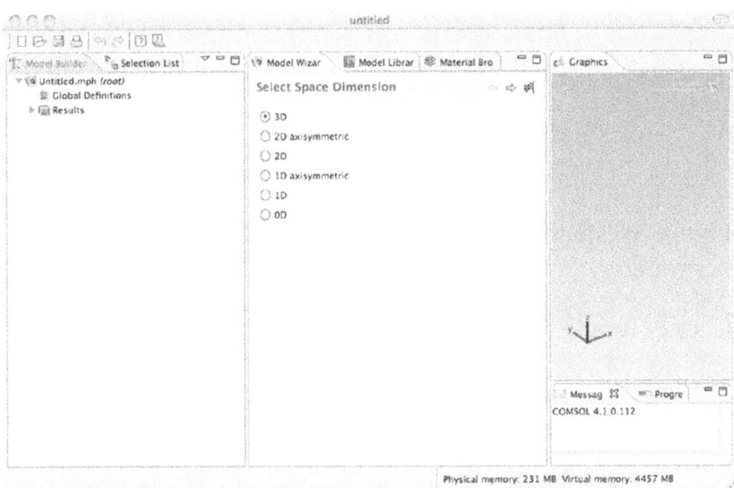

FIGURE 3.7 Default COMSOL Desktop Display.

The 0D Space Dimension will be used in this book only to explore electrical and/or electronic circuit model behavior through the use of Ohm's Law and Kirchhoff's Law calculations.

Select a Space Dimension

For this first demonstration model of the Engineering Circuit Interface, Select (Click) the 0D selection (radio) button.

The 0D selection (radio) button will show a Dot in the center once selected.

Then, Click the Next arrow (right pointing).

Once the Next arrow is clicked, the Desktop Display changes to show the Add Physics window as shown in Figure 3.8.

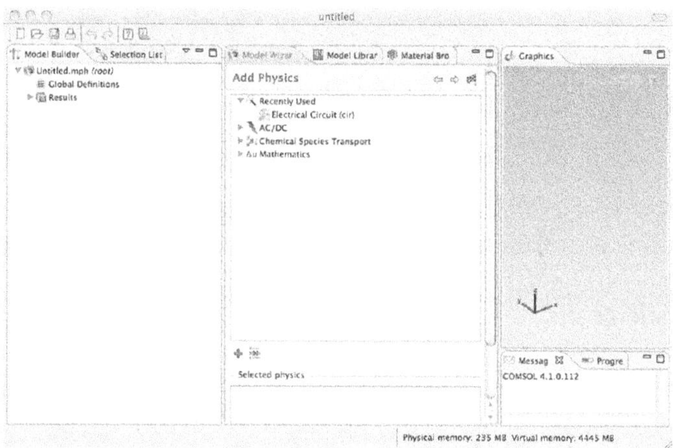

FIGURE 3.8 Desktop Display with the Add Physics Window.

Figure 3.8 shows the default COMSOL Desktop Display with the Add Physics window.

Click > Twistie of the AC/DC Physics Interface.

Select > Electrical Circuit (cir).

Click > Add Selected button (Plus).

Figure 3.9 shows the Desktop Display with the Add Physics window after the selection of Electrical Circuit (cir).

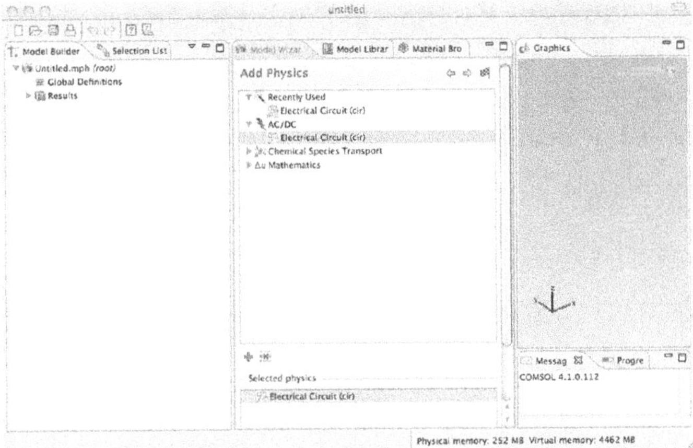

FIGURE 3.9 Electrical Circuit (cir) Selected.

Figure 3.9 shows the Electrical Circuit (cir) selected.

Click > Next.

Select > Time Dependent.

Click > Finish (Flag).

The modeler has now been shown fundamentally how to select the 0D Space Coordinates and how to access the Electrical Circuit (cir) Interface. See Figure 3.10.

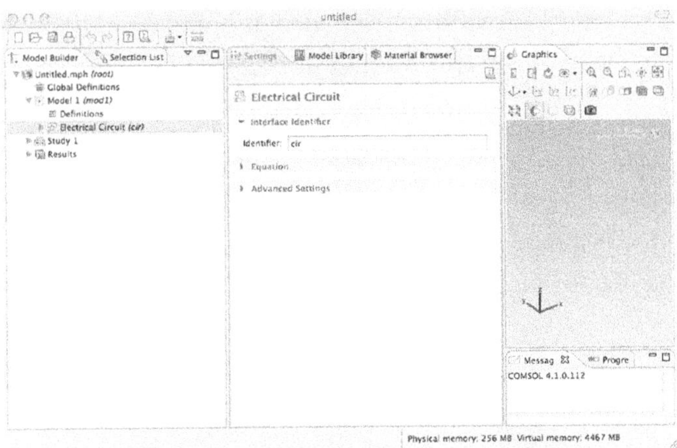

FIGURE 3.10 Desktop Display Prepared for an Electrical Circuit (cir) Model.

Figure 3.10 shows the Desktop Display prepared for an Electrical Circuit (cir) model.

Now that the initial steps have been demonstrated to start building a new model, the modeler can close down the 4.x application.

If the modeler is using a Macintosh,

Select > COMSOL Multiphysics menu > Quit COMSOL Multiphysics.

Select the Do Not Save option on closing the 4.x Macintosh application.

If the modeler is using a PC,

Select > File > Exit.

Select the Do Not Save option on closing the 4.x PC application.

Basic Problem Formulation and Implicit Assumptions

NOTE

A first-cut problem solution is the equivalent of a back-of-the-envelope or an on-a-napkin solution. Problem solutions of that type are more easily formulated, more quickly built, and typically provide a first estimate of whether the final solution to the full problem will be deemed to be (should possibly be) within reasonable time constraint or budgetary bounds. Creating first-cut solutions will often allow the 4.x modeler to easily decide whether or not it is worth the additional effort and cost needed to build a fully implemented higher dimensionality model.

The Electrical Circuit Interface models built herein (0D) will start with very simple models and then begin to explore more complicated models. As the modeler knows, there are an infinite number of variations of simple and complex circuits that could be explored. The models built in this chapter will demonstrate the use of the Electrical Circuit Interface as a means to explore the building of possible circuit configurations as a tutorial mechanism.

In later chapters, the Electrical Circuit Interface will be used to develop driver circuits for more complex field-based physical models.

NOTE

In a transient solution model, all of the appropriate variables in the model are a function of time. The model solution builds from a set of suitable initial conditions, through a set of incremental intermediate solutions, to a final solution.

0D BASIC CIRCUIT MODELS

Let us consider first the application of the 4.x Electrical Circuit Interface to some of the basic circuits just presented.

0D Resistor-Capacitor Series Circuit Model

Startup 4.x.

Select > 0D.

Click > Next.

Click > Twistie for AC/DC in the Add Physics window.

Select > Electrical Circuit (cir).

Click > Add Selected.

Click > Next.

Select > Time Dependent in the Select Study Type window.

Click > Finish (Flag).

Click > Save As > MMUC4_0D_SRC1.mph.

NOTE *In the Electrical Circuit (cir) Interface, the Node Connections specify exactly where each component is connected. Each circuit must have a Ground connection and that Node is 0. The completed circuit must tie back to the Ground Node for completion.*

The completed battery-powered series resistor-capacitor circuit will be connected as follows:

Ground (0) <> (0) Voltage Source (1) <> (1) Resistor (2) <> (2) Capacitor (0)

NOTE *In this book <> means is electrically connected to.*

See Figure 3.11.

Figure 3.11 shows the circuit diagram for the MMUC4_0D_SRC1.mph model.

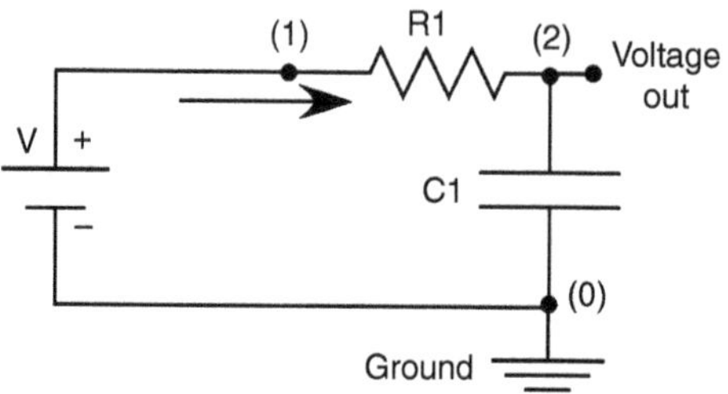

FIGURE 3.11 Circuit Diagram for the MMUC4_0D_SRC1.mph Model.

Right-Click in the Model Builder window on Electrical Circuit (cir).

Select > Voltage Source from the Pop-up menu.

See Figure 3.12.

Ground Node
Resistor
Capacitor
Inductor
Voltage Source
Current Source
Voltage-Controlled Voltage Source
Voltage-Controlled Current Source
Current-Controlled Voltage Source
Current-Controlled Current Source
Subcircuit Definition
Subcircuit Instance
NPN BJT
n-Channel MOSFET
Diode
External I Vs. U
External U Vs. I
External I-Terminal

Import SPICE Netlist

Delete
Disable
Rename F2

Properties

Dynamic Help F1

FIGURE 3.12 Electrical Circuit (cir) Pop-up Window.

Figure 3.12 shows the Electrical Circuit (cir) Pop-up window.

Enter 5 in the Voltage Source – Device Parameters – Electric potential entry window.

Press the Tab key.

See Figure 3.13.

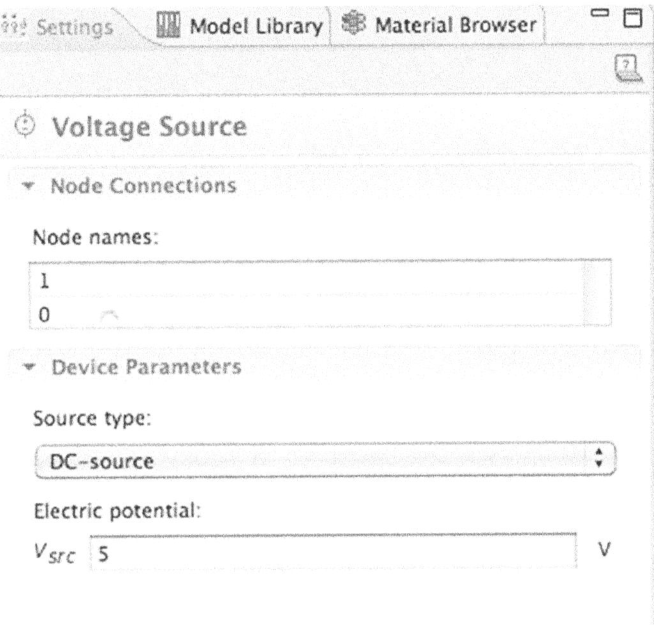

FIGURE 3.13 Desktop Display - Settings - Voltage Source - Device Parameters - Electric Potential Entry Window.

Figure 3.13 shows the Desktop Display – Settings – Voltage Source – Device Parameters – Electric potential entry window.

Right-Click in the Model Builder window on Electrical Circuit (cir).

Select > Resistor from the Pop-up menu.

This resistor is the first component in the series circuit.

Enter 2, 1 in the Settings – Resistor – Node Connections – Node names entry windows.

NOTE *When using 4.x on the Mac, the modeler will need to press the Tab key to see both entries.*

Leave the value in the Resistance entry window (1000[ohm]) set at the default value.

Figure 3.14 shows the Settings – Resistor – Node Connections – Node names.

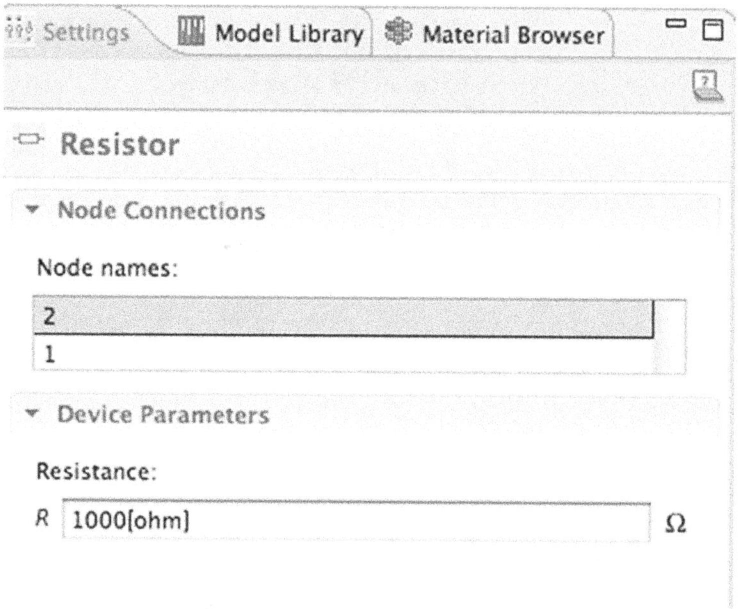

FIGURE 3.14 Settings - Resistor - Node Connections - Node names.

Right-Click in the Model Builder window on Electrical Circuit (cir).

Select > Capacitor from the Pop-up menu.

This capacitor is the second component in the series circuit.

Enter 2, 0 in the Settings – Capacitor – Node Connections – Node names entry windows.

Enter 1000[nF] in the Settings – Capacitor – Device Parameters – Capacitance entry window.

See Figure 3.15.

Figure 3.15 shows the Settings – Capacitor – Device Parameters – Capacitance.

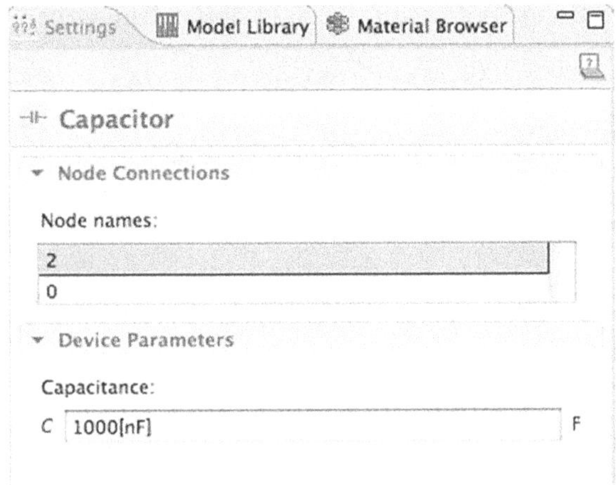

FIGURE 3.15 Settings - Capacitor - Device Parameters - Capacitance.

Now that the building of the series resistor-capacitor circuit model is complete, the next step is to prepare to solve the model.

NOTE *The default settings for the Time Dependent Study Type are not appropriate for the solution of this problem and thus need to be modified.*

Click > Study 1 Twistie.

Select > Step 1: Time Dependent. See Figure 3.16.

Figure 3.16 shows the Settings – Time Dependent window.

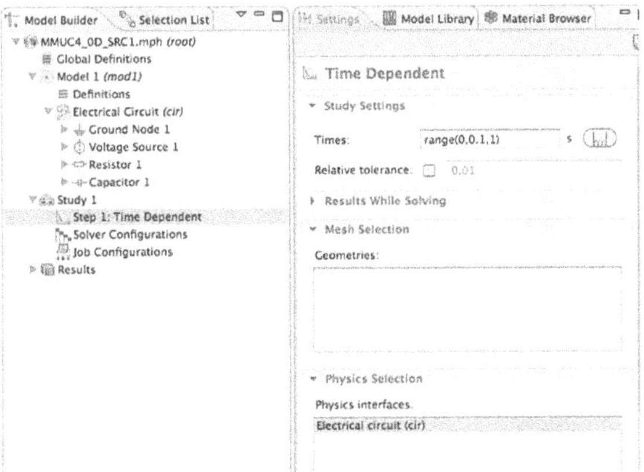

FIGURE 3.16 Settings - Time Dependent Window.

Click the Range button (at the right of the Times entry window).

Enter > Start = 0.0, Stop = 7e-3, Step = 1e-4.

Click > Replace.

Right-Click > Study 1.

Select > Compute.

NOTE *The default settings for the 1D Plot of the results of this model are not appropriate and thus the settings need to be modified.*

Click > 1D Plot Group 1 Twistie.

Click > Settings – 1D Plot Group 1 – Legend Twistie.

Select > Lower right from the Position Pull-down menu.

Click > 1D Plot Group 1 Twistie.

Select > 1 D Plot Group 1 – Global 1.

Click > Add Expression (Plus).

Select > Electrical Circuit > Voltage across device C1 (mod1.cir.C1.v).

Click > Plot.

See Figure 3.17 for the computed solution.

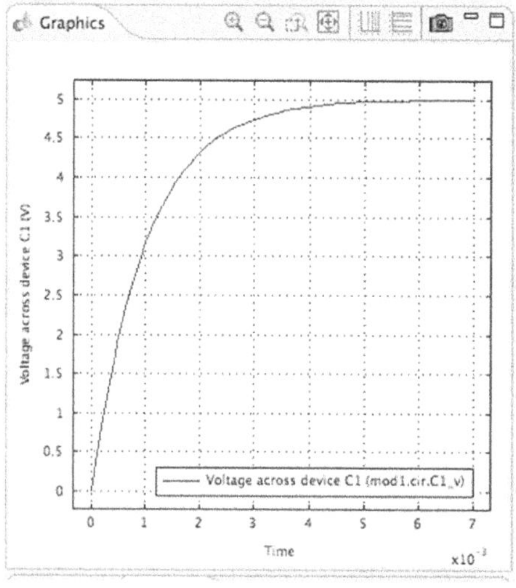

FIGURE 3.17 Resistor-Capacitor Transient Response Curve.

Figure 3.17 shows the resistor-capacitor transient response curve.

Save the completed model.

0D Inductor-Resistor Series Circuit Model

Startup 4.x.

Select > 0D.

Click > Next.

Click > Twistie for AC/DC in the Add Physics window.

Select > Electrical Circuit (cir).

Click > Add Selected.

Click > Next.

Select > Time Dependent in the Select Study Type window.

Click > Finish (Flag).

Click > Save As > MMUC4_0D_SLR1.mph.

NOTE *In the Electrical Circuit (cir) Interface, the Node Connections specify exactly where each component is connected. Each circuit must have a Ground connection and that Node is 0. The completed circuit must tie back to the Ground Node for completion.*

The completed battery-powered series inductor-resistor circuit will be connected as follows:

Ground (0) <> (0) Voltage Source (1) <> (1) Inductor (2) <> (2) Resistor (0)

NOTE *In this book <> means is electrically connected to.*

See Figure 3.18.

Figure 3.18 shows the circuit diagram for the MMUC4_0D_SLR1.mph model.

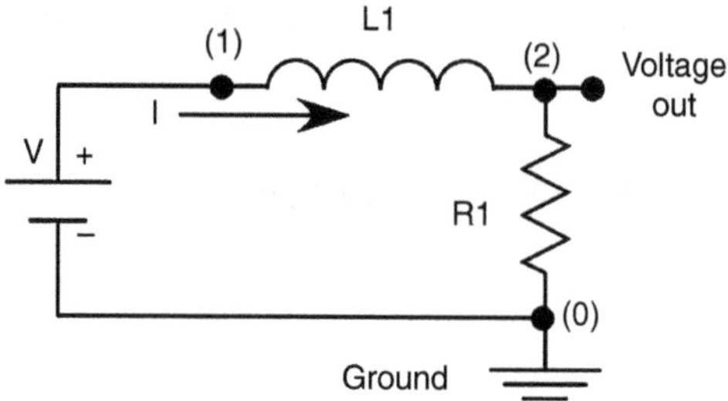

FIGURE 3.18 Circuit Diagram for the MMUC4_0D_SLR1.mph Model.

Right-Click in the Model Builder window on Electrical Circuit (cir).

Select > Voltage Source from the Pop-up menu.

See Figure 3.19.

Figure 3.19 shows the Electrical Circuit (cir) Pop-up window.

Ground Node
Resistor
Capacitor
Inductor
Voltage Source
Current Source
Voltage-Controlled Voltage Source
Voltage-Controlled Current Source
Current-Controlled Voltage Source
Current-Controlled Current Source
Subcircuit Definition
Subcircuit Instance
NPN BJT
n-Channel MOSFET
Diode
External I Vs. U
External U Vs. I
External I-Terminal
Import SPICE Netlist
Delete
Disable
Rename F2
Properties
Dynamic Help F1

FIGURE 3.19 Electrical Circuit (cir) Pop-up Window.

Enter 5 in the Voltage Source – Device Parameters – Electric potential entry window.

Press the Tab key.

See Figure 3.20.

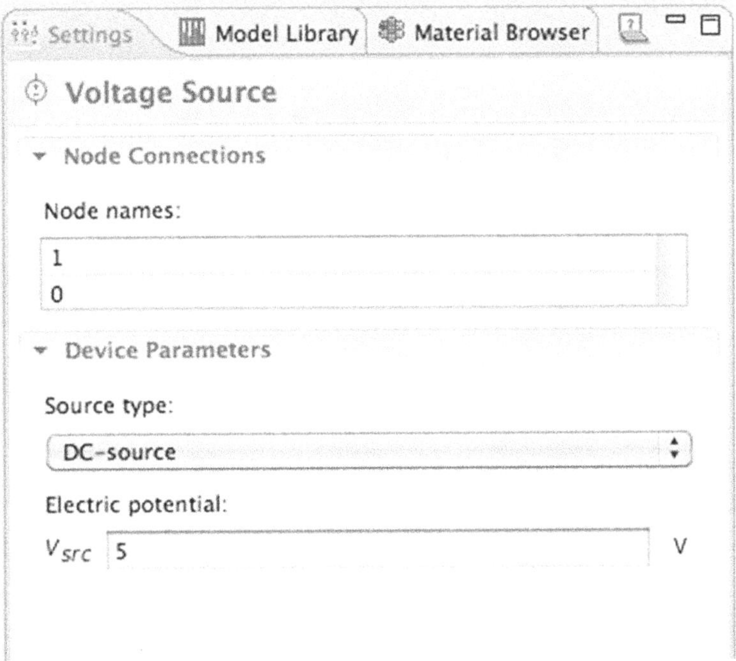

FIGURE 3.20 Desktop Display - Settings -Voltage Source - Device Parameters - Electric Potential Entry Window.

Figure 3.20 shows the Desktop Display – Settings – Voltage Source – Device Parameters – Electric potential entry window.

Right-Click in the Model Builder window on Electrical Circuit (cir).

Select > Inductor from the Pop-up menu.

This inductor is the first component in the series circuit.

Enter 2, 1 in the Settings – Inductor – Node Connections – Node names entry windows.

NOTE *When using 4.x on the Mac, the modeler will need to press the Tab key to see both entries.*

Leave the value in the Inductance entry window (1[mH]) set at the default value.

See Figure 3.21.

Figure 3.21 shows the Settings – Inductor – Node Connections – Node names.

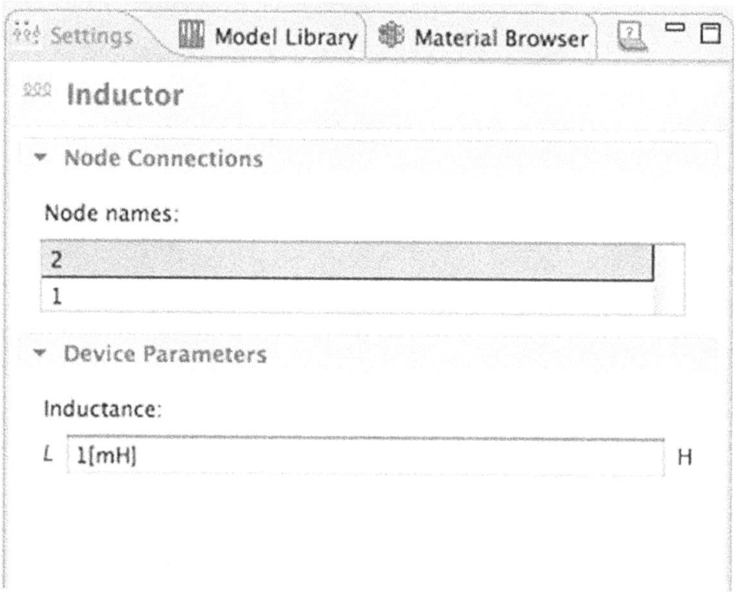

FIGURE 3.21 Settings - Inductor - Node Connections - Node Names.

Right-Click in the Model Builder window on Electrical Circuit (cir).

Select > Resistor from the Pop-up menu.

This resistor is the second component in the series circuit.

Enter 2, 0 in the Settings – Resistor – Node Connections – Node names entry windows.

Enter 1[ohm] in the Settings – Resistor – Resistance entry window.

See Figure 3.22.

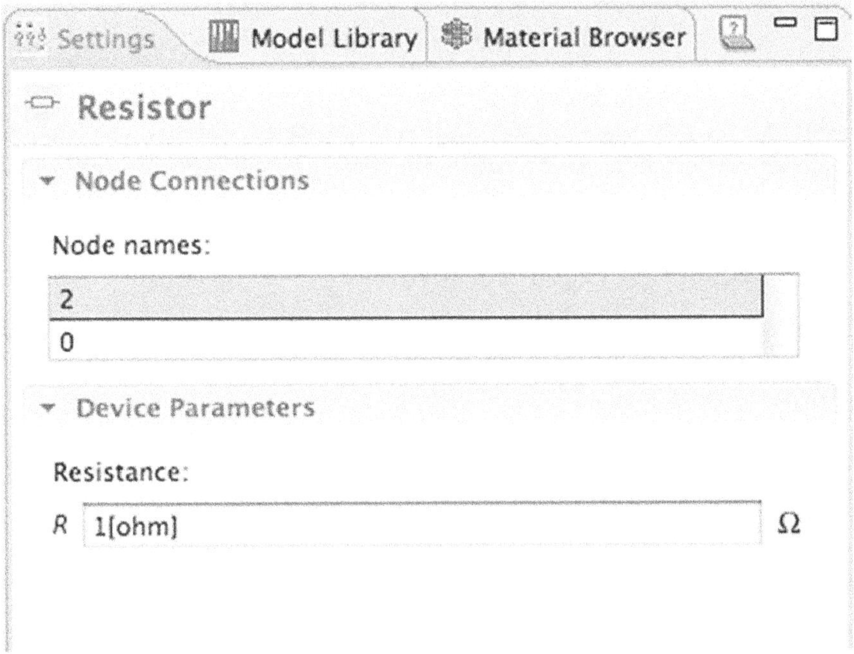

FIGURE 3.22 Settings - Resistor - Device Parameters - Resistance.

Figure 3.22 shows the Settings – Resistor – Device Parameters – Resistance.

Now that the building of the series resistor-capacitor circuit model is complete, the next step is to prepare to solve the model.

NOTE *The default settings for the Time Dependent Study Type are not appropriate for the solution of this problem and thus need to be modified.*

Click > Study 1 Twistie.

Select > Step 1: Time Dependent. See Figure 3.23.

Figure 3.23 shows the Settings – Time Dependent window.

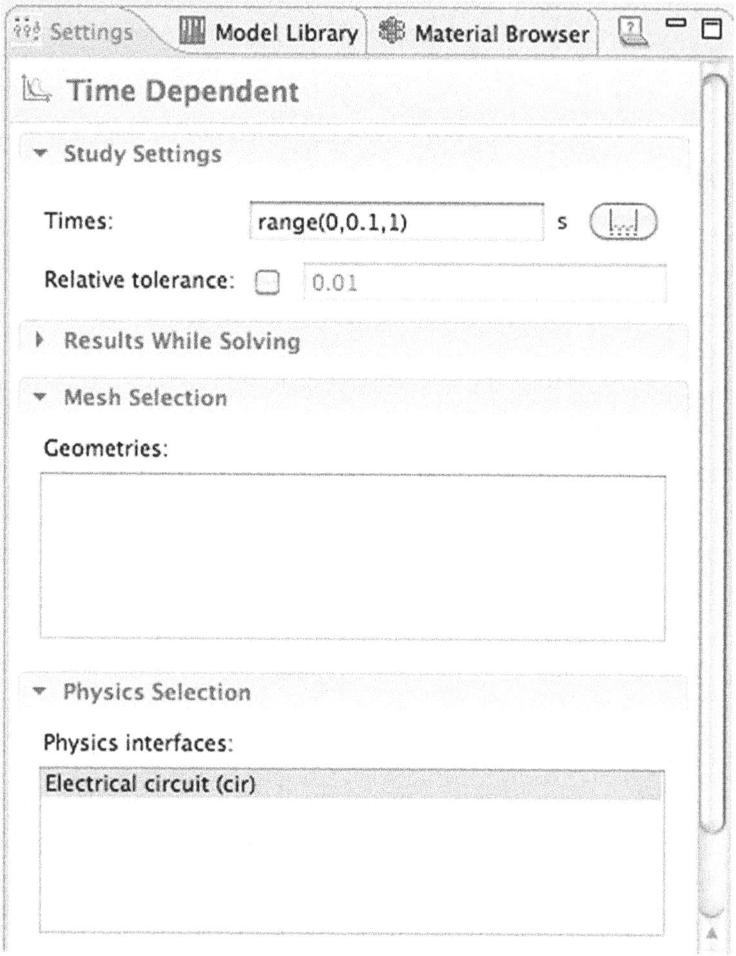

FIGURE 3.23 Settings - Time Dependent Window.

Click the Range button (at the right of the Times entry window).

Enter > Start = 0.0, Stop = 7e-3, Step = 1e-4.

Click > Replace.

Right-Click > Study 1.

Select > Compute.

NOTE *The default settings for the 1D Plot of the results of this model are not appropriate and thus the settings need to be modified.*

Click > 1D Plot Group 1 Twistie.

Click > Settings – 1D Plot Group 1 – Legend Twistie.

Select > Lower right from the Position Pull-down menu.

Select > 1 D Plot Group 1 – Global 1.

Click > Add Expression.

Select > Electrical Circuit > Voltage across device R1 (mod1.cir.R1.v).

Click > Plot.

See Figure 3.24 for the computed solution.

Figure 3.24 shows the inductor-resistor transient response curve.

Save the completed model.

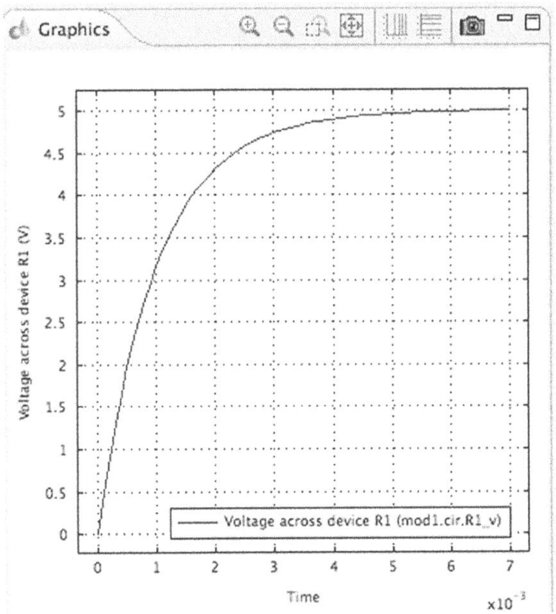

FIGURE 3.24 Inductor- Resistor Transient Response Curve.

0D Series-Resistor Parallel-Inductor-Capacitor Circuit Model

Startup 4.x.

Select > 0D.

Click > Next.

Click > Twistie for AC/DC in the Add Physics window.

Select > Electrical Circuit (cir).

Click > Add Selected.

Click > Next.

Select > Time Dependent in the Select Study Type window.

Click > Finish (Flag).

Click > Save As > MMUC4_0D_SRPLC1.mph.

NOTE *In the Electrical Circuit (cir) Interface, the Node Connections specify exactly where each component is connected. Each circuit must have a Ground connection and that Node is 0. The completed circuit must tie back to the Ground Node for completion.*

The completed battery-powered series resistor parallel inductor-capacitor circuit will be connected as follows:

Ground (0) <> (0) Voltage Source (1) <> (1) Resistor (2) <> (2) Inductor (0) <> (2) Capacitor (0)

NOTE *In this book <> means is electrically connected to.*

See Figure 3.25.

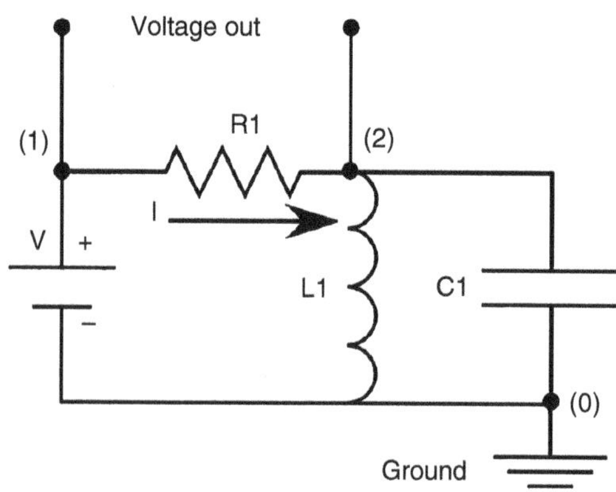

FIGURE 3.25 Circuit Diagram for the MMUC4_0D_SRPLC1.mph Model.

Figure 3.25 shows the Circuit Diagram for the MMUC4_0D_SRPLC1.mph model.

Right-Click in the Model Builder window on Electrical Circuit (cir).

Select > Voltage Source from the Pop-up menu.

See Figure 3.26.

Figure 3.26 shows the Electrical Circuit (cir) Pop-up window.

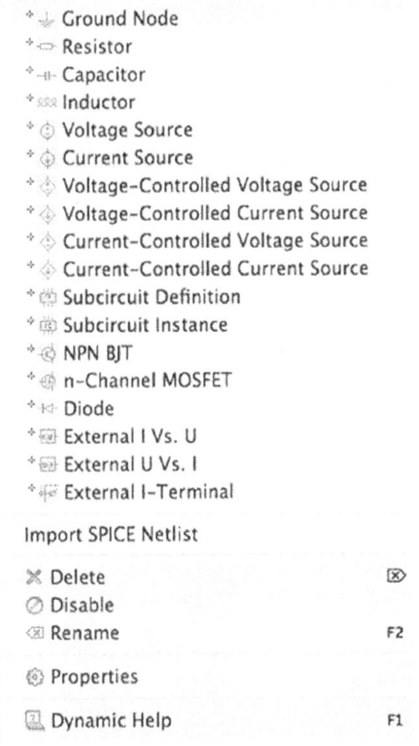

FIGURE 3.26 Electrical Circuit (cir) Pop-up Window.

Enter 5 in the Voltage Source – Device Parameters – Electric potential entry window.

Press the Tab key.

See Figure 3.27.

Figure 3.27 shows the Desktop Display – Settings – Voltage Source – Device Parameters – Electric potential entry window.

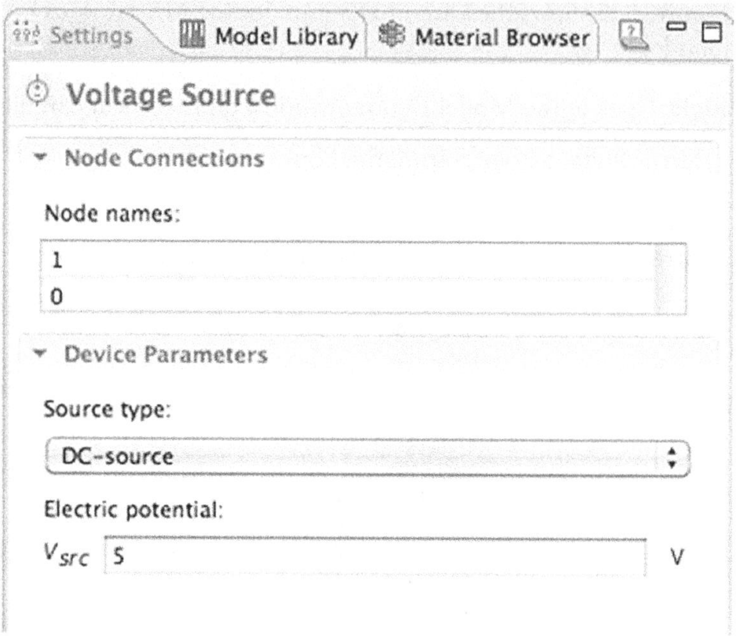

FIGURE 3.27 Desktop Display - Settings - Voltage Source - Device Parameters - Electric Potential Entry Window.

Right-Click in the Model Builder window on Electrical Circuit (cir).

Select > Resistor from the Pop-up menu.

This resistor is the first component in the series-parallel circuit.

Enter 2, 1 in the Settings – Resistor – Node Connections – Node names entry windows.

NOTE *When using 4.x on the Mac, the modeler will need to press the Tab key to see both entries.*

Enter the value in the Resistance entry window (3e3[ohm]).
See Figure 3.28.

Figure 3.28 shows the Settings – Resistor – Node Connections – Node names – Resistance.

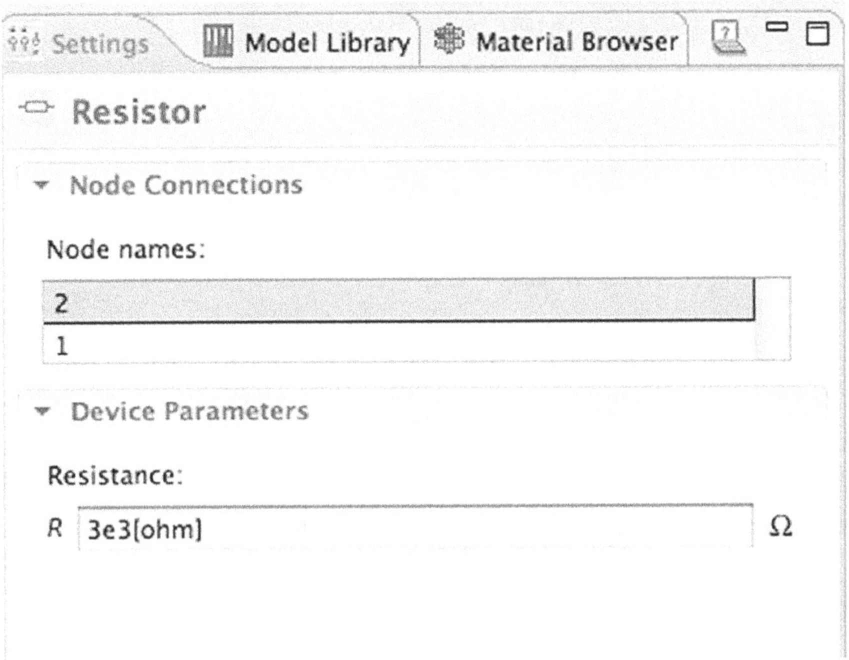

FIGURE 3.28 Settings – Resistor – Node Connections – Node names – Resistance.

Right-Click in the Model Builder window on Electrical Circuit (cir).

Select > Inductor from the Pop-up menu.

This inductor is the second component in the series circuit and the first component in the parallel circuit.

Enter 2, 0 in the Settings – Inductor – Node Connections – Node names entry windows.

Enter 100[mH] in the Settings – Inductor – Inductance entry window.

See Figure 3.29.

Figure 3.29 shows the Settings – Inductor – Device Parameters – Inductance.

Right-Click in the Model Builder window on Electrical Circuit (cir).

Select > Capacitor from the Pop-up menu.

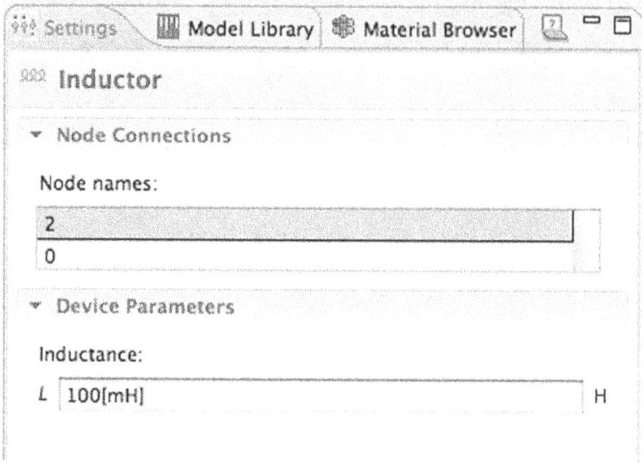

FIGURE 3.29 Settings - Inductor - Device Parameters - Inductance.

This capacitor is the third component in the series circuit and the second component in the parallel circuit.

Enter 2, 0 in the Settings – Capacitor – Node Connections – Node names entry windows.

Enter 1000[nF] in the Settings – Capacitor – Capacitance entry window.

See Figure 3.30.

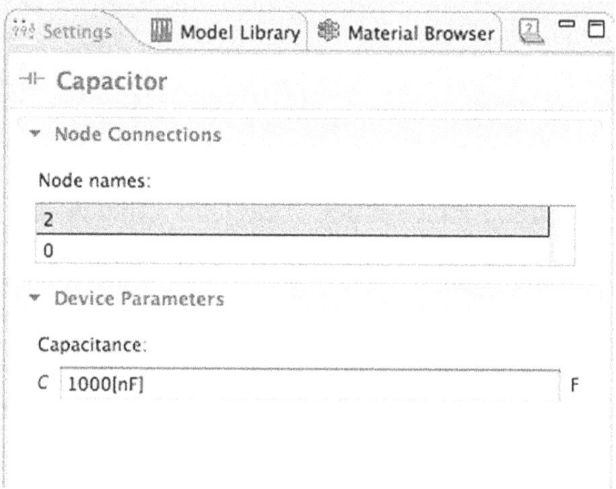

FIGURE 3.30 Settings - Capacitor - Device Parameters - Capacitance.

Now that the building of the series resistor-capacitor circuit model is complete, the next step is to prepare to solve the model.

NOTE *The default settings for the Time Dependent Study Type are not appropriate for the solution of this problem and thus need to be modified.*

Click > Study 1 Twistie.

Select > Step 1: Time Dependent. See Figure 3.31
Figure 3.31 shows the Settings – Time Dependent window.

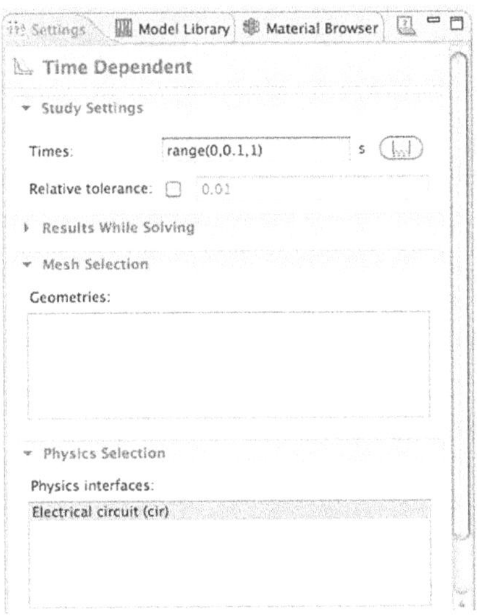

FIGURE 3.31 Settings - Time Dependent Window.

Click the Range button (at the right of the Times entry window).

Enter > Start = 0.0, Stop = 3.7e-2, Step = 1e-4.

Click > Replace.

Right-Click > Study 1.

Select > Compute.

NOTE *The default settings for the 1D Plot of the results of this model are not appropriate and thus the settings need to be modified.*

Click > 1D Plot Group 1 Twistie.

Click > Settings – 1D Plot Group 1 – Legend Twistie.

Select > Lower right from the Position Pull-down menu.

Select > 1 D Plot Group 1 - Global 1.

Click > Add Expression.

Select > Electrical Circuit > Voltage across device R1 (mod1.cir.R1.v).

Click > Plot.

See Figure 3.32 for the computed solution.

Figure 3.32 shows the series-resistor parallel-inductor-capacitor transient response curve.

Save the completed model.

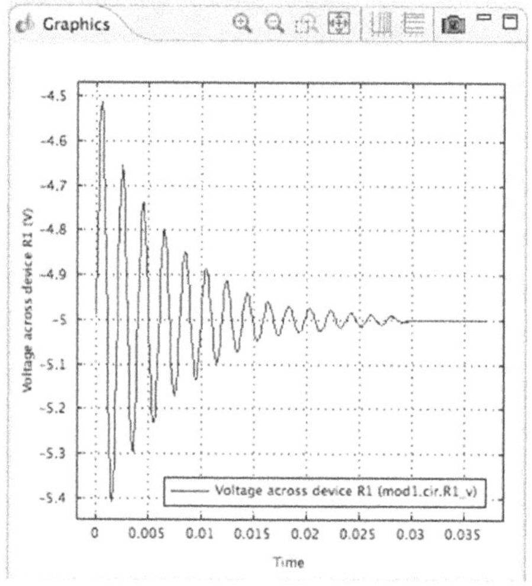

FIGURE 3.32 Series-Resistor Parallel-Inductor-Capacitor Transient Response Curve.

0D Basic Circuit Models Analysis and Conclusions

The 0D Resistor-Capacitor Series Circuit Model and the 0D Inductor-Resistor Series Circuit Model demonstrate that the addition of inductance or capacitance into a circuit along with resistance results in a finite transient response time. The transient response time of such circuits increases proportionally to the value of either the inductance or the capacitance {3.24}.

The 0D Series-Resistor Parallel-Inductor-Capacitor Circuit Model demonstrates that there is a non-linear interaction that may occur when both inductance and capacitance are incorporated into the same circuit. This potentially useful interaction is known as resonance {3.25}. Resonance can be very useful, somewhat annoying, or very destructive, depending upon whether it is purposely designed into the circuit or comes about inadvertently as a result of a lack of careful design.

NOTE

Useful resonance is found in designs for oscillators, clocks, mechanical vibrators, etc. Somewhat annoying resonance (ringing) can occur in almost any circuit due to unanticipated (parasitic) capacitances and/or inductances incorporated into the physical circuit inadvertently in the design process. Very destructive resonances are typically indicative of flaws incorporated into the artifact during the design process.

FIRST PRINCIPLES AS APPLIED TO 0D MODEL DEFINITION

First Principles Analysis derives from the fundamental laws of nature. In the case of models in this book or from any other source, the modeler should be able to demonstrate that the calculated results derived from the models are consistent with the laws of physics and the basic observed properties of materials. In the case of this Classical Physics Analysis, the laws of conservation in physics require that what goes in (as mass, energy, charge, etc.) must come out (as mass, energy, charge, etc.) or must accumulate within the boundaries of the model. To do otherwise violates fundamental principles.

NOTE

In the COMSOL Multiphysics software, the default interior boundary conditions are set to apply the conditions of continuity in the absence of sources (e.g. heat generation, charge generation, molecule generation, etc.) or sinks (e.g. heat loss, charge recombination, molecule loss, etc.).

The careful modeler must be able to determine by inspection of the model that the appropriate factors have been considered in the development of the specifications for the various geometries, for the material properties of each domain, and for the boundary conditions. He must also be knowledgeable of the implicit assumptions and default specifications that are normally incorporated into the COMSOL Multiphysics software model, when a model is built using the default settings.

Consider, for example, the three circuit models developed earlier in this chapter. By choosing to develop those models in the simplest 0D non-geometrical mode, the implicit assumption is that the circuit is isothermal (no heat generation or loss). That assumption essentially eliminates the consideration of thermally related changes (mechanical, electrical, etc.). It also assumes that the materials are homogeneous and isotropic and that there are no thin electrical contact barriers at the electrical junctions. None of these assumptions are typically true in the general case. However, by making such assumptions, it is possible to easily build a 0D First Approximation Model.

NOTE

A First Approximation Model is one that captures all the essential features of the problem that needs to be solved, without dwelling excessively on small details. A good First Approximation Model will yield an answer that enables the modeler to determine if he needs to invest the time and the resources required to build a more highly detailed model.

Also, the modeler needs to remember to name model parameters carefully as pointed out in Chapter 1.

REFERENCES

3.1 COMSOL Chemical Reaction Engineering Module Users Guide, pp. 18.

3.2 COMSOL Multiphysics Users Guide, pp. 584.

3.3 COMSOL Multiphysics Users Guide, pp. 43.

3.4 http://en.wikipedia.org/wiki/Electrical_circuit

3.5 http://en.wikipedia.org/wiki/Electronic_circuit

3.6 http://en.wikipedia.org/wiki/Georg_Ohm

3.7 http://en.wikipedia.org/wiki/Ohms_Law

3.8 http://en.wikipedia.org/wiki/James_Clerk_Maxwell

3.9 http://en.wikipedia.org/wiki/Electromotive_force

3.10 http://en.wikipedia.org/wiki/Charge_conservation

3.11 http://en.wikipedia.org/wiki/Gustav_Kirchhoff

3.12 http://en.wikipedia.org/wiki/Kirchhoff%27s_circuit_laws

3.13 http://en.wikipedia.org/wiki/Ewald_Georg_von_Kleist

3.14 http://en.wikipedia.org/wiki/Capacitor

3.15 http://en.wikipedia.org/wiki/Pieter_van_Musschenbroek

3.16 http://en.wikipedia.org/wiki/Dielectric

3.17 http://en.wikipedia.org/wiki/Michael_Faraday

3.18 http://en.wikipedia.org/wiki/Joseph_Henry

3.19 http://en.wikipedia.org/wiki/Electromagnetic_induction

3.20 http://en.wikipedia.org/wiki/Maxwell%27s_equations

3.21 E.U. Condon and H. Odishaw, "Handbook of Physics", McGraw-Hill, New York, 1958, pp. 4-28 – 4-46.

3.22 http://en.wikipedia.org/wiki/Inductance

3.23 http://en.wikipedia.org/wiki/Capacitance

3.24 J. J. Brophy, "Basic Electronics for Scientists", McGraw-Hill, New York, 1966, pp. 104 – 111.

3.25 http://en.wikipedia.org/wiki/Resonance

Suggested Modeling Exercises

1. Build, mesh, and solve the 0D Resistor-Capacitor Series Circuit Model problem presented earlier in this chapter.
2. Build, mesh, and solve the 0D Inductor-Resistor Series Circuit Model problem presented earlier in this chapter.
3. Build, mesh, and solve the 0D Series-Resistor Parallel-Inductor-Capacitor Circuit Model problem presented earlier in this chapter.
4. Change the values of the components comprising the 0D Resistor-Capacitor Series Circuit Model problem. Compute the new solution. Analyze, compare, and contrast the results with the new component values to the results found earlier.
5. Change the values of the components comprising the 0D Inductor-Resistor Series Circuit Model problem. Compute the new solution.

Analyze, compare, and contrast the results with the new component values to the results found earlier.

6. Change the values of the components comprising the 0D Series-Resistor Parallel-Inductor-Capacitor Circuit Model problem. Compute the new solution. Analyze, compare, and contrast the results with the new component values to the results found earlier.

CHAPTER 4

1D MODELING USING COMSOL MULTIPHYSICS 4.X

In This Chapter

- Guidelines for 1D Modeling in 4.x
  - 1D Modeling Considerations
- 1D Basic Models
  - 1D KdV Equation Model
  - 1D Telegraph Equation Model
  - 1D Spherically Symmetric Transport Model
- First Principles as Applied to 1D Model Definition
- References
- Suggested Modeling Exercises

GUIDELINES FOR 1D MODELING IN 4.X

NOTE

In this chapter, the modeler is introduced to the development and analysis of 1D models. 1D models have a single geometric dimension (x). 1D models are typically classed by level of difficulty as being of two types, introductory and advanced. In 4.x, there are two types of 1D geometries: 1D and 1D Axisymmetric.

In this text, only the introductory 1D models will be presented. Such 1D models have proven to be very valuable to the science and engineering communities, both in the past and currently, as first-cut evaluations of potential physical behavior under the influence of external stimuli. Those model responses and other such ancillary information are then gathered and screened early in a project for potential later use in building higher-dimensionality (2D, 3D, etc.) field-based (electrical, magnetic, etc.) models.

For information on the development of the more advanced models using the 1D Axisymmetric coordinate system, the modeler should consult the 4.x literature {4.1}.

1D Modeling Considerations

1D Modeling can potentially be both the least difficult and the most difficult type of model to build, no matter which modeling software is used. In a 1D model, the modeler can only have a single dimension (a single line or a sequence of line segments) as the modeling space. The potentially least difficult aspect of 1D model building arises from the fact that the geometry is dimensionally simple. However, the underlying physics in a 1D model can range from relatively easy (simple) to extremely complicated (complex).

The 1D model implicitly assumes that the modeling properties, such as the energy flow, the materials properties, the environment, and any other unspecified conditions and variables of interest are homogeneous, isotropic, and/or constant throughout the entire domain, both within the model and in the environs of the model. In other words, the properties assigned to the 1D model are representative of the properties of typical adjacent non-modeled regions. Considering that, the modeler needs to insure that all of the basic modeling conditions and associated parameters have been properly considered, defined, and set to the appropriate value.

NOTE

For any exploratory model built, the modeler should be able to anticipate reasonably accurately the expected numerical results of the model calculation. Calculated model solution values that widely deviate from the anticipated (estimated) values or from the comparison values measured in experimentally derived realistic models are probably indicative of one or more modeling errors either in the original model design, in the earlier model analysis, in the understanding of the underlying physics, or are simply due to incorporated human errors.

1D BASIC MODELS

1D KdV Equation Model

The KdV Equation {4.2} is a well-known example of a group of nonlinear partial differential equations {4.3} termed exactly solvable {4.4}. The exactly solvable type of equation has solutions that can be specified with exactness and precision.

NOTE

Nonlinear partial differential equations are extremely important in the formulation of an accurate mathematical description of physical systems {4.5}. Nonlinear partial differential equations are inherently difficult to solve and when solved often require a unique approach for the solution of each different type of equation.

Diederik Korteweg and Gustav de Vries solved the KdV equation in 1895. The KdV equation mathematically describes the propagation of a surface disturbance on a shallow canal. Their effort to solve this wave propagation problem was enabled as a result of the availability of earlier observations (data) by John Scott Russell in 1834 {4.6} and others. Subsequent activity in this mathematical area has led to soliton applications of the KdV equation in magnetics {4.7} and optics {4.8}. Work on using the KdV equation in the solution of soliton propagation problems is currently an active area of research.

The following numerical solution model (MMUC4_1D_KdV_1.mph) is based on a model (kdv_equation.mph) that was originally developed by COMSOL for distribution with 4.x as an Equation-Based Model. In the first part of this chapter, we will demonstrate how to build a model in 4.x for the solution of the KdV equation.

NOTE

It is important for the modeler to personally build each model presented within the text. There is no substitute for the hands-on experience of actually building, meshing, solving, and viewing the results of a solved model. The inexperienced modeler will many times make and subsequently correct errors, adding to his experience and his fund of modeling knowledge. Solving even the simplest model will expand the modeler's fund of knowledge.

The KdV Equation (as written in standard notation) is:

$$\partial_t u + \partial_x^3 u + 6u\partial_x u = 0 \tag{4.1}$$

In the 4.x Model Library kdv_equation.pdf documentation, the formula is shown as:

$$u_t + u_{xxx} = 6uu_x \quad in\,\Omega = [-8,8] \tag{4.2}$$

The difference between the two equations is that (4.2) is the negative form of (4.1), which will be adjusted during the results analysis.

The boundary conditions are periodic, as shown in (4.3)

$$u(-8,t) = u(8,t) \quad periodic \tag{4.3}$$

The initial condition for this model is:

$$u(x,0) = -6\,\mathrm{sech}^2(x) \tag{4.4}$$

Once the modeler builds and solves this model, it will be seen that the pulse immediately divides into two soliton pulses, with different width and propagation speeds.

NOTE *4.x does not evaluate third derivatives directly. Thus the original equation (4.2) needs to be rewritten as a system of two variables so that it can be solved in 4.x.*

Rewriting (4.2):

$$u_{1t} + u_{2x} = 6u_1 u_{1x} \tag{4.5}$$

And:

$$u_2 = u_{1xx} \tag{4.6}$$

And:

$$u(x,t) = \frac{1}{2}\lambda\,\mathrm{sech}^2\left(\frac{\sqrt{\lambda}}{2}(x - \lambda t - a)\right) \tag{4.7}$$

Where:

λ= phase velocity.

a = arbitrary constant.

x = position.

t = time.

Building the 1D KdV Equation Model

Startup 4.x.

Select > 1D.

Click > Next.

Click > Twistie for Mathematics in the Add Physics window.

Click > Twistie for PDE Interfaces > General Form PDE (g).

Click > Add Selected.

In the Model Wizard – Dependent variables – Number of dependent variables edit window.

Enter 2.

See Figure 4.1.

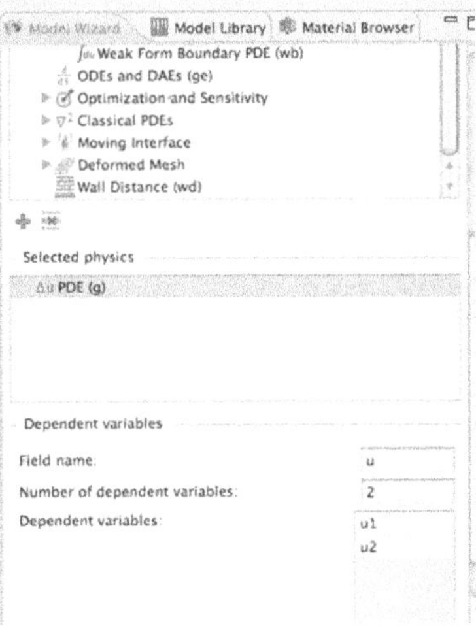

FIGURE 4.1 Desktop Display - Model Wizard - Dependent Variables - Number of Dependent Variables Edit Window.

Figure 4.1 shows the Desktop Display – Model Wizard – Dependent variables – Number of dependent variables edit window.

Click > Next.

Select > Time Dependent in the Select Study Type window.

Click > Finish (Flag).

Click > Save As > MMUC4_1D_KdV_1.mph.

See Figure 4.2.

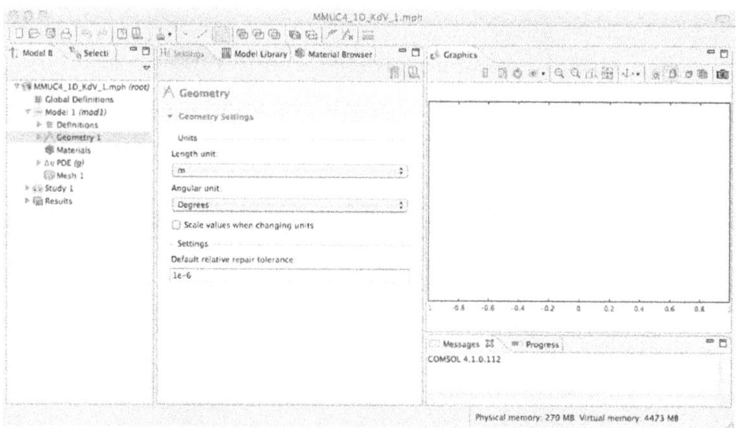

FIGURE 4.2 Desktop Display for the MMUC4_1D_KdV_1.mph Model.

Figure 4.2 shows the Desktop Display for the MMUC4_1D_KdV_1. mph model.

Right-Click in the Model Builder window on Geometry 1.

Select > Interval from the Pop-up menu.

Enter -8 in the Settings – Interval – Left endpoint entry window.

Enter 8 in the Settings – Interval – Right endpoint entry window.

Press the Tab key.

See Figure 4.3.

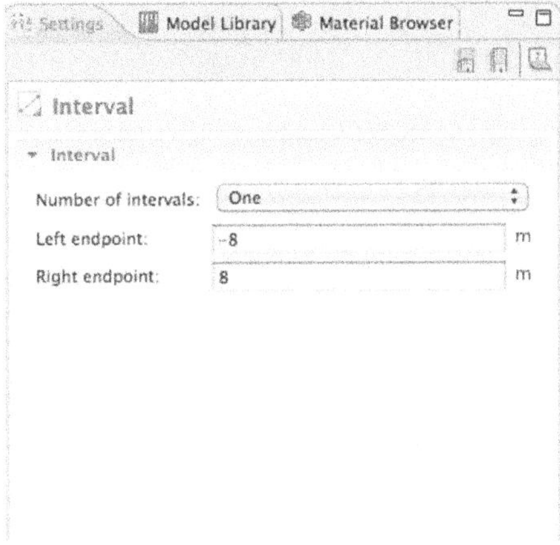

FIGURE 4.3 Desktop Display - Settings - Interval Entry Windows.

Figure 4.3 shows the Desktop Display – Settings – Interval entry windows.

Right-Click in the Model Builder window on Model 1 > Δu PDE (g).

Select > Periodic Condition from the Pop-up menu.

Select > Settings – Periodic Condition – Selection > All boundaries from the Pop-up menu.

See Figure 4.4.

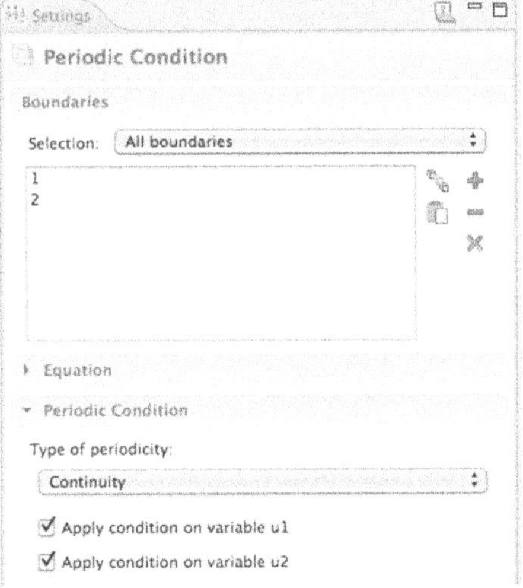

FIGURE 4.4 Desktop Display - Settings - Periodic Condition - Selection Window.

Figure 4.4 shows the Desktop Display – Settings – Periodic Condition – Selection window.

Click in the Model Builder window on Model 1 > Δu PDE (g) > General Form PDE 1.

Enter u2 in the first row of the Settings – General Form PDE – Conservative Flux – Γ edit-field array.

Enter u1x in the second row of the Settings – General Form PDE – Conservative Flux – Γ edit-field array.

Enter 6*u1*u1x in the first row of the Settings – General Form PDE – Source Term – *f* edit-field array.

Enter u2 in the second row of the Settings – General Form PDE – Source Term – *f* edit-field array.

Enter 0 in the second column - second row of the Settings – General Form PDE – Damping or Mass Coefficient – d_a edit-field array.

See Figure 4.5.

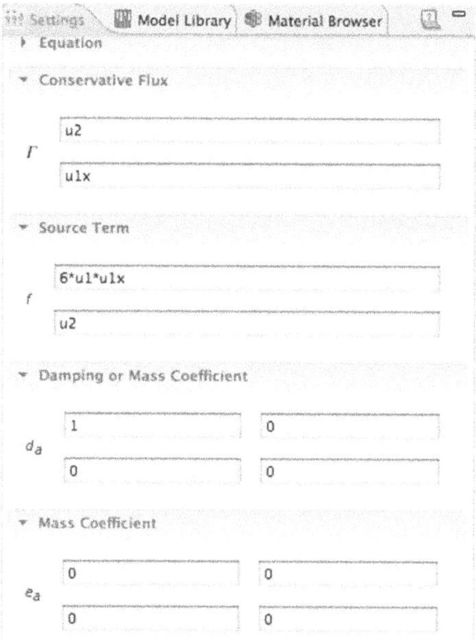

FIGURE 4.5 Desktop Display - Settings - General Form PDE Coefficients.

Figure 4.5 shows the Desktop Display – Settings – General Form PDE coefficients.

Click in the Model Builder window on Model 1 > Δu PDE (g) > Initial Values 1.

Enter -6*sech(x[1/m])^2 in the Settings – Initial Values – Initial value for *u*1.

Enter -24*sech(x[1/m])^2*tanh(x[1/m])^2+12*sech(x[1/m])^2*(1-tanh(x[1/m])^2) in the Settings – Initial Values – Initial value for *u*2.

NOTE *Inserting x[1/m] converts the dimensioned value of the x coordinate to a dimensionless number as required for its use in the hyperbolic trigonometric functions.*

See Figure 4.6.

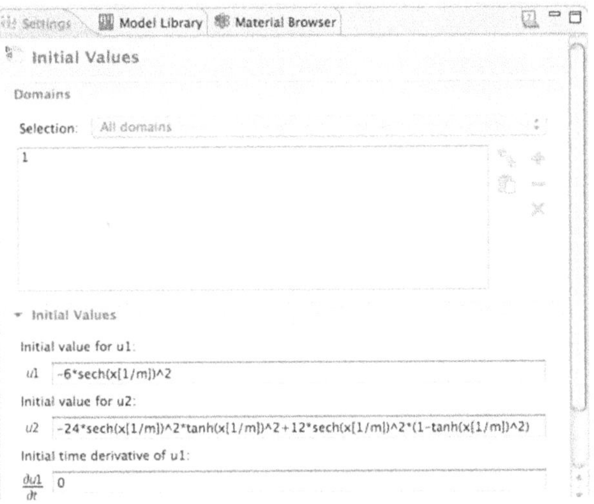

FIGURE 4.6 Desktop Display - Settings - Initial Values Coefficients.

Figure 4.6 shows the Desktop Display – Settings – Initial Values coefficients.

Mesh 1

NOTE *The default settings for the Mesh Type are not appropriate for the solution of this problem and thus need to be modified.*

Right-Click in the Model Builder window on Model 1 > Mesh 1.

Select > Edge.

Click Size.

Click > Element Size Parameters twistie.

Enter 0.1 in the Settings – Size – Element Size Parameters – Maximum element size edit window.

NOTE *The maximum element size setting (0.1) for the Mesh Type is chosen to ensure adequate resolution of the mesh for use with the hyperbolic trigonometric functions in the solution of this problem.*

See Figure 4.7.

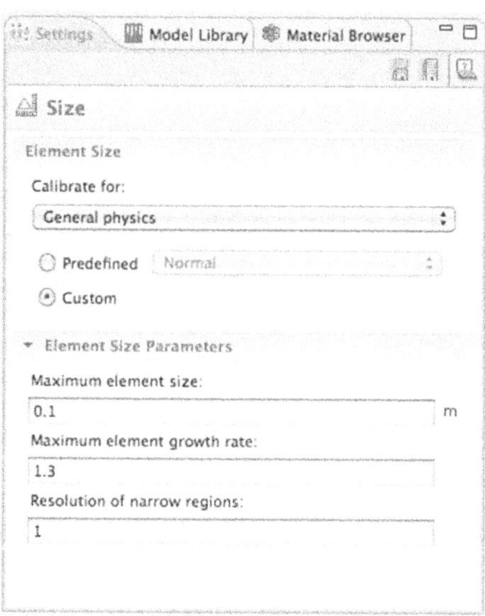

FIGURE 4.7 Desktop Display - Settings - Size - Element Size Parameters - Maximum Element Size Coefficient.

Figure 4.7 shows the Desktop Display – Settings – Size – Element Size Parameters – Maximum element size coefficient.

Click > Build All button.

After meshing, the modeler should see a message in the message window about the number of elements (160 elements) in the mesh.

See Figure 4.8.

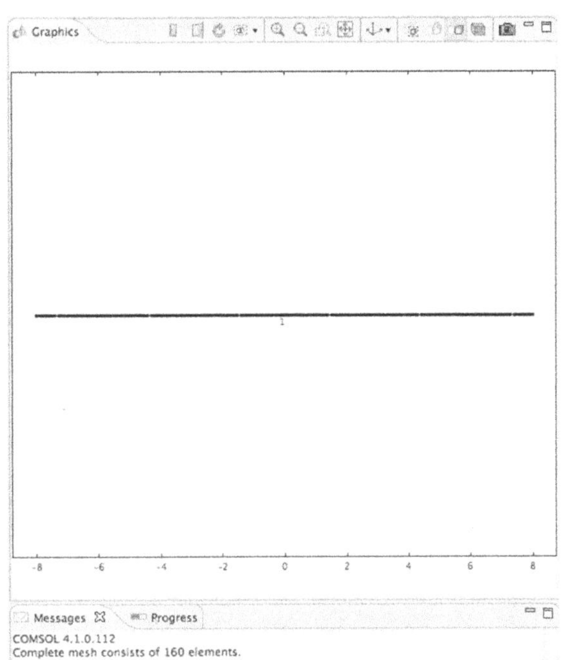

FIGURE 4.8 Desktop Display - Graphics - Meshed Domain.

Figure 4.8 shows the Desktop Display – Graphics – Meshed Domain.

Study 1

NOTE *The default settings for the Time Dependent Study Type are not appropriate for the solution of this problem and thus need to be modified.*

Click > Study 1 twistie.

Click > Step 1: Time Dependent.

Click the Range button (at the right of the Times entry window).

Enter > Start = 0.0, Stop = 2.0, Step = 2.5e-2.

Click > Replace.

Click > Relative tolerance check box.

Enter 1.0e-4 in the Relative tolerance edit window.

See Figure 4.9.

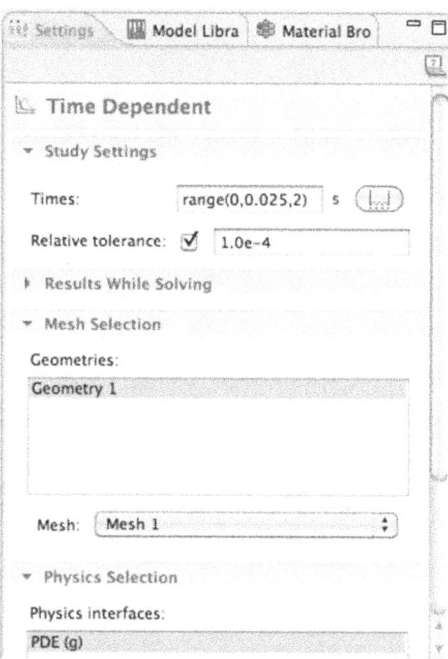

FIGURE 4.9 Desktop Display – Settings – Time Dependent – Study Settings.

Figure 4.9 shows the Desktop Display – Settings – Time Dependent – Study Settings.

Study 1 > Solver 1

In Model Builder,

Right-Click > Study 1.

Select > Show Default Solver.

NOTE *The default settings for the default solver for this model are not appropriate and thus the settings need to be modified.*

Click > Solver Configurations twistie.

Click > Solver 1 twistie.

Click > Time Dependent Solver 1.

In the Settings – Time Dependent Solver – Time Stepping section,

Select > Method > Generalized alpha from the Pull-down menu.

Click > Absolute Tolerance twistie.

Enter 1e-5 in the Absolute Tolerance – Tolerance edit window.

NOTE *The solver type and the absolute tolerance are chosen to ensure adequate resolution of the convergence process for use with the hyperbolic trigonometric functions in the solution of this problem.*

See Figure 4.10.

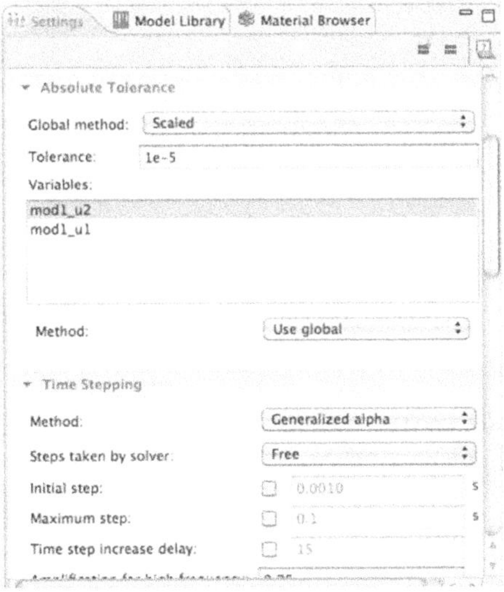

FIGURE 4.10 Desktop Display - Settings - Time Dependent - Solver Settings.

Figure 4.10 shows the Desktop Display – Settings – Time Dependent – Solver Settings.

In Model Builder, Right-Click Study 1 >Select > Compute.

Results

In Model Builder,

Click > Results > 1D Plot Group 1.

In Settings – 1D Plot Group – Data – Time selection,

Select > From list from the Pull-down menu.

In the Times window, locate and select 0.25.

See Figure 4.11.

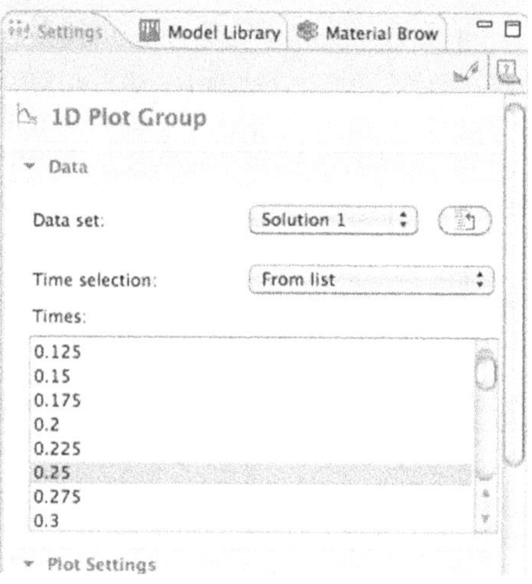

FIGURE 4.11 Desktop Display - Settings - 1D Plot Group - Time Data Settings.

Figure 4.11 shows the Desktop Display – Settings – 1D Plot Group – Time Data Settings.

Click > Plot button.

See Figure 4.12.

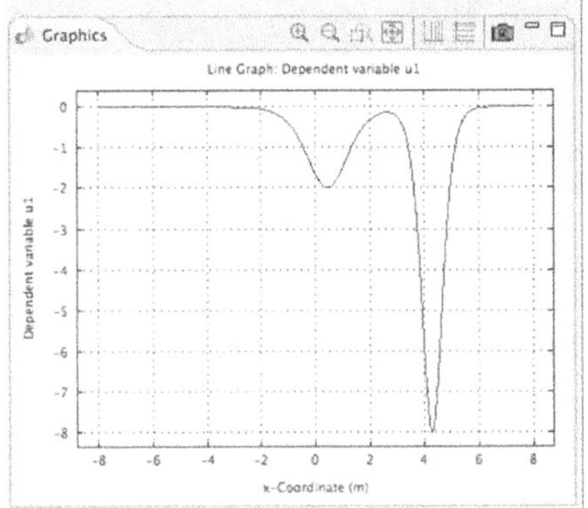

FIGURE 4.12 Desktop Display - Graphics - Initial KdV Solution Plot.

Figure 4.12 shows the Desktop Display – Graphics – Initial KdV Solution Plot.

As mentioned earlier, the plot needs to be inverted.

In Model Builder,

Click > Results > 1D Plot Group 1 twistie.

Click > Results > Line Graph 1.

In Settings – Line Graph – Y-Axis Data – Expression,

Enter -u1.

Click > Plot button.

See Figure 4.13.

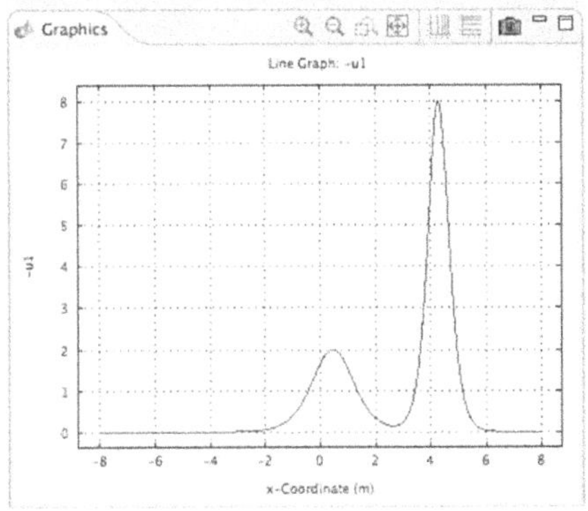

FIGURE 4.13 Desktop Display - Graphics - KdV Solution Plot.

Figure 4.13 shows the Desktop Display – Graphics – KdV Solution Plot.

KdV Equation Animation

To demonstrate the propagation of the soliton pulse or pulses in 4.x, the modeler can create an animation.

In Model Builder,

Right-Click > Results – Report.

Select > Animation.

In Settings – Scene,

Click > Go to Source button.

Select > All from the Time selection Pull-down menu.

Click > Results – Report – Animation 1.

In Settings – Animation – Output,

Select > Output type Movie from the Pull-down menu.

Select > Format GIF from the Pull-down menu.

NOTE *The detailed procedures for Animation in 4.x on the Macintosh and the PC are different. The Macintosh uses the GIF format. The PC uses the AVI format. If the modeler tries to use AVI on the Macintosh, an error results.*

In Settings – Animation – Frame Settings,

Click > Lock aspect ratio check box.

In Settings – Animation – Advanced,

Click > Antialiasing check box.

Click > Settings – Export.

NOTE *The movie will be exported to and run in the 4.x Graphics window, if there is no path designated in the Settings – Animation – Output Filename edit window.*

Saving the KdV Equation Animation Movie

The modeler can save the movie as a unique file.

Click > Settings – Animation – Output Browse button.

Select the desired location for saving the movie.

Enter the desired File Name in the Save-As edit window.

Click > Save.

Click > Save for the completed MMUC4_1D_KdV_1.mph KdV Equation model.

1D KdV Equation Model Summary and Conclusions

The 1D KdV Equation model is a powerful tool that can be used to explore soliton wave propagation in many different media (e.g. physical waves in liquids, electromagnetic waves in transparent media, etc.). As has been shown earlier in this chapter, the KdV Equation is easily and simply modeled with a 1D PDE mode model.

1D Telegraph Equation Model

The Telegraph Equation {4.9} was developed by Oliver Heaviside {4.10} and first published in the 1880s. The Telegraph Equation is based on a lumped-constant, four (4) terminal electrical component model, typically with earth (ground) as the return path, as shown in Figure 4.14.

See Figure 4.14.

Figure 4.14 shows the Telegraph Equation Lumped-Constant Circuit.

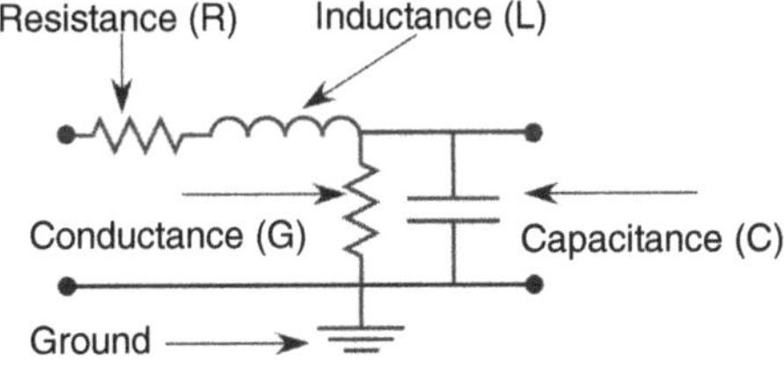

FIGURE 4.14 Telegraph Equation Lumped-Constant Circuit.

In this lumped-constant schematic model for telegraph wires (and similar transmission lines), there are four basic electrical components: R (Resistance) per unit of length, L (Inductance) per unit of length, G (Conductance) per unit of length, and C (Capacitance) per unit of length. The differential equations for Voltage (V) and Current (I) have the same form, as shown in equations 4.8 and 4.9.

Equation 4.8 shows the partial differential equation for voltage (V):

$$\frac{\partial^2}{\partial x^2}V = LC\frac{\partial^2}{\partial t^2}V + (RC+GL)\frac{\partial}{\partial t}V + GRV \tag{4.8}$$

Equation 4.9 shows the partial differential equation for current (I):

$$\frac{\partial^2}{\partial x^2}I = LC\frac{\partial^2}{\partial t^2}I + (RC+GL)\frac{\partial}{\partial t}I + GRI \tag{4.9}$$

Equations 4.8 and 4.9 are similar in form to the equation 4.10 as shown here for the COMSOL Multiphysics Telegraph Equation Model:

$$u_{tt} + (\alpha+\beta)u_t + \alpha\beta u = c^2 u_{xx} \tag{4.10}$$

Where α and β are positive constants, c is the transport velocity and u is the voltage.

Restating equation 4.8 in subscript notation:

$$u_{xx} = LC\,u_{tt} + (RC+GL)u_t + GR\,u \tag{4.11}$$

And rearranging the terms of equation 4.10:

$$u_{xx} = \frac{1}{c^2}u_{tt} + \frac{1}{c^2}(\alpha+\beta)u_t + \frac{1}{c^2}\alpha\beta\,u \tag{4.12}$$

Comparing equations 4.11 and 4.12 yields:

$$LC = \frac{1}{c^2} \tag{4.13}$$

And

$$\alpha+\beta = \frac{(RC+GL)}{LC} \tag{4.14}$$

Also:

$$\alpha\beta = \frac{GR}{LC} \tag{4.15}$$

Solving for α and β:

$$\alpha = \frac{CGL + C^2R - \sqrt{-4CGLR + (-CGL - C^2R)^2}}{2L}$$

$$and$$

$$\beta = \frac{CGL + C^2R + \sqrt{-4CGLR + (-CGL - C^2R)^2}}{2L} \tag{4.16}$$

Or:

$$\alpha = \frac{CGL + C^2R + \sqrt{-4CGLR + (-CGL - C^2R)^2}}{2L}$$

$$and$$

$$\beta = \frac{CGL + C^2R - \sqrt{-4CGLR + (-CGL - C^2R)^2}}{2L}$$

In the event that:

$$R = G == 0 \tag{4.17}$$

Then, the transmission line is considered lossless and the Telegraph Equation becomes:

$$u_{xx} = LC u_{tt} \tag{4.18}$$

Building the 1D Telegraph Equation Model

Startup 4.x.

Select > 1D.

Click > Next.

Click > Twistie for Mathematics in the Add Physics window.

Click > Twistie for PDE Interfaces > Coefficient Form PDE (c).

Click > Add Selected.

Click > Next.

Select > Time Dependent in the Select Study Type window.

Click > Finish (Flag).

Click > Save As > MMUC4_1D_TelE_1.mph.

See Figure 4.15.

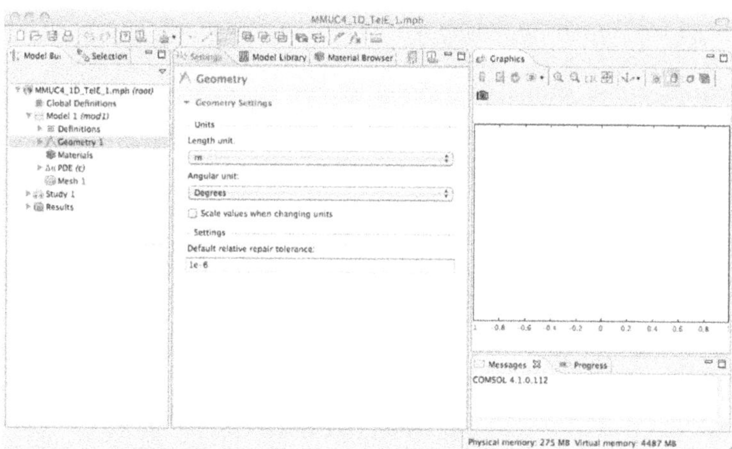

FIGURE 4.15 Desktop Display for the MMUC4_1D_TelE_1.mph Model.

Figure 4.15 shows the Desktop Display for the MMUC4_1D_TelE_1.mph model.

Right-Click in the Model Builder window on Global Definitions.

Select > Parameters from the Pop-up menu.

In the Settings – Parameters – Parameters edit window,

Enter the parameters as shown in Table 4.1.

TABLE 4.1 Parameters Window

Name	Expression	Description
c	1	Transport velocity
alpha	0.25	PDE coefficient parameter alpha
beta	0.25	PDE coefficient parameter beta

Click > Save.

NOTE *The modeler should save the parameters file at this time so that if the model needs to be recovered or the parameters need to be modified, they will be available to be changed or corrected without the need to reenter them.*

See Figure 4.16.

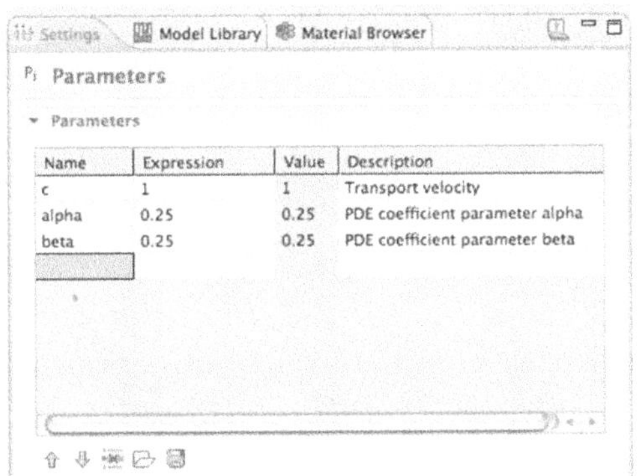

FIGURE 4.16 Desktop Display - Settings - Parameters Entry Windows.

Figure 4.16 shows the Desktop Display – Settings – Parameters entry windows.

Click > Settings – Parameter Save to File.

Enter > MMUC4_1D_TelE_1_Param.txt in the Save as edit window.

Geometry 1

Right-Click in the Model Builder window on Model 1 > Geometry 1.

Select > Interval from the Pop-up menu.

NOTE *The default settings for the specified interval are adequate for this model.*

Right-Click > Interval 1.

Select > Build Selected.

See Figure 4.17.

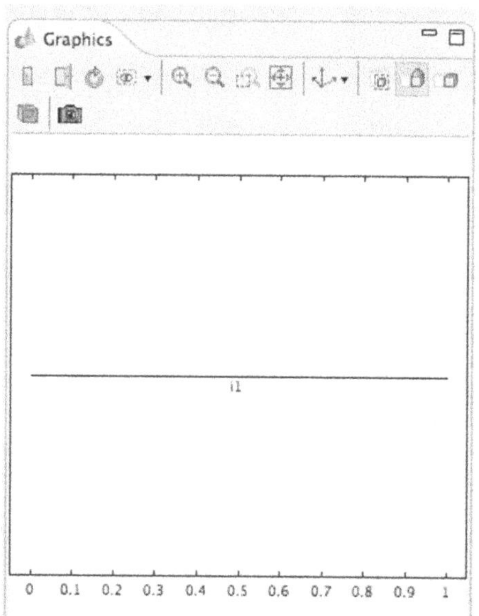

FIGURE 4.17 Desktop Display - Graphics - Geometry 1 Window.

Figure 4.17 shows the Desktop Display – Graphics – Geometry 1 Window.

PDE

Click in the Model Builder window on Model 1 > Δu PDE (c) twistie.

In the Model Builder window,

Click > Model 1 > Δu PDE (c) > Coefficient Form PDE 1.

NOTE

The default settings for the Coefficient Form PDE are not appropriate for the solution of this problem and thus need to be modified. The modeler should note the nominal equation (by Clicking on the Settings – Coefficient Form PDE – Equation twistie) and the manner in which the entered coefficients change that equation.

Enter c*c in Settings – Coefficient Form PDE – Diffusion Coefficient – *c* edit window.

Enter alpha*beta in Settings – Coefficient Form PDE – Absorption Coefficient – *a* edit window.

Enter -(alpha+beta)*ut in Settings – Coefficient Form PDE – Source Term – *f* edit window.

Enter 1 in Settings – Coefficient Form PDE – Mass Coefficient – e_a edit window.

Enter 0 in Settings – Coefficient Form PDE – Damping or Mass Coefficient – d_a edit window.

See Figure 4.18.

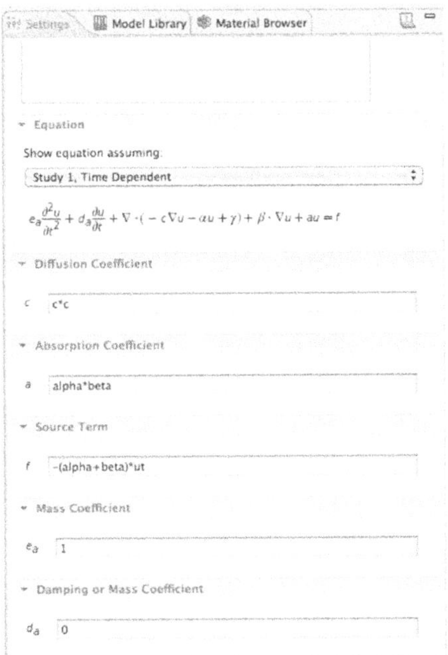

FIGURE 4.18 Desktop Display - Settings - Coefficient Form PDE Coefficients.

Figure 4.18 shows the Desktop Display – Settings – Coefficient Form PDE coefficients.

Zero Flux 1

NOTE *The default settings for the Zero Flux 1 set the Neumann Boundary Conditions by default and are appropriate for the solution of this problem. They thus do not need to be modified.*

Initial Values 1

Click in the Model Builder window on Model 1 > Δu PDE (c) > Initial Values 1.

NOTE *The default settings for the Initial Values are not appropriate for the solution of this problem and thus need to be modified. The entered equation describes a bell-shaped pulse applied to the transmission line.*

Enter exp(-3*((x[1/m]/0.2)-1)^2) in the Settings – Initial Values – Initial value for *u*.

NOTE *The [1/m] term applied to the x converts the dimensioned value of x to a dimensionless number for use in the exponential parametric equation.*

See Figure 4.19.

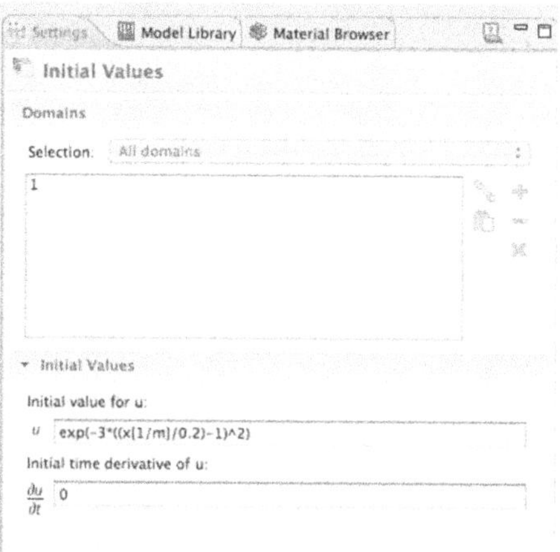

FIGURE 4.19 Desktop Display - Settings - Initial Values Coefficients.

Figure 4.19 shows the Desktop Display – Settings – Initial Values coefficients.

Mesh 1

NOTE *The default settings for the Mesh Type are not appropriate for the solution of this problem and thus need to be modified.*

Right-Click in the Model Builder window on Model 1 > Mesh 1.

Select > Edge.

Right-Click in the Model Builder window on Model 1 > Mesh 1.

Select > Refine.

Right-Click in the Model Builder window on Model 1 > Mesh 1.

Select > Build All.

After meshing, the modeler should see information in the message window about the number of elements (30 elements) in the mesh.

See Figure 4.20.

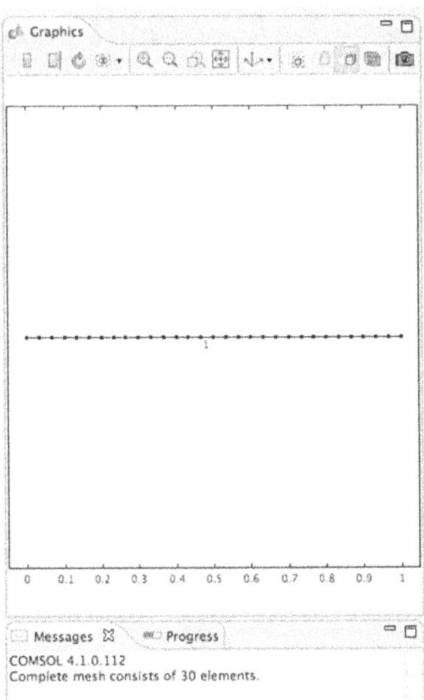

FIGURE 4.20 Desktop Display - Graphics - Meshed Domain.

Figure 4.20 shows the Desktop Display – Graphics – Meshed Domain.

Study 1

NOTE *A Parametric Sweep study step is added at this point to this model in order to observe the influence of changes in alpha and beta.*

Right-Click > Study 1.

Select > Parametric Sweep.

Click > Settings – Parameter names – Add.

NOTE *After clicking on the alpha parameter and before clicking on the beta parameter, hold down the Shift key and then click on the beta parameter to add both of the parameters.*

Select > PDE coefficient parameter alpha (alpha).

Select > PDE coefficient parameter beta (beta).

Click > OK button.

Enter > 0.25 0.25 0.5 0.5 1 1 2 2 in the Parameter values edit window.

NOTE *The individual parametric values for alpha and beta are entered as a string of numbers. Each number is separated from the next number by a single space (e.g. alpha1 beta1 alpha2 beta2 etc.).*

See Figure 4.21.

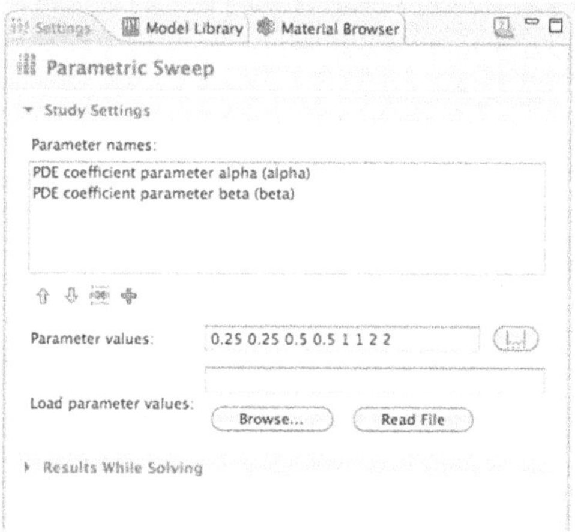

FIGURE 4.21 Desktop Display - Settings - Parametric Sweep - Study Settings Edit Windows.

Figure 4.21 shows the Desktop Display – Settings – Parametric Sweep – Study Settings edit windows.

Default Solver

Right-Click > Study 1.

Select > Show Default solver.

Click > in the Model Builder window on Model 1 > Study 1 > Solver Configurations twistie.

Click > in the Model Builder window on Model 1 > Study 1 > Solver Configurations > Solver 1 twistie.

Click > Time-Dependent Solver 1.

Click > Settings – Time-Dependent Solver – Time Stepping twistie.

Click > Settings – Time-Dependent Solver – Time Stepping – Method.

Select > Generalized alpha from the Method Pull-down menu.

NOTE *The Generalized alpha ODE solver is usually better for solving this type of wave equation model ($d_a = 0$, $e_a = 1$). It also avoids the need to manually specify time-step size values.*

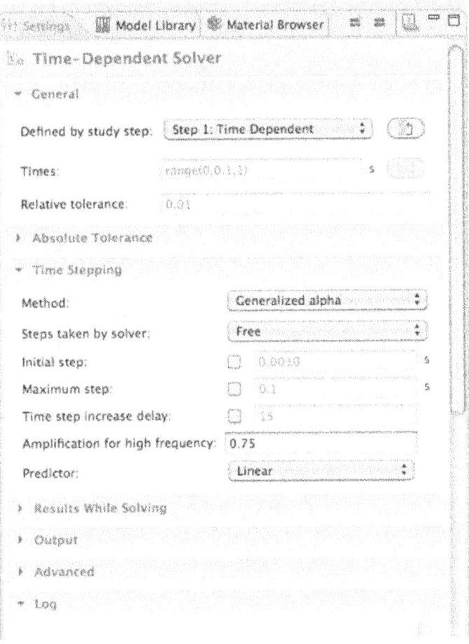

FIGURE 4.22 Desktop Display – Settings – Time-Dependent Solver – Time Stepping – Method Edit Windows.

See Figure 4.22.

Figure 4.22 shows the Desktop Display – Settings – Time-Dependent Solver – Time Stepping – Method edit windows.

In Model Builder, Right-Click Study 1 > Select > Compute.

Results

NOTE *The modeler should note that the shape and amplitude of the pulse change as a function of time and the values of the parameters alpha and beta.*

1D Plot Group 2

In Model Builder,

Right-Click > Results.

Select > 1D Plot Group from the Pull-down menu to create 1D Plot Group 2.

In Settings – 1D Plot Group – Data – Data set,

Select > Solution 2 from list from the Pull-down menu.

In Settings – 1D Plot Group – Data – Parameter selection (alpha, beta),

Select > From list from the Pull-down menu.

In Settings – 1D Plot Group – Data – Parameter values,

Select > 1:alpha=0.25,beta=0.25.

In Settings – 1D Plot Group – Data – Time selection,

Select > Interpolated from the Pull-down menu.

In the Times edit window,

Enter > 0 0.5 1.

See Figure 4.23.

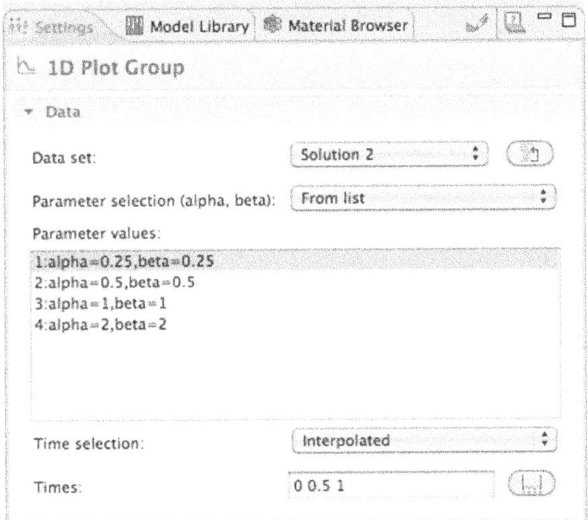

FIGURE 4.23 Desktop Display - Settings - 1D Plot Group - Data Edit Windows.

Figure 4.23 shows the Desktop Display – Settings – 1D Plot Group – Data edit windows.

Right-Click > Results > 1D Plot Group 2.

Select > Line Graph.

Click > Settings – Line Graph – Selection – Selection.

Select > All domains from the Pull-down menu.

Click > Settings – Line Graph – X-Axis Data – Replace Expression (green/orange arrows on the far right).

Select > Geometry and Mesh > Coordinate > x-Coordinate (x).

See Figure 4.24.

Figure 4.24 shows the Desktop Display – Settings – Line Graph – X-Axis Data settings.

Click > Plot button.

See Figure 4.25.

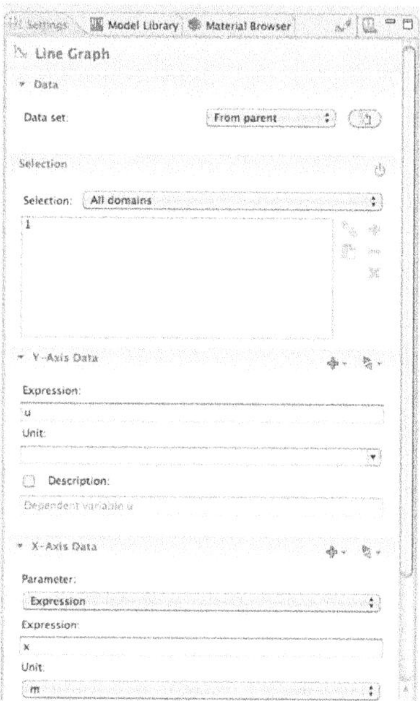

FIGURE 4.24 Desktop Display - Settings - Line Graph - X-Axis Data Settings.

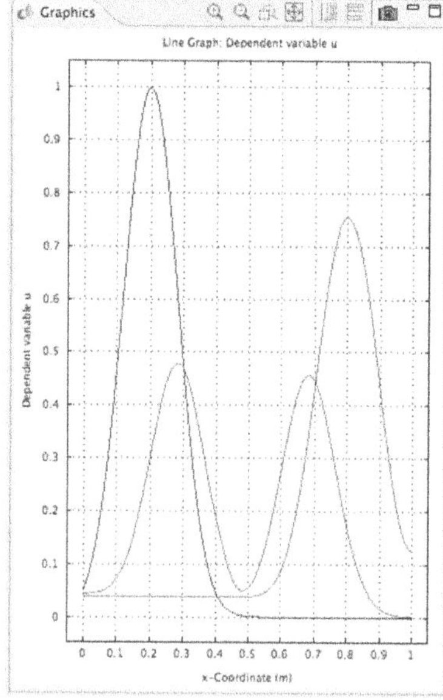

FIGURE 4.25 Desktop Display - Graphics - Initial Telegraph Equation Solution Plot.

Figure 4.25 shows the Desktop Display – Graphics – Initial Telegraph Equation Solution Plot.

NOTE *The modeler can now note that the shape and amplitude of the pulse change as a function of time and the values of the parameters alpha and beta.*

In Model Builder,

Click > Results > 1D Plot Group 2.

In Settings – 1D Plot Group – Data – Parameter values,

Click > 2:alpha=0.5,beta=0.5.

Click > Plot button.

See Figure 4.26.

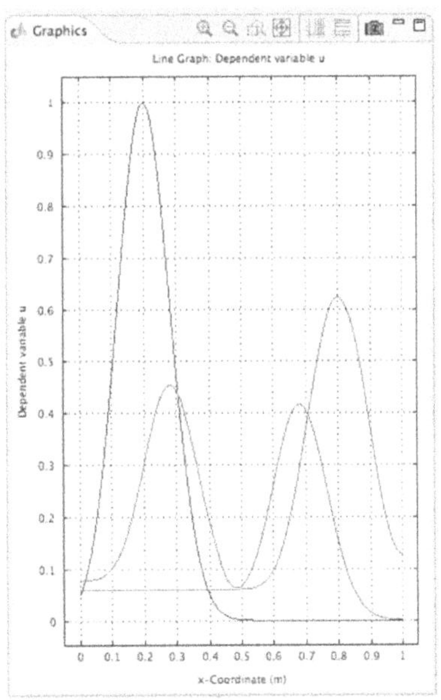

FIGURE 4.26 Desktop Display - Graphics - Telegraph Equation Solution Plot (alpha = beta = 0.5).

Figure 4.26 shows the Desktop Display – Graphics – Telegraph Equation Solution Plot (alpha = beta = 0.5).

In Model Builder,

Click > Results > 1D Plot Group 2.

In Settings – 1D Plot Group – Data – Parameter values,

Click > 3:alpha=1,beta=1.

Click > Plot button.

See Figure 4.27.

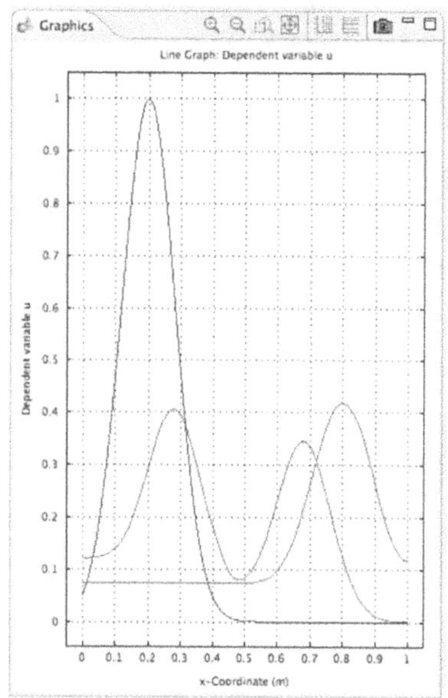

FIGURE 4.27 Desktop Display - Graphics - Telegraph Equation Solution Plot (alpha = beta = 1).

Figure 4.27 shows the Desktop Display – Graphics – Telegraph Equation Solution Plot (alpha = beta = 1).

In Model Builder,

Click > Results > 1D Plot Group 2.

In Settings – 1D Plot Group – Data – Parameter values,

Click > 4:alpha=2,beta=2.

Click > Plot button.

See Figure 4.28.

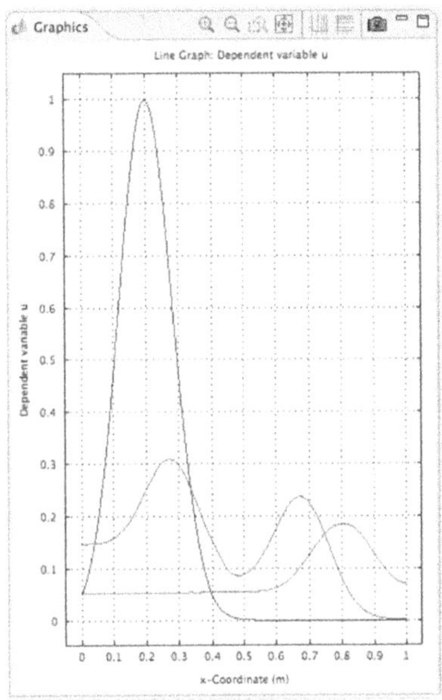

FIGURE 4.28 Desktop Display - Graphics - Telegraph Equation Solution Plot (alpha = beta = 2).

Figure 4.28 shows the Desktop Display – Graphics – Telegraph Equation Solution Plot (alpha = beta = 2).

Telegraph Equation Animation

To demonstrate the propagation of the Telegraph Equation pulse or pulses in 4.x, the modeler can create an animation.

In Model Builder,

Right-Click > Results – Report.

Select > Animation.

In Model Builder,

Click > Results – Report – Animation 1.

In Settings – Animation – Output,

Select > Output type Movie from the Pull-down menu.

Select > Format GIF from the Pull-down menu.

NOTE *The detailed procedures for Animation in 4.x on the Macintosh and the PC are different. The Macintosh uses the GIF format. The PC uses the AVI format. If the modeler tries to use AVI on the Macintosh, an error results.*

In Settings – Animation – Frame Settings,

Click > Lock aspect ratio check box.

In Settings – Animation – Parameter Sweep,

Click > Solutions.

Select > Outer solutions from the Pull-down menu.

In Settings – Animation – Parameter Sweep – Inner type,

Select > All from the Pull-down menu.

Click > Settings – Animation – Advanced twistie.

In Settings – Animation – Advanced,

Click > Antialiasing check box.

In Settings – Animation – Parameter Sweep – Parameter values,

Select > 1:alpha=0.25,beta=0.25 by double-clicking.

Click > Settings – Export.

NOTE *The movie will be exported to and run in the 4.x Graphics window, if there is no path designated in the Settings – Animation – Output Filename edit window. If the modeler wishes to run the Animation again, repeat the previous Selection and Export instructions.*

Saving the Telegraph Equation Animation Movie

The modeler can save the movie as a unique file.

Click > Settings – Animation – Output Browse button.

Select the desired location for saving the movie.

Enter the desired File Name in the Save-as edit window.

Click > Save.

In Settings – Animation – Parameter Sweep – Parameter values,

Select > 1:alpha=0.25,beta=0.25 by double-clicking.

Click > Settings – Export.

Click > Save for the completed MMUC4_1D_TelE_1.mph Telegraph Equation model.

1D Telegraph Equation Model Summary and Conclusions

The 1D Telegraph Equation model is a powerful tool that can be used to explore pulse wave propagation in electrical communication cables. As has been shown earlier in this chapter, the Telegraph Equation is easily and simply modeled with a 1D PDE mode model.

1D Spherically Symmetric Transport Model

The fabrication and use of small and microscopic spherical and spheroidal particles for the manufacturing of metal artifacts date to as early as 1200 B.C. {4.11}. In the present era, spherical and spheroidal particles are also employed in fluidized beds {4.12}, sintering (sintered artifact fabrication) {4.13}, lithium-ion batteries {4.14}, pharmaceuticals {4.15}, combustion {4.16}, and numerous other practical applications.

In this model, by making a few simple assumptions and by converting the model from Cartesian to Spherical coordinates, the modeling calculations are reduced from a 3D model to a 1D model. Figure 4.29 shows both a Cartesian and a Spherical coordinate system parametrically.

See Figure 4.29.

Figure 4.29 shows both the Cartesian and the Spherical coordinate parameters.

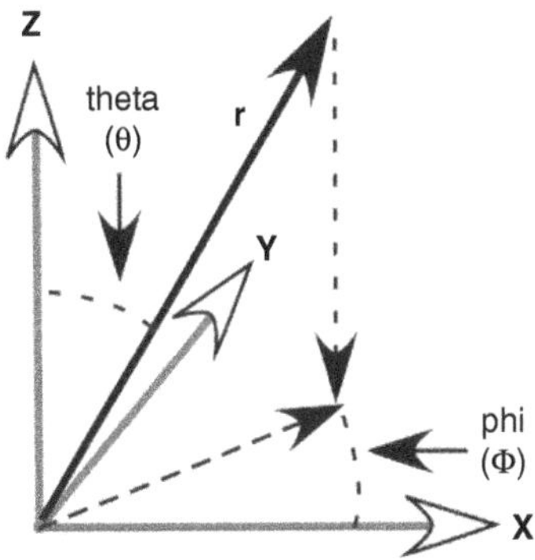

FIGURE 4.29 Both Cartesian and Spherical Coordinate Parameters.

A spherical radius vector with magnitude |r|, originating at coordinate (0,0,0) in Cartesian coordinates can be indicated as follows:

$$\vec{r} = \hat{i}x + \hat{j}y + \hat{k}z \tag{4.19}$$

Where:

$$r = \left(x^2 + y^2 + z^2\right)^{1/2} \tag{4.20}$$

And where:

$$\hat{i}, \hat{j}, \hat{k}$$

are unit vectors in the X, Y, Z directions, respectively.

The conversion factors for the magnitude of |r| to x, y, z values are {4.17}:

$$x = r \sin\Theta \cos\Phi \tag{4.21}$$

And:

$$y = r \sin\Theta \sin\Phi \tag{4.22}$$

And:

$$z = r \cos\Theta \tag{4.23}$$

Where: theta (Θ) is the angle between the radius vector (r) and the Z-axis (as shown in Figure 4.29).

phi (Φ) is the angle between the projection of the radius vector (r) into the X-Y plane and the X-axis (as shown in Figure 4.29)

NOTE *As mentioned previously, there are a large number of potential applications that are suitable for reduction from 3D to 1D through the use of this technique. The author recommends that the modeler explore those as the opportunity arises. In this text, however, we will explore the case of time-dependent heat transfer by conduction, since that type of model is so broadly applicable to numerous commonly observed different problems in its own right.*

Time-Dependent Heat Transfer by Conduction

The time-dependent heat conduction equation in Cartesian coordinates is expressed as follows:

$$\rho c_p \frac{\partial T}{\partial t} + \nabla \cdot (-k \nabla T) = Q \tag{4.24}$$

Where: ρ is the density of the material [kg/m^3].

c_p is the heat capacity of the material [J/(kg · K)].

T is the temperature [K].

k is the thermal conductivity [W/(m · K)].

Q is the internal heat source [W/m^3].

Conversion of the time-dependent heat conduction equation from Cartesian coordinates to Spherical coordinates requires the expression of the gradient operator in Spherical coordinates {4.18}, as shown in equation 4.25:

$$\nabla \cdot \nabla T = \frac{1}{r^2}\frac{\partial}{\partial r}\left[r^2 \frac{\partial T}{\partial r}\right] + \frac{1}{r^2}\left[\frac{1}{\sin\Theta}\frac{\partial}{\partial \Theta}\left[\sin\Theta \frac{\partial T}{\partial \Theta}\right] + \frac{1}{\sin^2\Theta}\frac{\partial^2 T}{\partial \Phi^2}\right] \tag{4.25}$$

Converting the Cartesian time-dependent heat conduction equation to a Spherical time-dependent heat conduction equation, yields:

$$\rho c_p \frac{\partial T}{\partial t} - k\left[\frac{1}{r^2}\frac{\partial}{\partial r}\left[r^2 \frac{\partial T}{\partial r}\right] + \frac{1}{r^2}\left[\frac{1}{\sin\Theta}\frac{\partial}{\partial \Theta}\left[\sin\Theta \frac{\partial T}{\partial \Theta}\right] + \frac{1}{\sin^2\Theta}\frac{\partial^2 T}{\partial \Phi^2}\right]\right] = Q \tag{4.26}$$

NOTE *In order to simplify the 3D problem to a 1D problem, it will be assumed that the sphere is nominally perfectly spherical, of radius* R_p*, and that T does not vary as a function of the angles* Θ *and/or* Φ*.*

Since T does not vary as a function of the angular variables, then:

$$\frac{\partial T}{\partial \Theta} = \frac{\partial T}{\partial \Phi} \equiv 0 \tag{4.27}$$

Based on these assumptions, the time-dependent heat conduction equation is modified as follows:

$$\rho c_p \frac{\partial T}{\partial t} - k\left[\frac{1}{r^2}\frac{\partial}{\partial r}\left[r^2 \frac{\partial T}{\partial r}\right]\right] = Q \tag{4.28}$$

In order to eliminate the problems associated with division by zero, the time-dependent heat conduction equation needs to be multiplied by r^2. The following equation results:

$$r^2 \rho c_p \frac{\partial T}{\partial t} + \frac{\partial}{\partial r}\left[-kr^2 \frac{\partial T}{\partial r}\right] = r^2 Q \tag{4.29}$$

NOTE

Since k is not a function of r, it can be moved inside the differentiation operation with no adverse effect in equation 4.29.

It is also convenient to solve the 1D problem for a dimensionless radial coordinate, because this allows the modeler to rapidly change the sphere size without changing the model geometry.

The conversion of the time-dependent heat conduction equation to a dimensionless form is as follows:

$$r_a = \frac{r}{R_p} \tag{4.30}$$

Where: R_p is the radius of the sphere.

Then:

$$\frac{\partial}{\partial r} = \frac{1}{R_p}\frac{\partial}{\partial r_a} \tag{4.31}$$

Substituting the dimensionless operator into equation 4.29, yields:

$$r_a^{\,2} \rho c_p \frac{\partial T}{\partial t} + \frac{\partial}{\partial r_a}\left[\frac{-kr_a^2}{R_p^2}\frac{\partial T}{\partial r_a}\right] = r_a^{\,2} Q \tag{4.32}$$

The dimensionless radius vector for a sphere of nominal size is shown in Figure 4.30.

Figure 4.30 shows a Radius Vector for a dimensionless nominal sphere.

Building the 1D Spherically Symmetric Transport Model

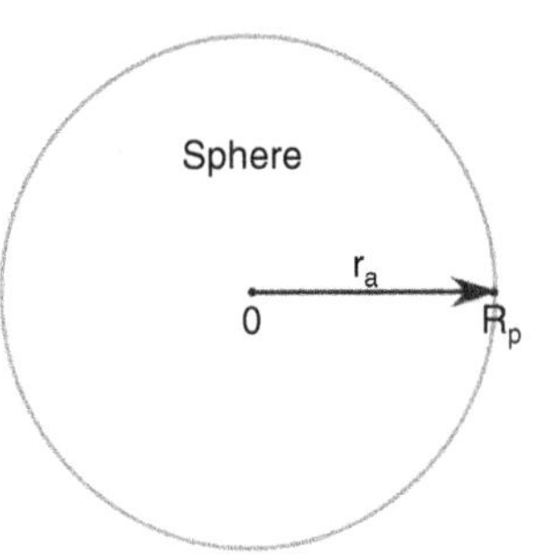

FIGURE 4.30 Radius Vector for a Dimensionless Nominal Sphere.

Startup 4.x.

Select > 1D.

Click > Next.

Click > Twistie for Mathematics in the Add Physics window.

Click > Twistie for PDE Interfaces > General Form PDE (g).

Click > Add Selected.

In the Model Wizard – Dependent variables – Dependent variables entry window,

Enter > T.

See Figure 4.31.

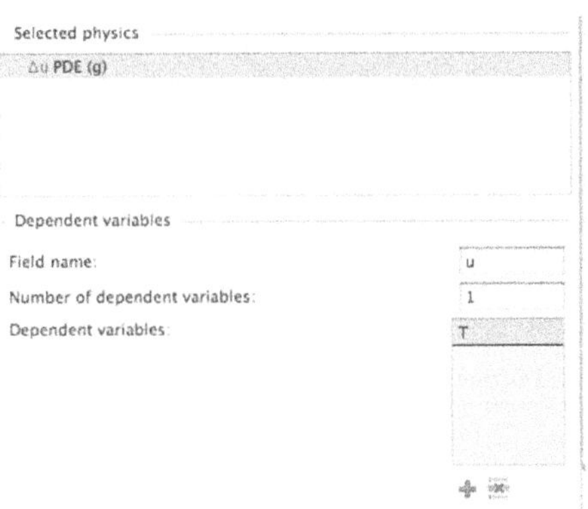

FIGURE 4.31 Dependent Variables Entry Window.

Figure 4.31 shows the dependent variables entry window.

Click > Next.

Select > Time Dependent in the Select Study Type window.

Click > Finish (Flag).

Click > Save As > MMUC4_1D_SST_1.mph.

See Figure 4.32.

Figure 4.32 shows the Desktop Display for the MMUC4_1D_SST_1.mph model.

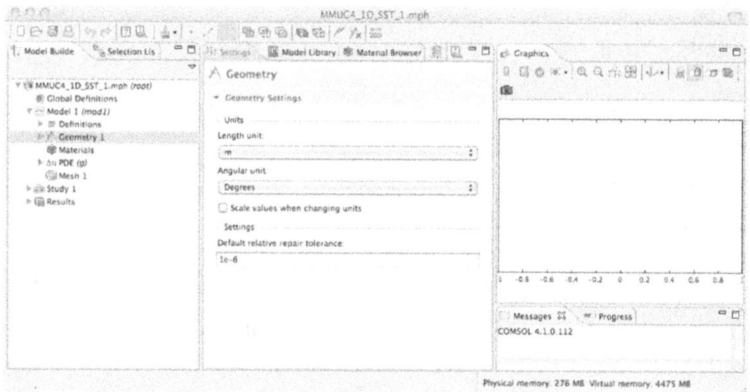

FIGURE 4.32 Desktop Display for the MMUC4_1D_SST_1.mph Model.

Global Definitions

Right-Click in the Model Builder window on Global Definitions.

Select > Parameters from the Pop-up menu.

In the Settings – Parameters – Parameters edit window,

Enter the parameters as shown in Table 4.2.

TABLE 4.2 Parameters Window

Name	Expression	Description
rho	2000[kg/m^3]	Density
cp	300[J/(kg*K)]	Heat capacity
k	0.5[W/(m*K)]	Thermal conductivity
Rp	0.005[m]	Sphere radius
Qs	0[W/m^3]	Heat source
hs	1000[W/(m^2*K)]	Heat transfer coefficient
Text	368[K]	External temperature
Tinit	298[K]	Initial temperature

Click > Save (to save the model).

NOTE *The modeler should also save the parameters file at this time so that if the model needs to be recovered or the parameters need to be modified, they will be available to be changed or corrected without the need to reenter them.*

See Figure 4.33.

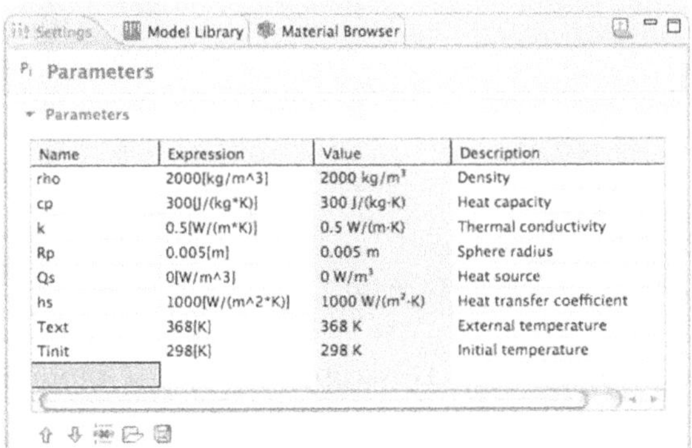

Parameters

Name	Expression	Value	Description
rho	2000[kg/m^3]	2000 kg/m³	Density
cp	300[J/(kg*K)]	300 J/(kg·K)	Heat capacity
k	0.5[W/(m*K)]	0.5 W/(m·K)	Thermal conductivity
Rp	0.005[m]	0.005 m	Sphere radius
Qs	0[W/m^3]	0 W/m³	Heat source
hs	1000[W/(m^2*K)]	1000 W/(m²·K)	Heat transfer coefficient
Text	368[K]	368 K	External temperature
Tinit	298[K]	298 K	Initial temperature

FIGURE 4.33 Desktop Display - Settings - Parameters Entry Windows.

Figure 4.33 shows the Desktop Display – Settings – Parameters entry windows.

Click > Settings – Parameter Save to File.

Enter > MMUC4_1D_SST_1_Param.txt in the Save to File window.

Geometry 1

Right-Click in the Model Builder window on Model 1 > Geometry 1.

Select > Interval from the Pop-up menu.

NOTE *The default settings for the specified interval are adequate for this model.*

Right-Click > Interval 1.

Select > Build Selected.

See Figure 4.34.

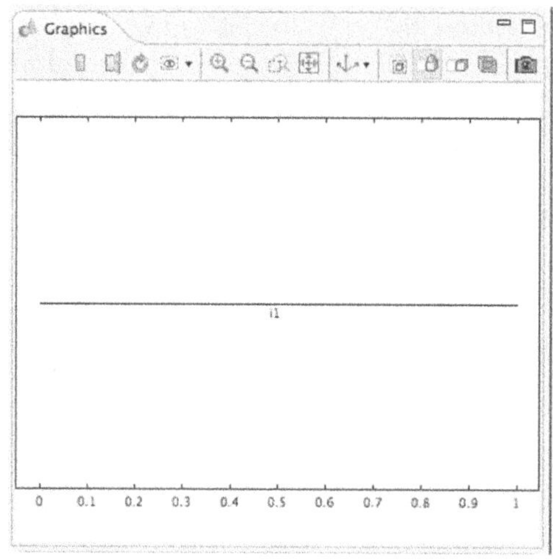

FIGURE 4.34 Desktop Display - Graphics - Geometry 1 Window.

Figure 4.34 shows the Desktop Display – Graphics – Geometry 1 window.

PDE

Click in the Model Builder window on Model 1 > Δu PDE (g) twistie.

In the Model Builder window,

Click > Model 1 > Δu PDE (g) > General Form PDE 1.

NOTE *The default settings for the General Form PDE are not appropriate for the solution of this problem and thus need to be modified. The modeler should note the nominal equation (by Clicking on the Settings – General Form PDE – Equation twistie) and the manner in which the entered coefficients change that equation.*

Enter –k*(x^2/Rp^2)*Tx in Settings – General Form PDE – Conservative Flux – Γ edit window.

Enter x^2*Qs in Settings – General Form PDE – Source Term – f edit window.

Enter x^2*rho*cp in Settings – General Form PDE – Damping or Mass Coefficient – d_a edit window.

See Figure 4.35.

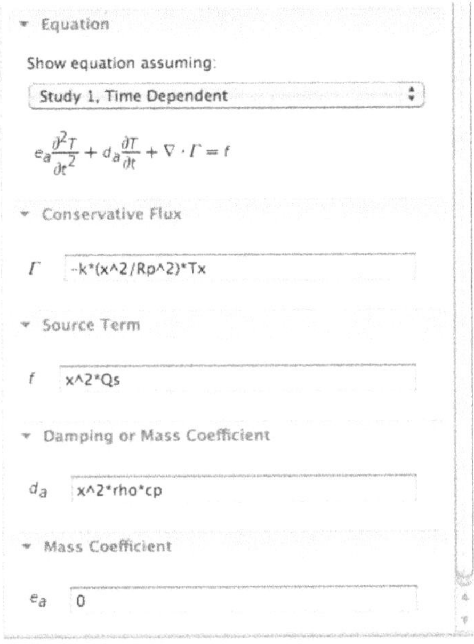

FIGURE 4.35 Desktop Display - Settings - General Form PDE Coefficients.

Figure 4.35 shows the Desktop Display – Settings – General Form PDE coefficients.

Initial Values 1

Click in the Model Builder window on Model 1 > Δu PDE (c) > Initial Values 1.

NOTE *The default settings for the Initial Values are not appropriate for the solution of this problem and thus need to be modified.*

Enter Tinit in the Settings – Initial Values – Initial value for T.

See Figure 4.36.

Figure 4.36 shows the Desktop Display – Settings – Initial Values coefficients.

NOTE *The default settings for the Boundary Conditions are not appropriate for the solution of this problem and thus need to be modified.*

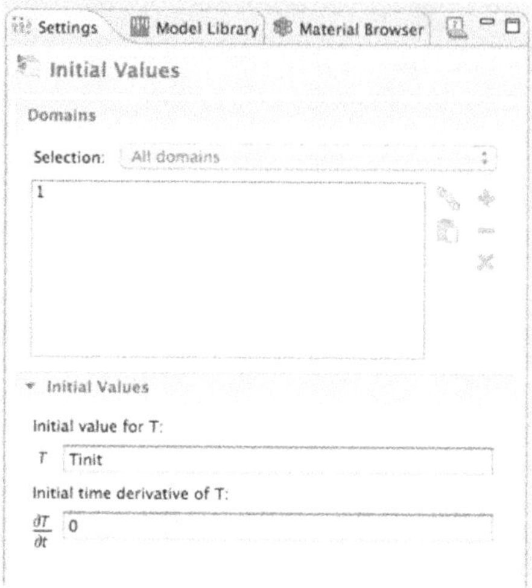

FIGURE 4.36 Desktop Display – Settings – Initial Values Coefficients.

In Model Builder,

Right-Click > Model 1 > Δu PDE (g).

Select > Flux/Source.

Click > Boundary 1 in the Graphics window.

In Settings – Flux/Source – Boundaries,

Click > Add to Selection (Plus).

In Model Builder,

Right-Click > Model 1 > Δu PDE (g).

Select > Flux/Source.

Click > Boundary 2 in the Graphics window.

In Settings – Flux/Source – Boundaries,

Click > Add to Selection (Plus).

In Settings – Flux/Source – Boundary Flux/Source – g edit window,

Enter (x^2/Rp)*hs*(Text-T).

See Figure 4.37.

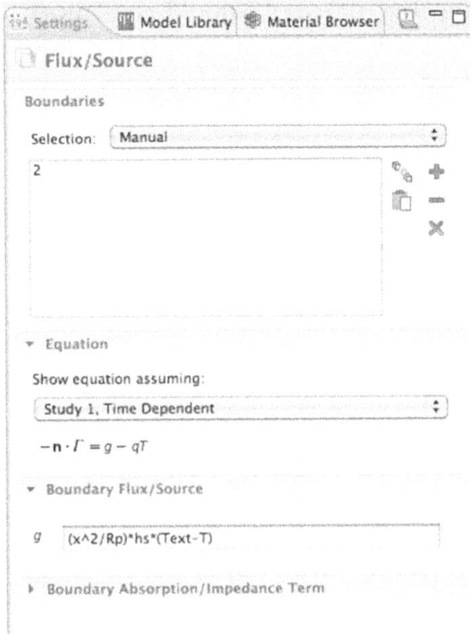

FIGURE 4.37 Desktop Display - Settings - Boundary Condition.

Figure 4.37 shows the Desktop Display – Settings – boundary condition.

Mesh 1

The default settings for the Mesh Type are not appropriate for the solution of this problem and thus need to be modified.

Right-Click in the Model Builder window on Model 1 > Mesh 1.

Select > Scale.

In Settings – Scale – Scale edit window,

Enter 0.4.

Right-Click in the Model Builder window on Model 1 > Mesh 1.

Select > Edge.

Right-Click in the Model Builder window on Model 1 > Mesh 1.

Select > Build All.

After meshing, the modeler should see information in the message window about the number of elements (38 elements) in the mesh.

See Figure 4.38.

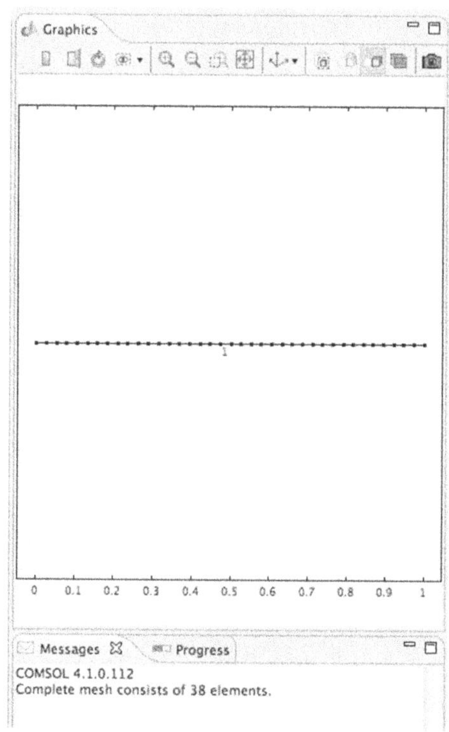

FIGURE 4.38 Desktop Display - Graphics - Meshed Domain.

Figure 4.38 shows the meshed domain in the Desktop Display – Graphics window.

Study 1

Click > Study 1 twistie.

Select > Step 1: Time Dependent.

In Settings – Time Dependent – Study Settings,

Click > Times Range button.

Enter > Start = 0, Stop = 10, Step = 0.25.

Click > Replace.

In Model Builder, Right-Click Study 1 >Select > Compute.

Results

The as-computed results are shown in Figure 4.39 using the default plot parameters.

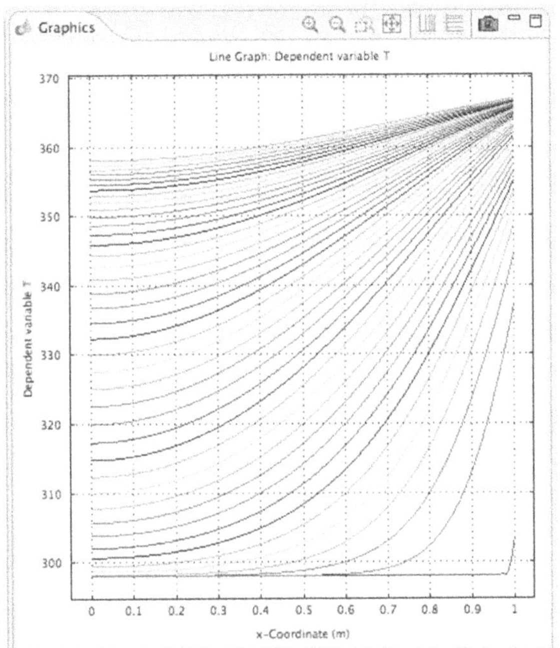

FIGURE 4.39 Desktop Display - Graphics - Initial Computed Radial Temperature Profile.

Figure 4.39 shows the initial computed radial temperature profile in the Desktop Display – Graphics window.

NOTE *The computed radial temperature profile is plotted using the default plot parameters in Figure 4.39. The modeler can adjust the output plot to better satisfy his particular information display needs as shown below.*

1D Plot Group 1

In Model Builder,

Click > Results – 1D Plot Group 1.

In Settings – 1D Plot Group – Plot Settings,

Select > Title checkbox.

Enter Temperature in the Title edit window.

Select > x-axis label checkbox,

Enter Dimensionless Radius in x-axis label edit window.

Select > y-axis label checkbox,

Enter T (K) in the y-axis label edit window.

In Model Builder – Results,

Click > 1D Plot Group 1 twistie.

In Model Builder – Results – 1D Plot Group 1,

Click > Line Graph 1.

In Settings – Line Graph,

Click > Legends twistie.

Uncheck, as needed, the Show legends checkbox.

On the Graphics Toolbar.

Click > Zoom Extents.

See Figure 4.40.

Figure 4.40 shows the Desktop Display – Graphics – Adjusted Plot Parameters Computed Radial Temperature Profile Graph.

1D Plot Group 2

In Model Builder,

Right-Click > Results.

Select > 1D Plot Group from the Pull-down menu to create 1D Plot Group 2.

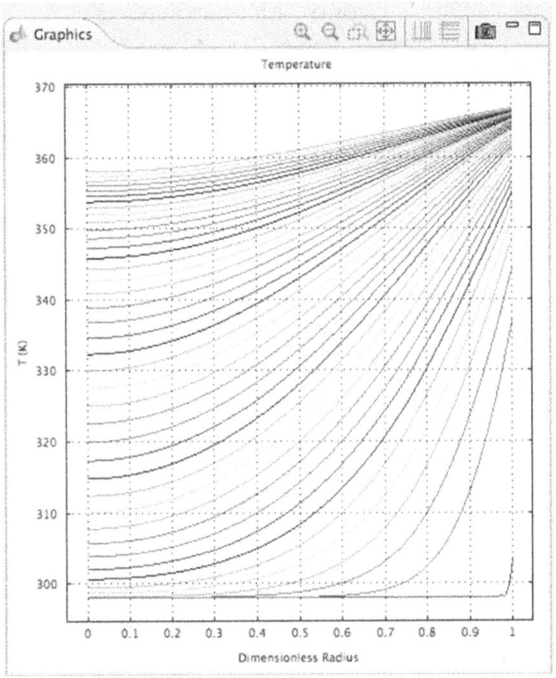

FIGURE 4.40 Desktop Display - Graphics -Adjusted Plot Parameters Computed Radial Temperature Profile Graph.

Right-Click> 1D Plot Group 2.

Select > Point Graph.

In the Graphics window,

Select > Boundary 1 only.

In Settings – 1D Plot Group – Point Graph – Selection,

Click > Add to Selection button (Plus).

In Settings – Point Graph,

Click > Legends twistie.

Uncheck, as needed, the Show legends checkbox.

In Model Builder,

Click > 1D Plot Group 2.

In Settings – 1D Plot Group – Plot Settings,

Select > Title checkbox.

Enter Temperature Response at Sphere Center in the Title edit window.

Select > x-axis label checkbox.

Enter Time (s) in the x-axis label edit window.

Select > y-axis label checkbox.

Enter T (K) in the y-axis label edit window.

Click > Plot.

See Figure 4.41.

Figure 4.41 shows the Desktop Display – Graphics – Temperature Response at Sphere Center.

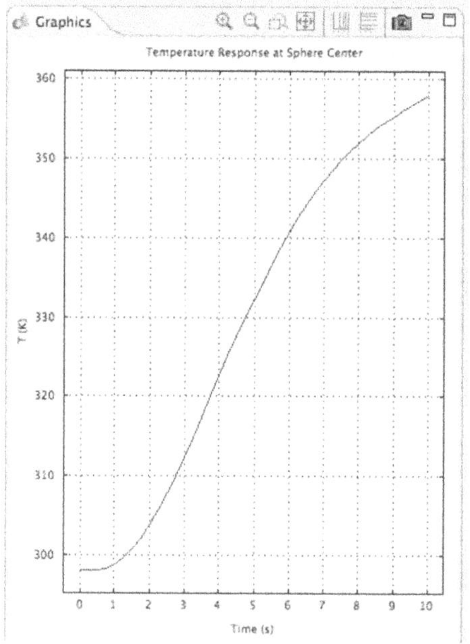

FIGURE 4.41 Desktop Display - Graphics - Temperature Response at Sphere Center.

1D Spherically Symmetric Transport Model Animation

To demonstrate the propagation of the heat transfer by conduction in 4.x, the modeler can create an animation.

In Model Builder,

Right-Click > Results – Report.

Select > Animation.

In Model Builder,

Click > Results – Report – Animation 1.

In Settings – Animation – Output,

Select > Output type Movie from the Pull-down menu.

Select > Format GIF (for the Macintosh or AVI for the PC) from the pull-down menu.

NOTE *The detailed procedures for Animation in 4.x on the Macintosh and the PC are different. The Macintosh uses the GIF format. The PC uses the AVI format. If the modeler tries to use AVI on the Macintosh, an error results.*

In Settings – Animation – Frame Settings,

Click > Lock aspect ratio check box.

In Settings – Animation – Parameter Sweep,

Click > Sweep Type.

Select > Stored Solutions from the Pull-down menu.

Click > Select via.

Select > Stored output times from the Pull-down menu.

In Settings – Animation – Advanced,

Click > Antialiasing check box.

Click > Settings – Export.

NOTE *The movie will be exported to and run in the 4.x Graphics window, if there is no path designated in the Settings – Animation – Output Filename edit window.*

1D Spherically Symmetric Transport Model Movie

The modeler can save the movie as a unique file.

Click > Settings – Animation – Output Browse button.

Select the desired location for saving the movie.

Enter the desired File Name in the Export Animation window.

Click > Save in the Export Animation window.

Click > Save for the completed MMUC4_1D_SST_1.mph 1D Spherically Symmetric Transport Model.

1D Spherically Symmetric Transport Model Summary and Conclusions

The 1D Spherically Symmetric Transport Model is a powerful tool that can be used to reduce the dimensionality of models from 3D to 1D through the assumption of homogeneous, isotropic materials properties and spherical symmetry. As has been shown earlier in this chapter, the 1D Spherically Symmetric Transport Model is easily and simply built with a 1D PDE mode model.

FIRST PRINCIPLES AS APPLIED TO 1D MODEL DEFINITION

First Principles Analysis derives from the fundamental laws of nature. In the case of models in this book or from any other source, the modeler should be able to demonstrate the calculated results derived from the models are consistent with the laws of physics and the basic observed properties of materials. In the case of models using this Classical Physics Analysis approach, the laws of conservation in physics require that what goes in (as mass, energy, charge, etc.) must come out (as mass, energy, charge, etc.) or must accumulate within the boundaries of the model. To do otherwise violates fundamental principles.

NOTE *In the COMSOL Multiphysics software, the default interior boundary conditions are set to apply the conditions of continuity in the absence of sources (e.g. heat generation, charge generation, molecule generation, etc.) or sinks (e.g. heat loss, charge recombination, molecule loss, etc.).*

The careful modeler must be able to determine by inspection of the model that the appropriate factors have been considered in the development of the specifications for the various geometries, for the material properties of each domain, and for the boundary conditions. He must also be knowledgeable of the implicit assumptions and default specifications that are normally incorporated into the COMSOL Multiphysics software model when a model is built using the default settings.

Consider, for example, the three 1D models developed earlier in this chapter. In these models, it is implicitly assumed there are no thermally related changes (mechanical, electrical, etc.). It is also assumed the materials are homogeneous and isotropic and there are no thin electrical contact barriers at the electrical junctions. None of these assumptions are typically true in the general case. However, by making such assumptions, it is possible to easily build a 1D First Approximation Model.

NOTE *A First Approximation Model is one that captures all the essential features of the problem that needs to be solved, without dwelling excessively on small details. A good First Approximation Model will yield an answer that enables the modeler to determine if he needs to invest the time and the resources required to build a more highly detailed model.*

Also, the modeler needs to remember to name model parameters carefully as pointed out in Chapter 1.

REFERENCES

4.1 COMSOL Multiphysics Users Guide, pp. 334

4.2 http://en.wikipedia.org/wiki/KdV

4.3 http://en.wikipedia.org/wiki/Nonlinear_partial_differential_equation

4.4 http://en.wikipedia.org/wiki/Exactly_solvable_model

4.5 http://en.wikipedia.org/wiki/Nonlinear_system

4.6 http://en.wikipedia.org/wiki/John_Scott_Russell

4.7 http://en.wikipedia.org/wiki/Soliton

4.8 http://en.wikipedia.org/wiki/Soliton_(optics)

4.9 http://en.wikipedia.org/wiki/Telegraph_equation

4.10 http://en.wikipedia.org/wiki/Oliver_Heaviside

4.11 http://en.wikipedia.org/wiki/Powder_metallurgy

4.12 http://en.wikipedia.org/wiki/Fluidized_bed

4.13 http://en.wikipedia.org/wiki/Sintering

4.14 http://en.wikipedia.org/wiki/Nanoarchitectures_for_lithium-ion_batteries

4.15 http://en.wikipedia.org/wiki/Spheronisation

4.16 http://en.wikipedia.org/wiki/Fluidized_bed_combustion

4.17 http://en.wikipedia.org/wiki/Spherical_coordinates

4.18 E.U. Condon and H. Odishaw, "Handbook of Physics", McGraw-Hill, New York, 1958, pp. 1–109

Suggested Modeling Exercises

1. Build, mesh, and solve the 1D KdV Equation Model problem presented earlier in this chapter.
2. Build, mesh, and solve the 1D Telegraph Equation Model problem presented earlier in this chapter.
3. Build, mesh, and solve the 1D Spherically Symmetric Transport Model problem presented earlier in this chapter.
4. Change the values of the parameters in the 1D KdV Equation Model problem. Compute the new solution. Analyze, compare, and contrast the results with the new parameter values to the results found earlier.
5. Change the values of the parameters in the 1D Telegraph Equation Model problem. Compute the new solution. Analyze, compare, and contrast the results with the new parameter values to the results found earlier.
6. Change the values of the parameters in the 1D Spherically Symmetric Transport Model problem. Compute the new solution. Analyze, compare, and contrast the results with the new parameter values to the results found earlier.

CHAPTER 5

2D MODELING USING COMSOL MULTIPHYSICS 4.X

In This Chapter

- Guidelines for 2D Modeling in 4.x
    - 2D Modeling Considerations
- 2D Basic Models
    - 2D Electrochemical Polishing Model
    - 2D Hall Effect Model
- First Principles as Applied to 2D Model Definition
- References
- Suggested Modeling Exercises

GUIDELINES FOR 2D MODELING IN 4.X

NOTE

In this chapter, the modeler is introduced to the development and analysis of 2D models. 2D models have two geometric dimensions (x, y). In 4.x, there are fundamentally two types of 2D geometries: 2D and 2D Axisymmetric. 2D models, for ease of calculation, are assumed to be homogeneous and isotropic in the third (non-planar) dimension. 2D Axisymmetric models are planar representations of cylindrical 3D objects that are rotationally homogeneous and isotropic. 2D models are typically classed by level of difficulty as being of two types, introductory and advanced.

In this chapter, introductory 2D models will be presented. Such 2D models are more geometrically complex than 1D models.

2D models have proven to be very valuable to the science and engineering communities, both in the past and currently, as first-cut evaluations of potential physical behavior under the influence of external stimuli. The 2D model responses and other such ancillary information are then gathered and screened early in a project for initial evaluation and potentially for later use in building higher-dimensionality (3D) field-based (electrical, magnetic, etc.) models.

Since the models in this and subsequent chapters are more complex to build and more difficult to solve than the models presented thus far, it is important that the modeler have available the tools necessary to most easily utilize the powerful capabilities of the 4.x software. In order to do that, the modeler should go to the main 4.x toolbar, Click > Options – Preferences – Model builder. When the Preferences – Model builder edit window is shown, Select > Show equation view checkbox and Show more options checkbox. Click > Apply {5.1}.

For information on the development of the more advanced models using the 2D Axisymmetric coordinate system, the modeler should consult Chapter 6 of this text.

2D Modeling Considerations

2D Modeling can in some cases be less difficult than 1D modeling, having fewer implicit assumptions, and yet potentially, it can still be a very challenging type of model to build, depending on the underlying physics involved. The least difficult aspect of 2D model creation arises from the fact that the geometry is relatively simple. (In a 2D model, the modeler has only a single plane as the modeling space.) However, the physics in a 2D model can range from relatively easy to extremely complex.

In compliance with the laws of physics, a 2D model implicitly assumes that energy flow, materials properties, environment, and all other conditions and variables that are of interest are homogeneous, isotropic, and/or constant, unless otherwise specified, throughout the entire domain of interest both within the model and through the boundary conditions and in the environs of the model.

The modeler needs to bear the above stated conditions in mind and insure that all of the modeling conditions and associated parameters (default settings) in each model created are properly considered, defined, verified, and/or set to the appropriate values.

It is always mandatory that the modeler be able to accurately anticipate the expected results of the model and accurately specify the manner in which those results will be presented. Never assume that any of the default values that are present when the model is created necessarily satisfy the needs or conditions of a particular model.

NOTE *Always verify that any parameters employed in the model are the correct value needed for that model. Calculated solutions that significantly deviate from the anticipated solution or from a comparison of values measured in an experimentally derived realistic model are probably indicative of one or more modeling errors either in the original model design, in the earlier model analysis, in the understanding of the underlying physics, or are simply due to human error.*

Coordinate System

In 2D models, there are two geometric coordinates, space (x), space (y), and the temporal coordinate, time (t). In a steady-state solution to a 2D model, parameters can only vary as a function of position in space (x) and space (y) coordinates. Such a 2D model represents the parametric condition of the model in a time-independent mode (quasi-static). In a transient solution model, parameters can vary both by position in space (x), space (y), and in time (t).

The space coordinates (x) and (y) typically represent a distance coordinate throughout which the model is to calculate the change of the specified observables (i.e. temperature, heat flow, pressure, voltage, current, etc.) over the range of coordinate values ($x_{min} <= x <= x_{max}$) and ($y_{min} <= y <= y_{max}$). The time coordinate (t) represents the range of temporal values ($t_{min} <= t <= t_{max}$) from the beginning of observation period (t_{min}) to the end of observation period (t_{max}).

Electrochemical Polishing (Electropolishing) Theory

Electrochemical {5.2} polishing (electropolishing {5.3}) is a well-known process in the metal finishing industry. This process allows the finished surface smoothness of a conducting material to be cleanly controlled to a high degree of precision, using relatively simple processing equipment. The electrochemical polishing technique eliminates the abrasive residue typically present on the polished surface from a mechanical polishing process and also eliminates the need for complex, mechanical polishing machinery.

NOTE *It is presently understood that the science of electricity and consequently that of electrochemistry started with the work of William Gilbert through the study of magnetism. He first published his studies in the year 1600 {5.4}. Charles-Augustin de Coulomb {5.5}, Joseph Priestley {5.6}, Georg Ohm {5.7}, and others made additional independent later contributions that furthered that basic understanding of the nature of electricity and electrochemistry. Those contributions led to the discovery and disclosure by Michael Faraday {5.8} of his two laws of electrochemistry in the year 1832.*

The following numerical solution model (Electrochemical Polishing) was originally developed by COMSOL for distribution with earlier versions of the Multiphysics software as a COMSOL Multiphysics Electromagnetics Model. This model introduces two important basic concepts, the first concept in applied physics, and the second concept in applied modeling: (1) Electropolishing and the (2) Moving Mesh (ALE = Arbitrary Lagrangian-Eulerian {5.9}). The Electrochemical Polishing model built below has been transformed into an Electromagnetics model that is functional in 4.x.

NOTE *It is important for the new modeler to personally build each model presented within this text. The best method of building an understanding of the modeling process is by obtaining the hands-on creation experience of actually building, meshing, solving, and plotting the results of a model. Many times the inexperienced modeler will make and subsequently correct errors, adding to his experience and fund of modeling knowledge. Even building the simplest model will expand the modeler's fund of knowledge.*

The polishing (smoothing) of a material surface, either mechanical or electrochemical, results from the reduction of asperities (bumps), thus achieving a nominally smooth surface profile (uniform thickness $\pm \Delta$ thickness). In a mechanical polishing technique, the reduction of asperities occurs through the use of finer (smaller) and finer grit (abrasive) sizes. The mechanical polishing of many non-uniform surfaces is difficult, if not impossible, due to the complexity and/or physical size of such surfaces. Figure 5.1 shows a simple asperity, as will be modeled in this section of the chapter.

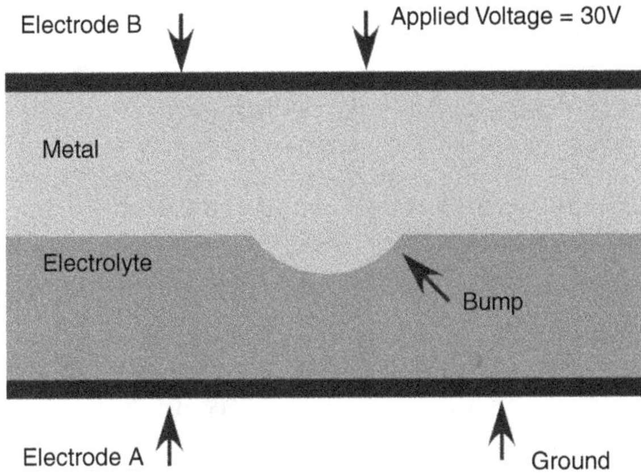

FIGURE 5.1 An Asperity (Bump) on a Metal Electrode.

The surface of the metal electrode, using the electrochemical method, is polished by the differential removal of material from local asperities in selected areas, by the immersion of the nominally rough metal electrode in an electrolyte, and the application of a current (electron bombardment). A first order approximation to the experimentally observed material removal process is that the rate (velocity) of material removal (U) from the electrode surface is proportional to the amplitude of the current and direction of the current **J**, relative to the local surface normal vector **n** (See Figure 5.2).

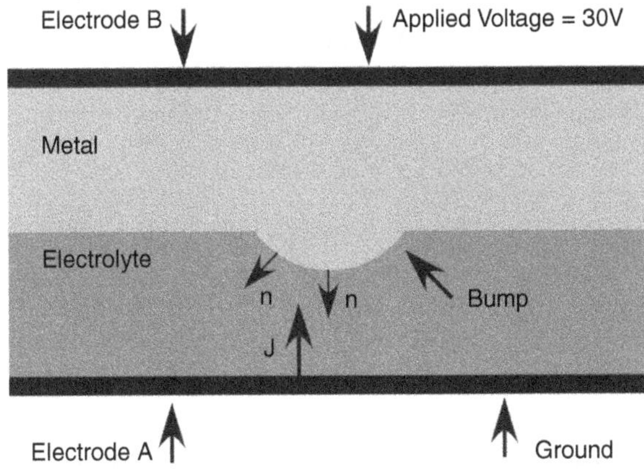

FIGURE 5.2 Surface Normal Vector **n** and the Current Vector **J**.

$$U = -K * \mathbf{J} \cdot \mathbf{n} = -K * J_n \tag{5.1}$$

NOTE

The electropolishing technique is substantially the inverse of the electroplating process and as a result the rate of removal of material (velocity = U) from the nominally rough surface of the positive electrode is proportional to the normal current density at the positive electrode surface, as shown in equation 5.1.

The exact value of the proportionality constant (K), to first order, in physical applications (e.g. research experiments, processing, etc.), is determined by the electrode material, the electrolyte, and the temperature.

For this model, the proportionality constant is chosen to be:

$$K = 1.0x10^{-11}\ m^3/(A * s) \tag{5.2}$$

Where: K is the proportionality constant in [m^3/(A*s)].

m is in meters [m].

A is amperes [A].

s is seconds [s].

Obviously, since material is removed from the positive electrode during the electropolishing process, the spacing between the upper and lower electrodes will increase. The time rate of change of the model geometry (electrode spacing) needs to be accommodated somewhere within the model. The Moving Mesh (ALE = Arbitrary Lagrangian-Eulerian) Application Mode accommodates that time rate of change, resulting from the normal current (J_n) flowing in the electrolyte during the use of the Electric Currents (ec) Interface {5.10}.

NOTE

The Moving Mesh Interface allows the modeler to create models in which the physics of the process introduces and controls geometric changes in the model. However, the modeler must be sure to know and to work carefully within the limits of the modeling system. The Moving Mesh Interface is a powerful tool. However, the calculated Mesh parameters can drift, as the mesh is deformed, and ultimately lead to non-physical, non-convergent results. Avoidance of such non-physical results requires the modeler to understand the basic physics of the modeled problem and to choose the meshing method that yields the best overall results.

2D BASIC MODELS

2D Electrochemical Polishing Model

Building the 2D Electrochemical Polishing Model

Startup 4.x.

Select > 2D.

Click > Next.

Click > Twistie for AC/DC in the Add Physics window.

Click > Electric Currents (ec).

Click > Add Selected.

Click > Twistie for Δu Mathematics > Twistie for Deformed Mesh in the Add Physics window.

Click > Moving Mesh (ale).

Click > Add Selected.

Click > Next.

Select > Time Dependent in the Select Study Type window.

Click > Finish (Flag).

Click > Save As.

Enter MMUC4_2D_EP_1.mph.

See Figure 5.3.

Figure 5.3 shows the Desktop Display for the MMUC4_2D_EP_1.mph model.

Right-Click > Global Definitions.

Select > Parameters from the Pop-up menu.

In the Settings – Parameters – Parameters edit window,

Enter > Name = K, Expression = 1.0e-11[m^3/(A*s)], Description = Proportionality constant.

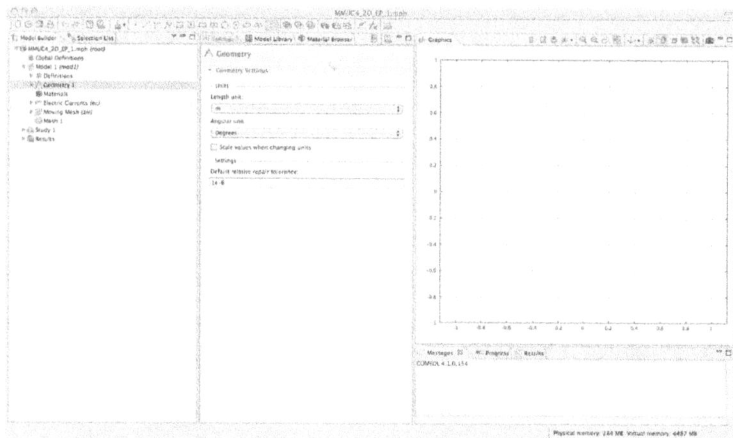

FIGURE 5.3 Desktop Display for the MMUC4_2D_EP_1.mph Model.

See Figure 5.4.

Figure 5.4 shows the Settings – Parameters – Parameters edit window filled.

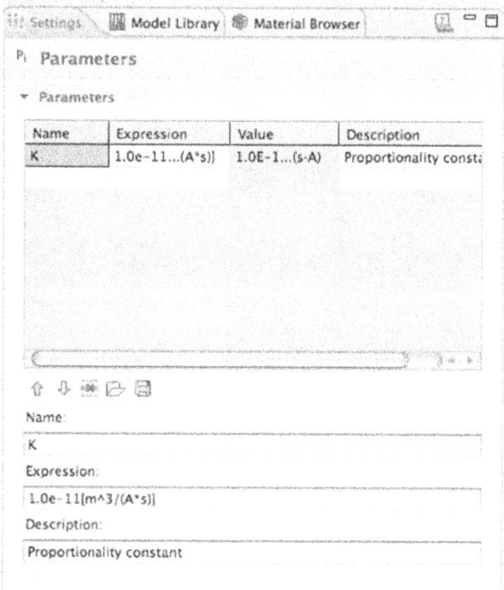

FIGURE 5.4 Settings - Parameters - Parameters Edit Window Filled.

NOTE *The Moving Mesh Interface {5.11} requires the specification of a set of coordinate systems to enable the calculation of the changing parameters within the model. The Deformed configuration and the Reference configuration are specified in the modeling steps that follow.*

In Model Builder – Model 1 *(mod1)*,

Click > Twistie for Definitions.

Right-Click > Boundary System 1 *(sys1)*.

Select > Duplicate from the Pop-up menu.

See Figure 5.5.

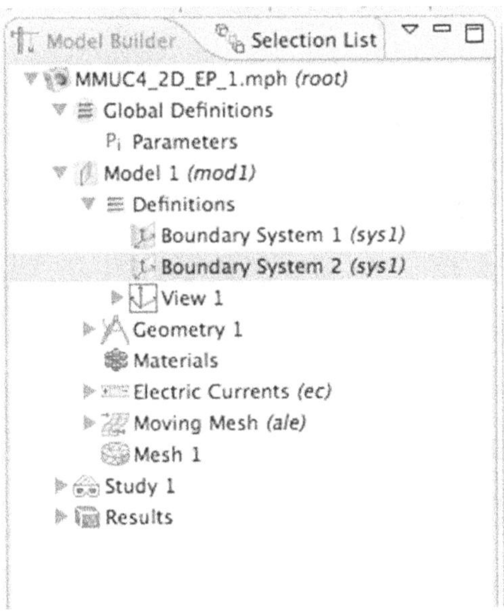

FIGURE 5.5 Model Builder - Model 1 *(mod1)* - Definitions - Boundary Systems Duplicate.

Figure 5.5 shows the Model Builder – Model 1 *(mod1)* – Definitions – Boundary Systems Duplicate.

Right-Click > Boundary System 2 *(sys1)*.

Select > Rename from the Pop-up menu.

Enter > Material Boundary System 2 in the Rename Boundary System edit window.

Click > OK.

In the Settings – Boundary System – Coordinate System Identifier – Identifier edit window,

Enter > sys2.

Click > Settings – Boundary System – Settings – Frame type.

Select > Reference configuration from the Pull-down menu.

See Figure 5.6.

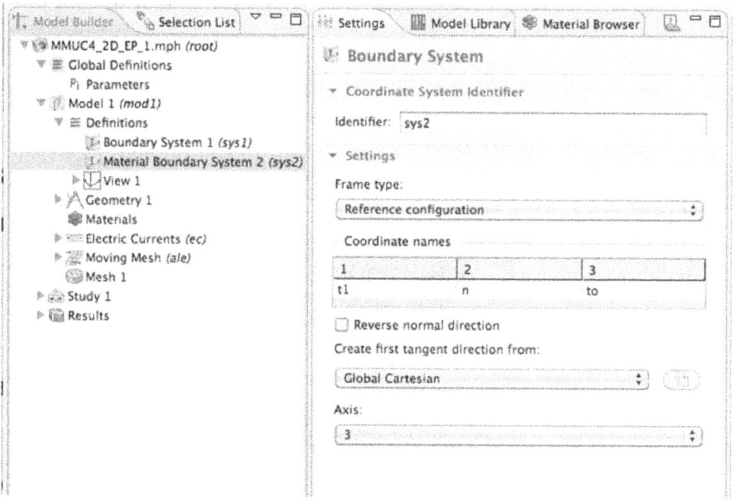

FIGURE 5.6 Model Builder - Model 1 *(mod1)* - Definitions - Boundary Systems Duplicate Configured.

Figure 5.6 shows the Model Builder – Model 1 *(mod1)* – Definitions – Boundary Systems Duplicate configured.

Right-Click > Model Builder – Geometry 1.

Select > Rectangle from the Pop-up menu.

Enter > 2.8e-3 in the Settings – Rectangle – Size – Width entry window.

Enter > 0.4e-3 in the Settings – Rectangle – Size – Height entry window.

Enter > –1.4e-3 in the Settings – Rectangle – Position – x entry window.

Enter > 0e-3 in the Settings – Rectangle – Position – y entry window.

Press the Tab key.

See Figure 5.7.

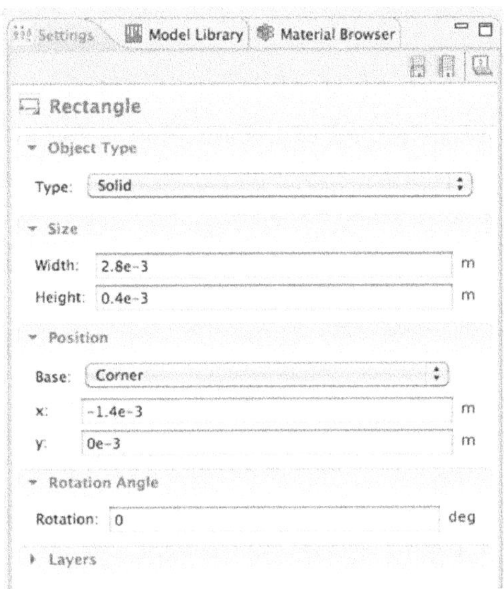

FIGURE 5.7 Settings - Rectangle Entry Windows.

Figure 5.7 shows the Settings – Rectangle Entry Windows.

Click > Build All.

Right-Click > Model Builder – Geometry 1.

Select > Circle from the Pop-up menu.

Enter > 0.3e-3 in the Settings – Circle – Size – Radius entry window.

Verify > Settings – Circle – Position – Base Pull-down menu is set to Center.

Enter > 0 in the Settings – Circle – Position – x entry window.

Enter > 0.6e-3 in the Settings – Circle – Position – y entry window.

Press the Tab key.

Click > Build All.

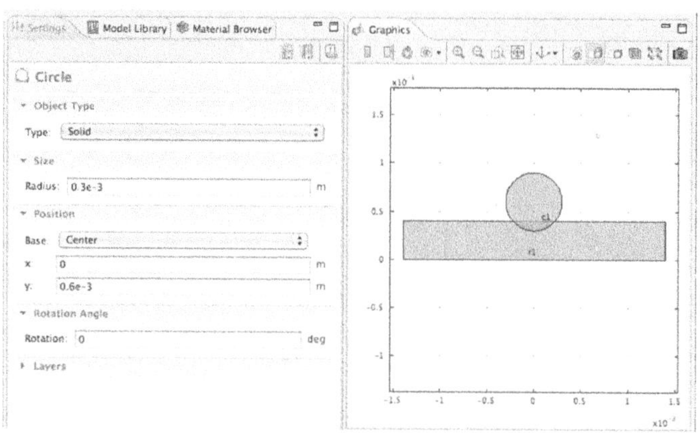

FIGURE 5.8 Settings - Circle Entry Windows and Graphics Window.

See Figure 5.8.

Figure 5.8 shows the Settings – Circle Entry Windows and Graphics Window.

Right-Click > Model Builder – Geometry 1.

Select > Boolean Operations – Difference from the Pop-up menu.

Verify > Settings – Difference – Difference – Objects to add entry window is selected.

NOTE *The Objects to add entry window is selected when the topmost button (Activate Selection button) on the right side of the window is surrounded by a grayed square indicator.*

Click > r1 in the Graphics window.

Click > Add to Selection button.

Click > Activate Selection button for the Objects to subtract entry window.

Click > c1 in the Graphics window.

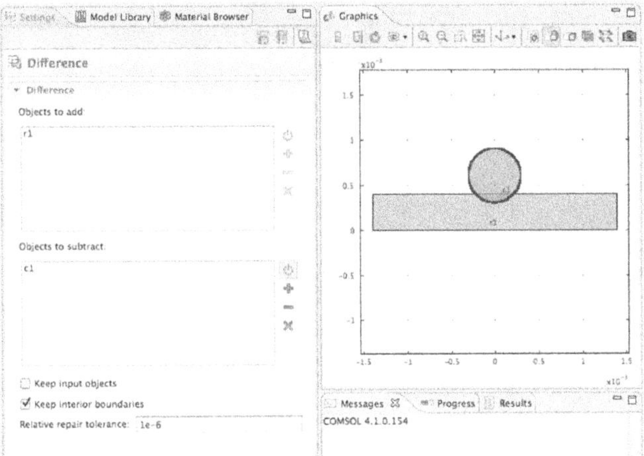

FIGURE 5.9 Settings - Difference - Difference Windows.

Click > Add to Selection button.

Figure 5.9 shows the Settings – Difference – Difference windows.

Click > Build All.

See Figure 5.10.

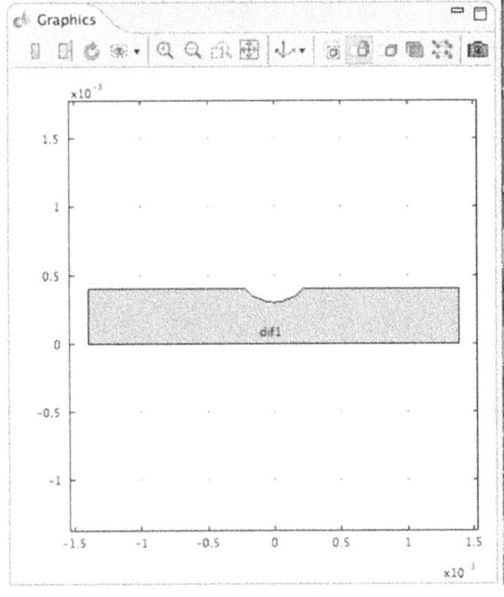

FIGURE 5.10 Model Geometry in the Graphics Window.

Figure 5.10 shows the Model Geometry in the Graphics window.

Electric Currents (ec) Interface

Click > Twistie for the Electric Currents (ec) interface.

Click > Electric Currents (ec).

Enter > 2.8e-3 in the Settings – Electric Currents – Out-of-Plane-Thickness – Thickness edit window.
See Figure 5.11.

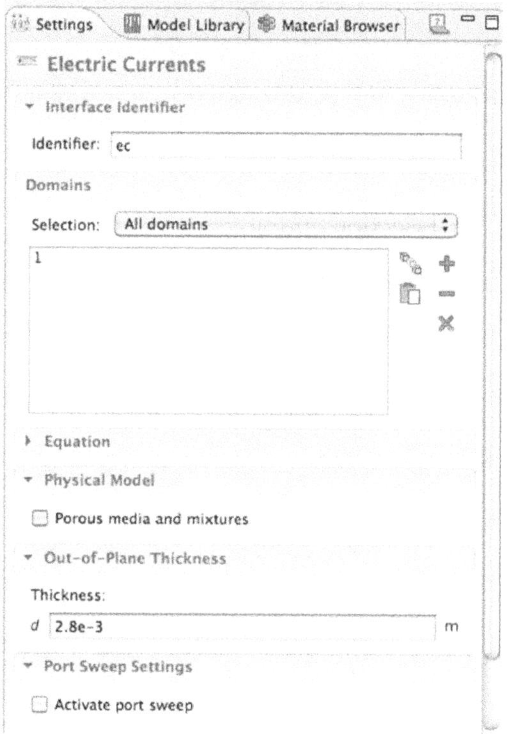

FIGURE 5.11 Model Geometry Out-of-Plane-Thickness Thickness Edit Window.

Figure 5.11 shows the Model Geometry Out-of-Plane-Thickness Thickness edit window.

Click > Model Builder – Model 1 – Electric Currents – Current Conservation 1.

Select > User defined from the Pull-down menu in Settings – Current Conservation – Conduction Current – Electric conductivity.

Enter > 10 in the Settings – Current Conservation – Conduction Current – Electric conductivity edit window.

Scroll-down > Find the Settings – Current Conservation – Conduction Current – Electric Field – Relative permittivity Pull-down menu.

Select > User defined from the Pull-down menu in Settings – Current Conservation – Conduction Current – Electric Field – Relative permittivity Pull-down menu.

Either use the default value of 1 that appears or enter the value 1 in the edit window.

In this case, a relative permittivity of 1 is appropriate. If the electrolyte had a different relative permittivity, then the modeler would need to enter that value here.

Right-Click > Model Builder – Model 1 – Electric Currents.

Select > Electric Insulation from the Pop-up menu.

Click > Model Builder – Model 1 – Electric Currents – Electric Insulation 2.

Click > Boundary 1 in the Graphics window.

Click > Add to Selection in the Settings – Electric Insulation – Boundaries window.

Click > Boundary 5 in the Graphics window.

Click > Add to Selection in the Settings – Electric Insulation – Boundaries window.

See Figure 5.12.

Figure 5.12 shows the Model Geometry Electrically Insulation 2 Boundaries.

Right-Click > Model Builder – Model 1 – Electric Currents.

Select > Ground from the Pop-up menu.

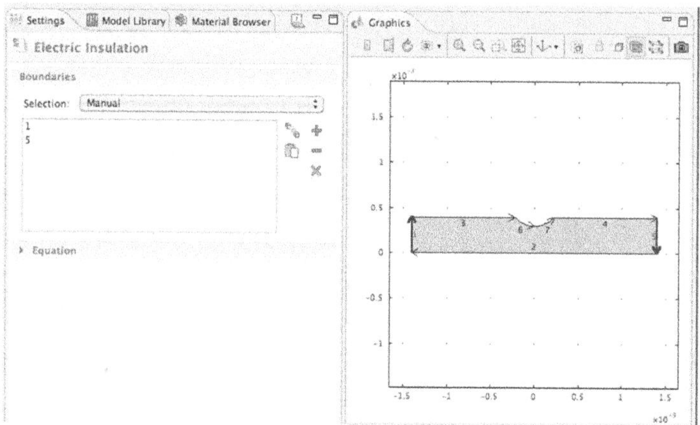

FIGURE 5.12 Model Geometry Electrically Insulation 2 Boundaries.

Click > Model Builder – Model 1 – Electric Currents – Ground 1.

Click > Boundary 2 in the Graphics window.

Click > Add to Selection in the Settings – Ground – Boundaries window.

See Figure 5.13.

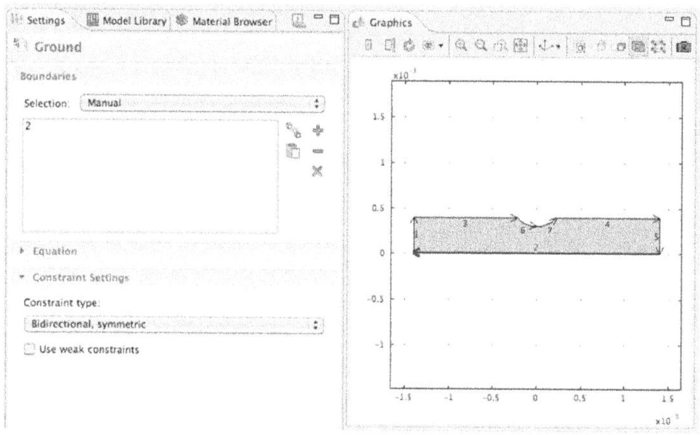

FIGURE 5.13 Model Geometry Ground 1 Boundaries.

Figure 5.13 shows the Model Geometry Ground 1 Boundaries.

Right-Click > Model Builder – Model 1 – Electric Currents.

Select > Electric Potential from the Pop-up menu.

Click > Model Builder – Model 1 – Electric Currents – Electrical Potential 1.

Shift-Click > Boundaries 3, 4, 6, 7 in the Graphics window.

Click > Add to Selection in the Settings – Electric Potential – Boundaries window.

Enter > 30 in the Settings – Electric Potential – Electric Potential – Voltage edit window.

See Figure 5.14.

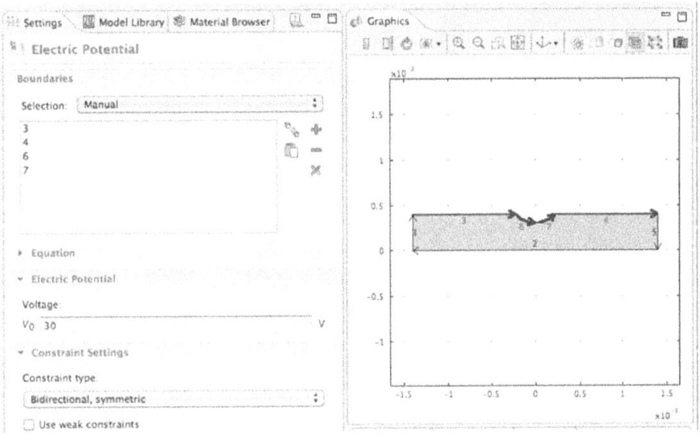

FIGURE 5.14 Model Geometry Electric Potential 1 Boundaries.

Figure 5.14 shows the Model Geometry Electric Potential 1 Boundaries.

Moving Mesh (ale) Interface

Click > Twistie for the Moving Mesh (ale) interface.

Click > Moving Mesh (ale).

Verify > Domain 1 is selected.

Right-Click > Model Builder – Model 1 – Moving Mesh (ale).

Select > Free Deformation from the Pop-up menu.

Click > Model Builder – Model 1 – Moving Mesh (ale) – Free Deformation 1.

Click > Domain 1 in the Graphics window.

Click > Add to Selection in the Settings – Free Deformation – Domains window.

NOTE *In this case, the default settings for the Initial mesh displacement are correct. If however the mesh were initially displaced, then the modeler would need to enter different values in the Initial mesh displacement edit windows.*

See Figure 5.15.

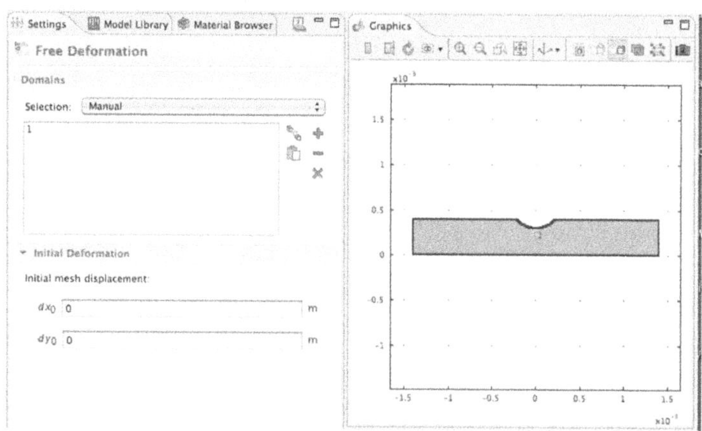

FIGURE 5.15 Settings - Free Deformation - Domains and Graphics Windows.

Figure 5.15 shows the Settings – Free Deformation – Domains and Graphics windows.

Right-Click > Model Builder – Model 1 – Moving Mesh (ale).

Select > Prescribed Mesh Velocity from the Pop-up menu.

Click > Model Builder – Model 1 – Moving Mesh (ale) – Prescribed Mesh Velocity 1.

Shift-Click > Boundaries 1 and 5 in the Graphics window.

Click > Add to Selection in the Settings – Prescribed Mesh Velocity – Boundaries edit window.

Select > Prescribed x velocity checkbox in Settings – Prescribed Mesh Velocity – Prescribed Mesh Velocity.

Uncheck > Prescribed y velocity checkbox in Settings – Prescribed Mesh Velocity – Prescribed Mesh Velocity.

NOTE

In this case, the default settings (0) for the Prescribed Mesh Velocity are correct. If however the Prescribed Mesh Velocity were different, then the modeler would need to enter other values in the Prescribed Mesh Velocity edit windows.

Select > Use weak constraints checkbox in Settings – Prescribed Mesh Velocity – Constraint Settings.

NOTE

In 4.x, the use of "weak constraints" implements constraints by using finite elements on the constrained domain for the Lagrange multipliers, and by solving for the Lagrange multipliers along with the original problem {5.12}.

See Figure 5.16.

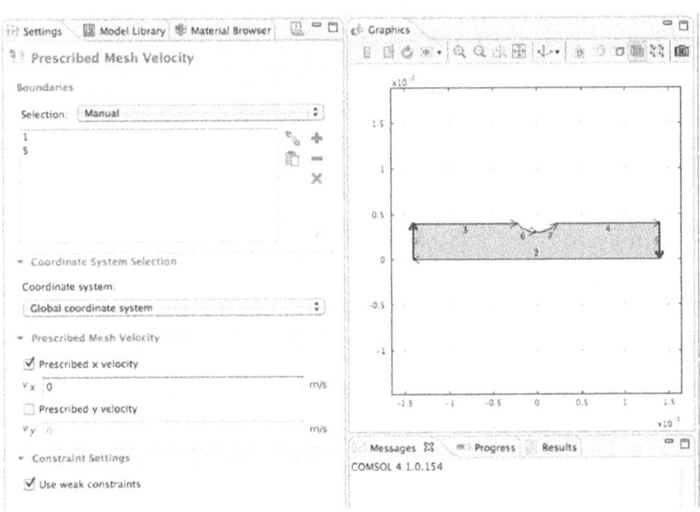

FIGURE 5.16 Settings – Prescribed Mesh Velocity – Prescribed Mesh Velocity and Constraint Settings.

Figure 5.16 shows the Settings – Prescribed Mesh Velocity – Prescribed Mesh Velocity and Constraint Settings.

Right-Click > Model Builder – Model 1 – Moving Mesh (ale).

Select > Prescribed Mesh Velocity from the Pop-up menu.

Click > Model Builder – Model 1 – Moving Mesh (ale) – Prescribed Mesh Velocity 2.

Click > Boundaries 3, 4, 6, and 7 in the Graphics window.

Click > Add to Selection in the Settings – Prescribed Mesh Velocity – Boundaries edit window.

Click > Pull-down menu in Settings – Prescribed Mesh Velocity – Coordinate System Selection.

Select > Boundary System 1 (sys1).

Uncheck > Prescribed t1 velocity checkbox in Settings – Prescribed Mesh Velocity – Prescribed Mesh Velocity.

Select > Prescribed n velocity checkbox in Settings – Prescribed Mesh Velocity – Prescribed Mesh Velocity.

NOTE *The modeler should note the referenced coordinate system variables change to match the Coordinate system selected from the Pull-down menu. In this case, the default settings for both the t1 and the n Prescribed Mesh Velocity are not correct.*

Enter > -K*(-ec.normJ) in the Prescribed n velocity edit window in Settings – Prescribed Mesh Velocity – Prescribed Mesh Velocity – Prescribed n velocity v_y edit window.

NOTE *The names of the 4.x Predefined Physics Interface Variables may be determined as follows: First, build the geometry of the desired model; Second, expand the Physics Interface to show Equation View; Third, Click > Equation View; Fourth, Click > Refresh Equations; Fifth, Scroll until you find the desired variable Name.*

Select > Use weak constraints checkbox in Settings – Prescribed Mesh Velocity – Constraint Settings.

See Figure 5.17.

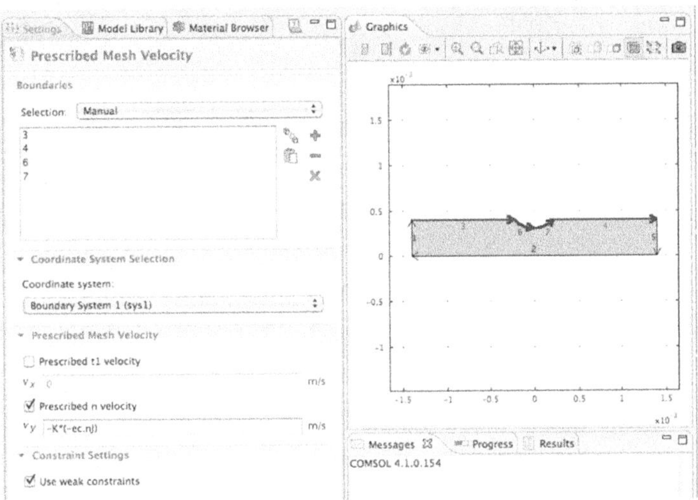

FIGURE 5.17 Settings - Prescribed Mesh Velocity - Prescribed Mesh Velocity, Coordinate System, and Constraint Settings.

Figure 5.17 shows the Settings – Prescribed Mesh Velocity – Prescribed Mesh Velocity, Coordinate System, and Constraint Settings.

Mesh 1

NOTE

The default settings for the Mesh Type are not appropriate for the solution of this problem and thus need to be modified.

Click > Model Builder – Model 1 – Mesh 1.

Click > Sequence type Pull-down menu in Settings – Mesh – Mesh Settings – Sequence type.

Select > Physics-controlled mesh.

Click > Element size Pull-down menu in Settings – Mesh – Mesh Settings – Element size.

Select > Extra fine.

Click > Build All button.

After meshing, the modeler should see a message in the message window about the number of elements (920 elements) in the mesh.

See Figure 5.18.

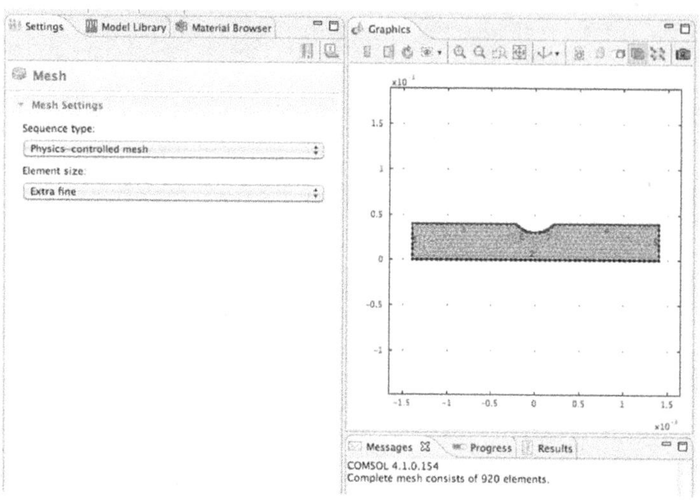

FIGURE 5.18 Desktop Display - Graphics - Meshed Domain.

Figure 5.18 shows the Desktop Display – Graphics – Meshed Domain.

Study 1

NOTE *The default settings for the Time Dependent Study Type are not appropriate for the solution of this problem and thus need to be modified.*

Click > Study 1 twistie.

Click > Step 1: Time Dependent.

Click the Range button (at the right of the Times entry window).

Enter > Start = 0, Stop = 10, Step = 1.

Click > Replace.

See Figure 5.19.

Figure 5.19 shows the Desktop Display – Settings – Time Dependent – Study Settings.

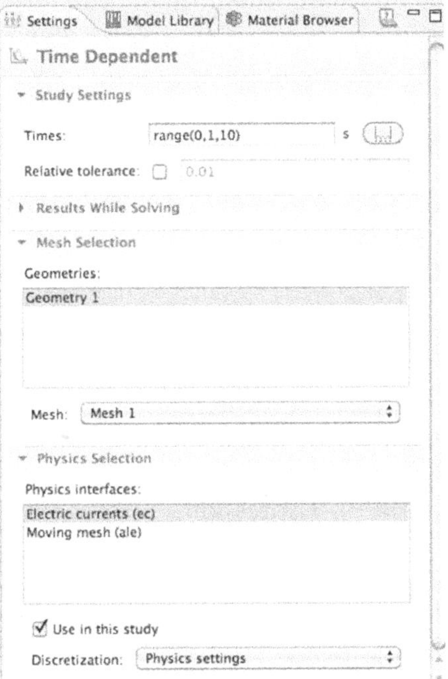

FIGURE 5.19 Desktop Display - Settings - Time Dependent - Study Settings.

Study 1 > Solver 1

In Model Builder,

Right-Click Study 1 >Select > Compute.

Computed results, using the default display settings, are shown in Figure 5.20.

Figure 5.20 shows the computed results, using the default display settings.

Results

In Model Builder,

Click > Results > 2D Plot Group 1 Twistie.

Click > Model Builder – Results – 2D Plot Group 1 – Surface 1.

In Settings – Surface – Data – Data set,

Select > Solution 1 from the Pull-down menu.

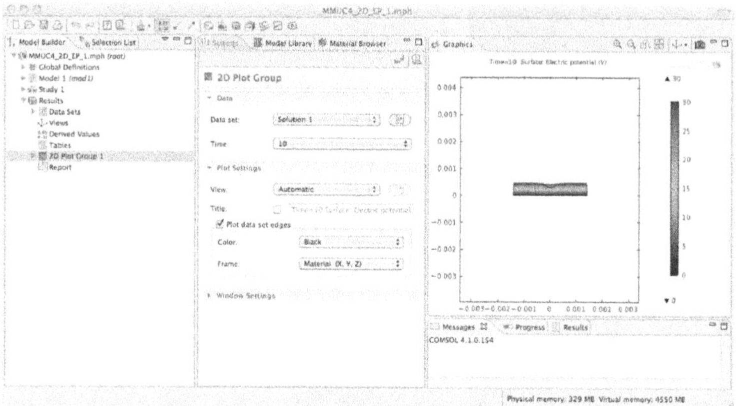

FIGURE 5.20 Computed Results, Using the Default Display Settings.

In Settings – Surface – Expression,

Click > Replace Expression.

Select > Electric Currents – Current density norm (ec.normJ) from the Pop-up menus.

Click > Plot.

See Figure 5.21.

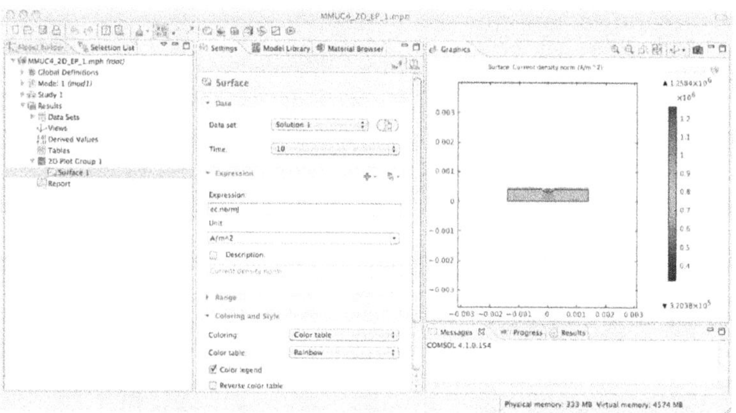

FIGURE 5.21 Desktop Display - Computed Solution.

Figure 5.21 shows the Desktop Display – Computed Solution.

2D Electrochemical Polishing Model Animation

This animation shows the electropolishing process as it progresses in time.

In Model Builder,

Right-Click > Results – Report.

Select > Animation.

In Settings – Scene,

Verify > 2D Plot Group 1 is the Subject shown on the Pull-down menu.

In Settings – Animation – Output,

Select > Output type Movie from the Pull-down menu.

Select > Format GIF from the Pull-down menu.

NOTE *The detailed procedures for Animation in 4.x on the Macintosh and the PC are different. The Macintosh uses the GIF format. The PC uses the AVI format. If the modeler tries to use AVI on the Macintosh, an error results.*

In Settings – Animation – Frame Settings,

Click > Lock aspect ratio check box.

In Settings – Animation – Advanced,

Click > Antialiasing check box.

Click > Settings – Export.

NOTE *The movie will be exported to and run in the 4.x Graphics window, if there is no path designated in the Settings – Animation – Output Filename edit window.*

2D Electrochemical Polishing Model Movie

The modeler can save the movie as a unique file.

Click > Settings – Animation – Output Browse button.

Select the desired location for saving the movie.

Enter the desired File Name in the Save-as edit window.

Click > Save.

Click > Save for the completed MMUC4_2D_EP_1_An.gif EP movie.

2D Electrochemical Polishing Model Summary and Conclusions

The 2D Electrochemical Polishing Model is a powerful tool that can be used to explore electrochemical polishing techniques in many different conductive media (e.g. metals, semimetals, alloys, graphene films, graphite films, etc.). As has been shown earlier in this chapter, the 2D Electrochemical Polishing Model is easily and simply modeled with a combination of a 2D Electrical Currents Interface and a Moving Mesh Interface model.

2D Hall Effect Model Considerations

In 1827, Georg Ohm published {5.13} his now fundamental and famous Ohm's Law:

$$I = \frac{V}{R} \tag{5.3}$$

Where: I = Current in Amperes.

V = Potential Difference in Volts.

R = Resistance in Ohms.

See Figure 5.22.

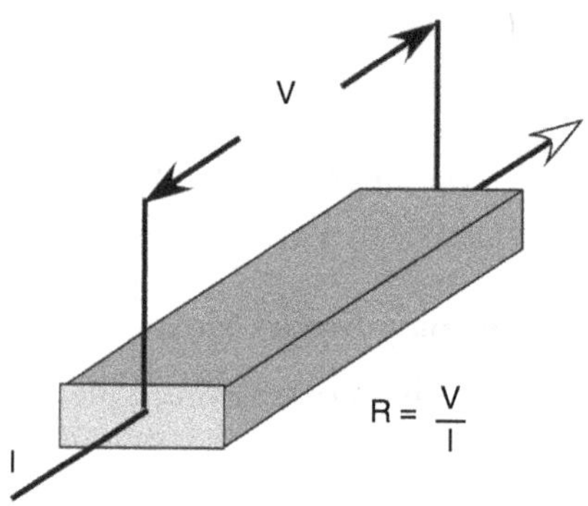

FIGURE 5.22 Ohm's Law.

Ohm's Law is an extremely useful phenomenological law. However, in order to better understand conduction in homogeneous, isotropic solid materials, the calculations used herein need to be expanded until they represent the physical behavior (motion) of the fundamental charged particles (electrons, holes).

NOTE

In solid materials (e.g. metals, semiconductors, etc.), there are three potential mobile carriers of charge: electrons (–), holes (+), and ions (charge sign can be either + or –, depending upon the type of ion). Ions in a solid typically have a very low mobility (pinned in position) and thus contribute little to the observed current flow in most solids. Ion flow will not be considered herein.

In metals, due to the underlying physical and electronic structure, electrons are the only carrier. In semiconductors (e.g. Si, Ge, GaAs, InP), either electrons and/or holes (the absence of an electron) can exist as the primary carrier types. The density of each carrier type (electrons, holes) is determined by the electronic structure of the host material (e.g. Si, Ge, SiGe) and the density and distribution of foreign impurity atoms (e.g. As, P, N, Al) within the host solid material. For further information on the nature of solids and the behavior of impurity atoms in a host matrix see works by Kittel {5.14} and/or Sze {5.15}.

Hall Effect sensors are widely available in a large number of different geometric configurations. They are typically applied in sensing fluid flow, rotating, and/or linear motion, proximity, current, pressure, orientation, etc. In the 2D model presented in the remainder of this chapter, several simplifying assumptions will be made that allow the basic physics principles to be demonstrated without excess complexity.

The resistance of a homogeneous, isotropic solid material R is defined:

$$R = \frac{\rho L}{A} \tag{5.4}$$

Where: ρ = resistivity in Ohm-meters (Ω-m).

L = Length of Sample in meters (m).

A= Cross-sectional Area of Sample in meters squared (m^2).

See Figure 5.23.

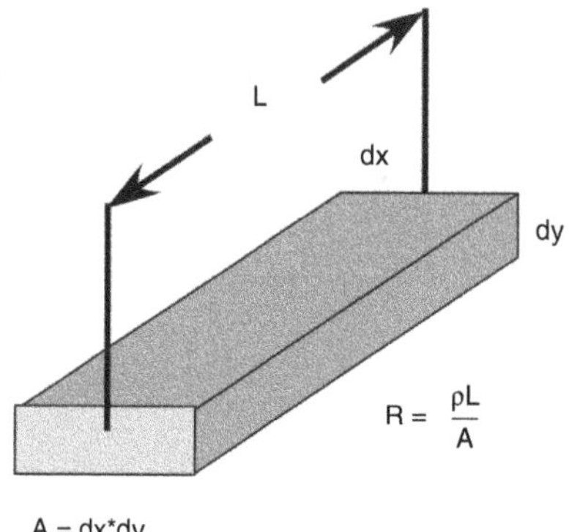

FIGURE 5.23 Resistance.

The resistivity of a homogeneous, isotropic solid material is defined {5.16}:

$$\sigma \equiv \frac{1}{\rho} = n_e |e| \mu_e + n_h |e| \mu_h \tag{5.5}$$

Where:
- ρ = resistivity Ohm-meters (Ω-m).
- σ = conductivity in Siemens per meter (S/m).
- n_e = electron density in electrons per cubic meter (N_e/m^3).
- n_h = hole density in holes per cubic meter (N_h/m^3).
- $|e|$ = absolute value of the charge on an electron (hole) in Coulombs (C).
- μ_e = electron mobility in meters squared per volt-second ($m^2/(V*s)$).
- μ_h = hole mobility in meters squared per volt-second ($m^2/(V*s)$).

The Hall Effect {5.17} was discovered by Edwin Hall in 1879 {5.18}. His discovery was made through making measurements on the behavior of electrical currents in thin gold foils, in the presence of a magnetic field. By introducing a magnetic field into the current flow region of the solid, the measurer effectively adds an anisotropic term into the conductivity of a nominally homogeneous, isotropic solid material. The magnetic field causes the anisotropic conductivity by acting on the carriers through the Lorentz

Force {5.19}. The Lorentz Force produces a proportional, differential voltage/charge accumulation between two surfaces or edges of a conducting material orthogonal to the current flow.

The Lorentz Force is:

$$\mathbf{F} = q\,(\mathbf{E} + (\mathbf{v} \times \mathbf{B})) \tag{5.6}$$

Where: **F** = Force Vector on the charged particle (electron and/or hole).

q = charge on the particle (electron and/or hole).

E = Electric Field Vector.

v = Instantaneous velocity vector of the particle.

B = Magnetic Field Vector.

The Hall Voltage {5.20} is:

$$V_H = \frac{R_H * I * B}{t} \tag{5.7}$$

Where: V_H = Hall Voltage.

R_H = Hall coefficient.

I = Current.

B = Magnetic Field.

t = thickness of sample.

The Hall coefficient (R_H) is:

$$R_H = -\frac{r}{n_e e} \tag{5.8}$$

Where: R_H = Hall coefficient.

$r = 1 \leq x \leq 2$.

n_e = density of electrons.

e = charge on the electron.

NOTE

In the Hall Effect model presented herein, it is assumed that r = 1. That assumption is a valid first approximation. For applied development models, the modeler will need to determine experimentally the best approximation for the value of r for the particular material and physical conditions being modeled.

For example, in the case that the charge carrier is a "hole," the minus sign (–) in the equation for the Hall coefficient changes to a plus sign (+). In the case of mixed electron/ hole flow, R_H can become zero (0).

The differential voltage/charge accumulation (Hall Voltage (V_H)) that results from the Lorentz Force interaction between any currents (electron and/or hole) flowing through that conducting material and the local magnetic field is shown in Figure 5.24.

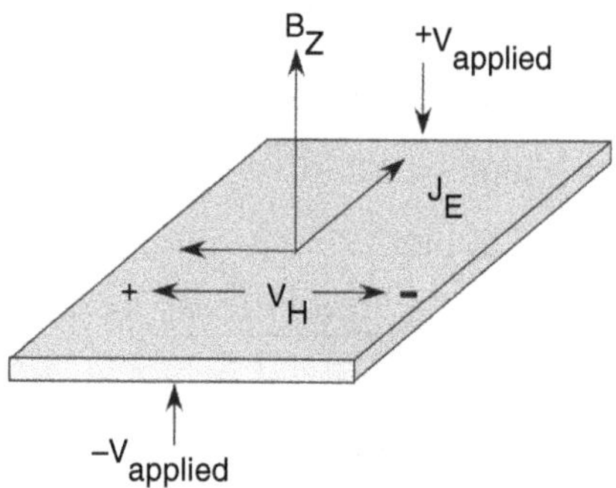

FIGURE 5.24 Hall Effect Sensor Geometry, Electron Flow.

As can be seen from the introductory material, depending upon the characteristics of the material being modeled, the calculation of the Hall Effect can be very complex. The Hall coefficient (R_H) varies for different materials and has a predominant functional dependence that involves temperature, carrier type, carrier concentration, carrier mobility, carrier lifetime, and carrier velocity. In a dual carrier system, such as semiconducting materials (electrons and holes), under the proper conditions, R_H can become equal to zero. Semiconductor sensors, however, are among the most sensitive magnetic field Hall sensors currently manufactured.

Due to the underlying complexity of the Hall Effect, the model in this section of Chapter 5 requires the use of either the AC/DC module or the MEMS Module, in addition to the basic COMSOL Multiphysics Software. In this model, only a single carrier conduction system, electrons, is employed. For ease of modeling, it is assumed that the system is quasi-static. This model introduces the COMSOL modeling concepts of Point-wise Constraints and Floating Potential {5.21, 5.22}.

2D Hall Effect Model

Building the 2D Hall Effect Model

Startup 4.x.

Select > 2D.

Click > Next.

Click > Twistie for AC/DC in the Add Physics window.

Click > Electric Currents (ec).

Click > Add Selected.

Click > Next.

Select > Stationary in the Select Study Type window.

Click > Finish (Flag).

Click > Save As.

Enter MMUC4_2D_HE_1.mph.

See Figure 5.25.

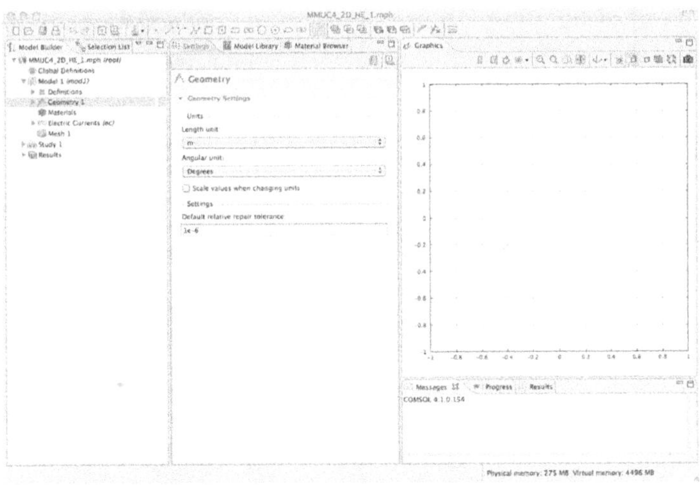

FIGURE 5.25 Desktop Display for the MMUC4_2D_HE_1.mph Model.

Figure 5.25 shows the Desktop Display for the MMUC4_2D_HE_1.mph model.

Right-Click > Global Definitions.

Select > Parameters from the Pop-up menu.

Enter > In the Settings – Parameters – Parameters edit window the parameters as shown in Table 5.1.

TABLE 5.1 Parameters Edit Window

Name	Expression	Description
sigma0	1.04e3[S/m]	Silicon conductivity
Rh	1.25e-4[m^3/C]	Hall coefficient
Bz	0.1[T]	Magnetic field
coeff0	sigma0/(1+(sigma0*Rh*Bz)^2)	Conductivity anisotropy 2
V0	5.0[V]	Applied voltage
t_Si	1.0e-3[m]	Silicon thickness
coeff1	sigma0*Rh*Bz	Conductivity anisotropy 1
s11	coeff0	Conductivity matrix term 11
s12	coeff0*coeff1	Conductivity matrix term 12
s21	-coeff0*coeff1	Conductivity matrix term 21
s22	coeff0	Conductivity matrix term 22
epsilon_Si	11.68	Relative permeability silicon

See Figure 5.26.

Figure 5.26 shows the Settings – Parameters – Parameters edit window filled.

In Model Builder – Model 1 (mod1),

Right-Click > Model Builder – Geometry 1.

Select > Rectangle from the Pop-up menu.

Enter > 1.8e-2 in the Settings – Rectangle – Size – Width entry window.

Enter > 6e-3 in the Settings – Rectangle – Size – Height entry window.

Enter > -9e-3 in the Settings – Rectangle – Position – x entry window.

Settings | Model Library | Material Browser

Parameters

Parameters

Name	Expression	Value	Description
Rh	1.25e-4[m^3/C]	1.25E-4 $m^3/(s\cdot A)$	Hall coefficient
Bz	0.1[T]	0.1 T	Magnetic field
coeff0	sigma0/(1+(sigma0*Rh*Bz)^2)	1040 S/m	Conductivity anisotropy 2
V0	5.0[V]	5 V	Applied voltage
t_Si	1.0e-3[m]	0.001 m	Silicon thickness
coeff1	sigma0*Rh*Bz	0.013	Conductivity anisotropy 1
s11	coeff0	1040 S/m	Conductivity matrix term 11
s12	coeff0*coeff1	13.52 S/m	Conductivity matrix term 12
s21	-coeff0*coeff1	-13.52 S/m	Conductivity matrix term 21
s22	coeff0	1040 S/m	Conductivity matrix term 22

Name:

Expression:

Description:

FIGURE 5.26 Settings - Parameters - Parameters Edit Window Filled.

Enter > -3e-3 in the Settings – Rectangle – Position – y entry window.

Click > Build All.

See Figure 5.27.

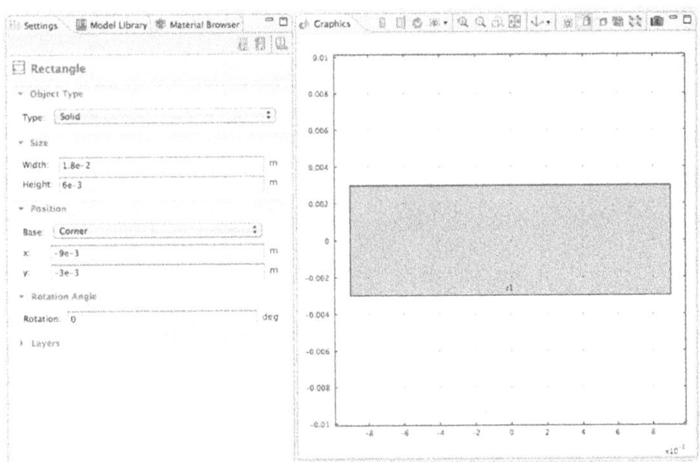

FIGURE 5.27 Settings - Rectangle Entry Windows.

Figure 5.27 shows the Settings – Rectangle Entry Windows.

Create points at the locations shown in Table 5.2:

TABLE 5.2 Points List

Point #	x Coordinate	y Coordinate
1	−1e−3	3e−3
2	1e−3	3e−3
3	−1e−3	-3e−3
4	1e−3	-3e−3

Right-Click > Model Builder – Geometry 1.

Select > Point from the Pop-up menu.

Enter > Point 1 coordinates.

Click > Build All.

Select > Point from the Pop-up menu.

Enter > Point 2 coordinates.

Click > Build All.

Select > Point from the Pop-up menu.

Enter > Point 3 coordinates.

Click > Build All.

Select > Point from the Pop-up menu.

Enter > Point 4 coordinates.

Click > Build All.

See Figure 5.28.

Figure 5.28 shows the Rectangle with Points in the Graphics Window.

Electric Currents (ec) Interface

Click > Twistie for the Electric Currents (ec).

Click > Electric Currents (ec),

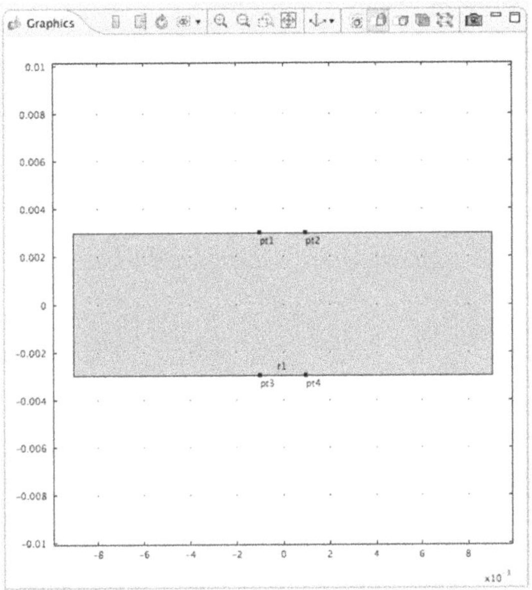

FIGURE 5.28 Rectangle with Points in the Graphics Window.

Enter > t_Si in the Settings – Electric Currents – Out-of-Plane-Thickness – Thickness edit window.

See Figure 5.29.

Figure 5.29 shows the Model Geometry Out-of-Plane-Thickness Thickness edit window.

Click > Model Builder – Model 1 – Electric Currents – Current Conservation 1.

Select > User defined from the Pull-down menu in Settings – Current Conservation – Conduction Current – Electric conductivity.

Select > Anisotropic from the Pull-down menu in Settings – Current Conservation – Conduction Current – Electric conductivity.

Enter the parameter values as indicated in Table 5.3 in the Settings – Current Conservation – Conduction Current – Electric conductivity edit window.

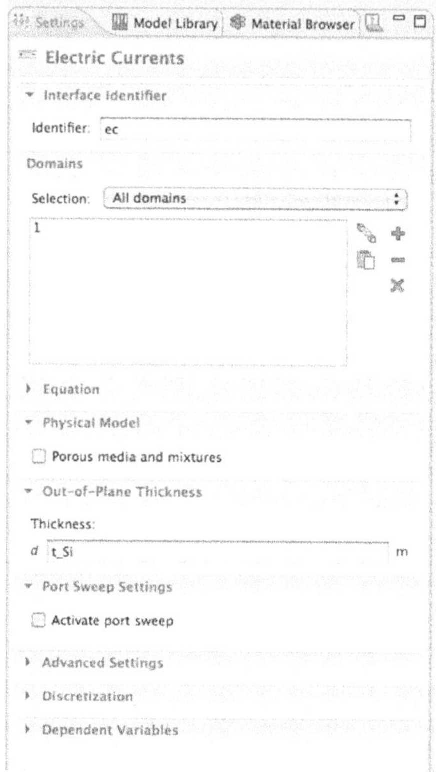

FIGURE 5.29 Model Geometry Out-of-Plane-Thickness Thickness Edit Window.

TABLE 5.3 Silicon Conductivity Parameters

Window #	Expression
11	s11
12	s12
21	s21
22	s22

Scroll-down > Find the Settings – Current Conservation – Conduction Current – Electric Field – Relative permittivity Pull-down menu.

Select > User defined from the Pull-down menu in Settings – Current Conservation – Conduction Current – Electric Field – Relative permittivity Pull-down menu.

Enter > epsilon_Si in the Settings – Current Conservation – Conduction Current – Electric Field – Relative permittivity edit window.

NOTE *In this case, a relative permittivity of 1 is not appropriate. Since silicon has a different relative permittivity, then the modeler needs to enter that value as indicated.*

See Figure 5.30.

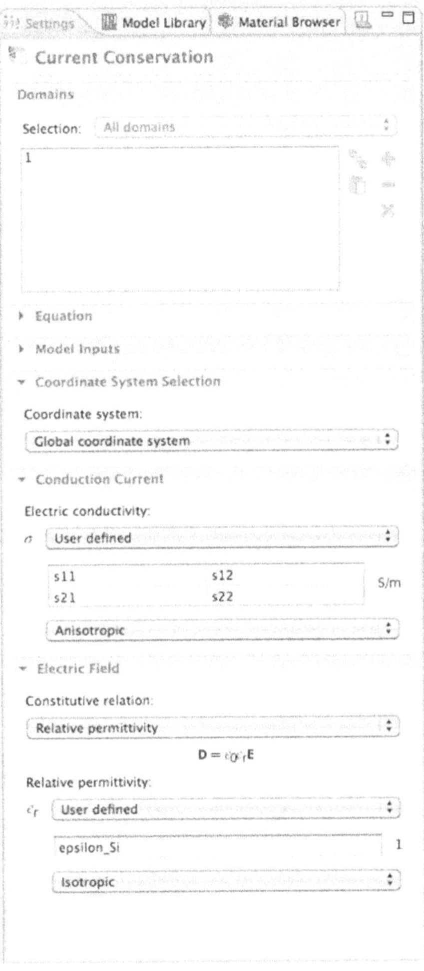

FIGURE 5.30 Settings - Current Conservation Edit Windows.

Figure 5.30 shows the Settings – Current Conservation edit windows.

Right-Click > Model Builder – Model 1 – Electric Currents.

Select > Electric Insulation from the Pop-up menu.

Click > Model Builder – Model 1 – Electric Currents – Electric Insulation 2.

Shift-Click > Boundaries 2, 3, 6, 7 in the Graphics window.

Click > Add to Selection (plus) in Settings – Electric Insulation – Boundaries.

See Figure 5.31.

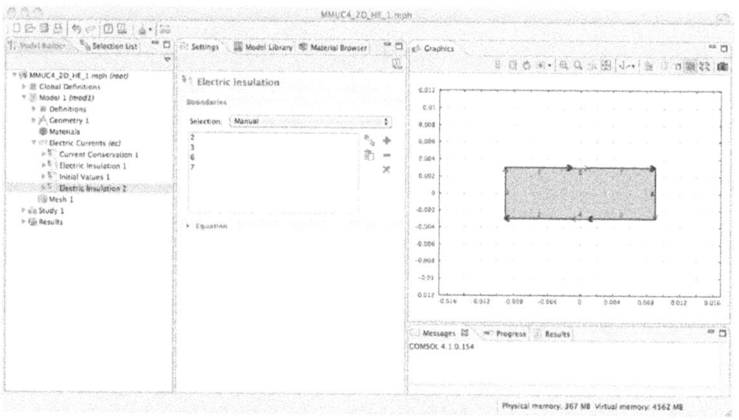

FIGURE 5.31 Model Geometry Electrical Insulation 2 Boundaries.

Figure 5.31 shows the Model Geometry Electrical Insulation 2 boundaries.

Right-Click > Model Builder – Model 1 – Electric Currents.

Select > Ground from the Pop-up menu.

Click > Model Builder – Model 1 – Electric Currents – Ground 1.

Click > Boundary 1 in the Graphics window.

Click > Add to Selection in the Settings – Ground – Boundaries window.

Select > Use weak constraints checkbox in Settings – Ground – Constraint Settings edit panel.

See Figure 5.32.

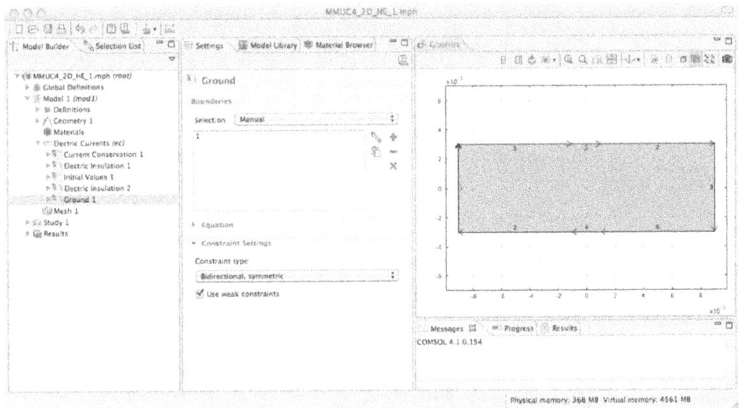

FIGURE 5.32 Model Geometry Ground 1 Boundary and Constraint Settings.

Figure 5.32 shows the Model Geometry Ground 1 Boundary and Constraint settings.

Right-Click > Model Builder – Model 1 – Electric Currents.

Select > Electric Potential from the Pop-up menu.

Click > Model Builder – Model 1 – Electric Currents – Electrical Potential 1.

Click > Boundary 8 in the Graphics window.

Click > Add to Selection in the Settings – Electric Potential – Boundaries window.

Enter > V0 in the Settings – Electric Potential – Electric Potential – Voltage edit window.

Select > Use weak constraints checkbox in Settings – Electric Potential – Constraint Settings edit panel.

See Figure 5.33.

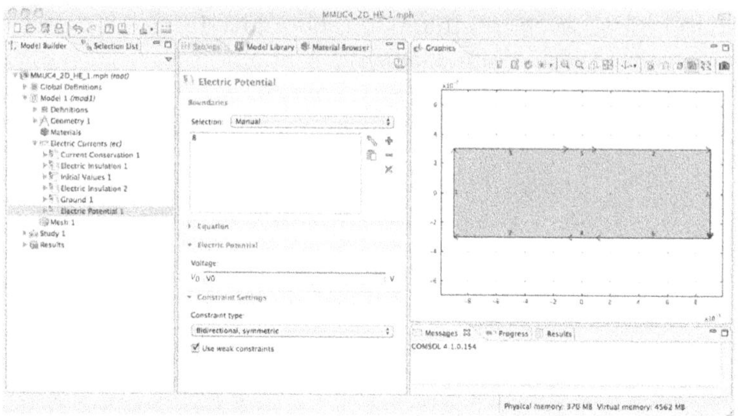

FIGURE 5.33 Model Geometry Electric Potential 1 and Constraint Settings.

Figure 5.33 shows the Model Geometry Electric Potential 1 and Constraint Settings.

Right-Click > Model Builder – Model 1 – Electric Currents.

Select > Floating Potential from the Pop-up menu.

Click > Model Builder – Model 1 – Electric Currents – Floating Potential 1.

Click > Boundary 5 in the Graphics window.

Click > Add to Selection in the Settings – Floating Potential – Boundaries window.

NOTE *In this case, a Terminal current of zero (0) is appropriate. If a different value of current were to be drawn, then the modeler would need to enter that value as indicated.*

See Figure 5.34.

Figure 5.34 shows the Model Geometry Floating Potential 1 Setting.

Right-Click > Model Builder – Model 1 – Electric Currents.

Select > Floating Potential from the Pop-up menu.

Click > Model Builder – Model 1 – Electric Currents – Floating Potential 2.

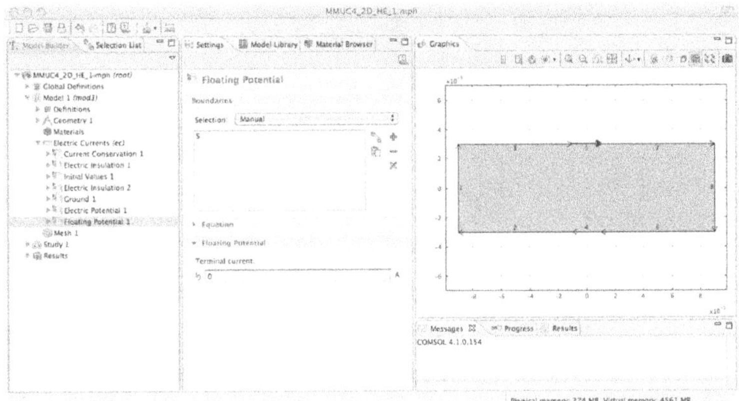

FIGURE 5.34 Model Geometry Floating Potential 1 Setting.

Click > Boundary 4 in the Graphics window.

Click > Add to Selection in the Settings – Floating Potential – Boundaries window.

NOTE *In this case, a Terminal current of zero (0) is appropriate. If a different value of current were to be drawn, then the modeler would need to enter that value as indicated.*

See Figure 5.35.

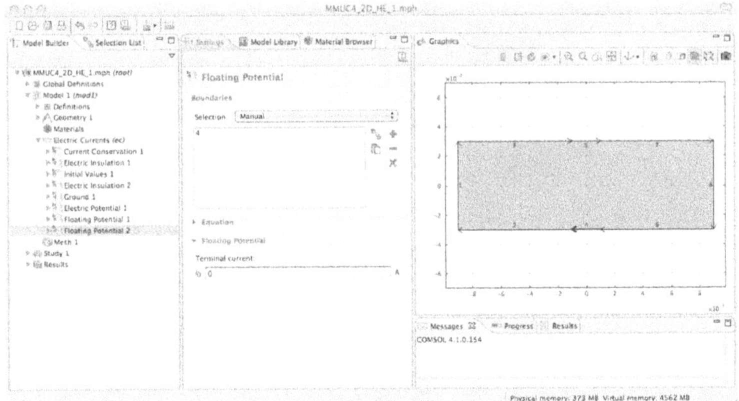

FIGURE 5.35 Model Geometry Floating Potential 2 Setting.

Figure 5.35 shows the Model Geometry Floating Potential 2 Setting.

Mesh 1

NOTE *The default settings for the Mesh Type are not appropriate for the solution of this problem and thus need to be modified.*

Click > Model Builder – Model 1 – Mesh 1.

Click > Sequence type Pull-down menu in Settings – Mesh – Mesh Settings – Sequence type.

Select > Physics-controlled mesh.

Click > Element size Pull-down menu in Settings – Mesh – Mesh Settings – Element size.

Select > Extra fine.

Click > Build All button.

After meshing, the modeler should see a message in the message window about the number of elements (2222 elements) in the mesh.

See Figure 5.36.

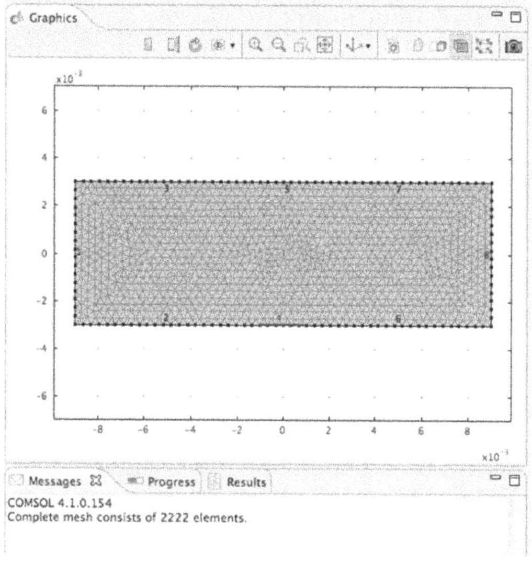

FIGURE 5.36 Desktop Display - Graphics - Meshed Domain.

Figure 5.36 shows the Desktop Display – Graphics – Meshed Domain.

Study 1

Right-Click > Model Builder – Study 1.

Select > Parametric Sweep from the Pop-up menu.

Click > Model Builder – Study 1 – Parametric Sweep.

Click > Add in Settings – Parametric Sweep – Study Settings.

Select > Magnetic field (Bz) from the Pop-up menu.

Click > OK.

Click > Range button in Settings – Parametric Sweep – Study Settings.

Enter > Start = 0, Stop = 2, Step = 0.1 in the Range Pop-up edit window.

Click > Replace button in the Range Pop-up edit window.

See Figure 5.37.

Figure 5.37 shows the Settings – Parametric Sweep – Study Settings.

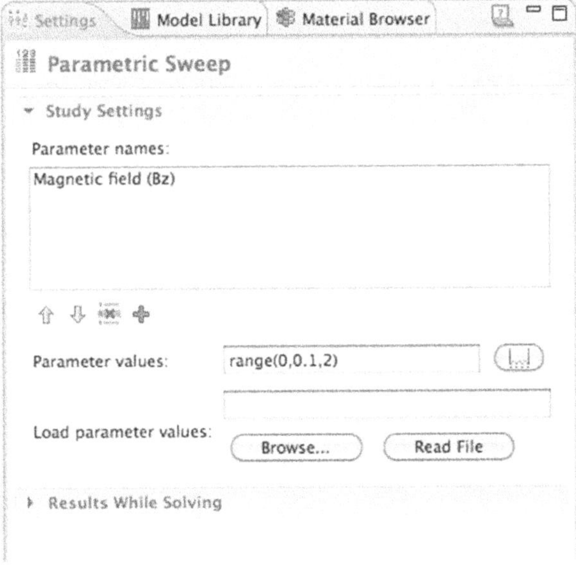

FIGURE 5.37 Settings - Parametric Sweep - Study Settings.

Study 1

In Model Builder, Right-Click Study 1 > Select > Compute.

Computed results, using the default display settings, are shown in Figure 5.38.

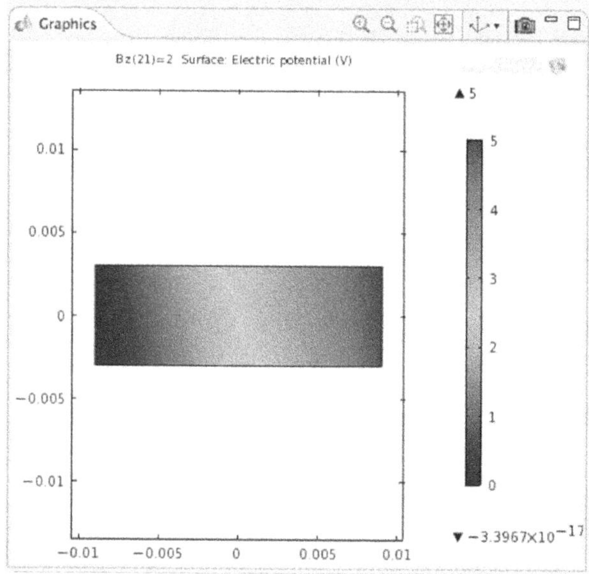

FIGURE 5.38 Hall Effect Model Computed Results, Using the Default Display Settings.

Figure 5.38 shows the Hall Effect Model Computed results, using the default display settings.

Results

In Model Builder,

Click > Results > 2D Plot Group 1 twistie.

Right-Click > 2D Plot Group 1.

Select > +Contour from the pop-up menu.

Click > Contour 1 in Model Builder – Results – 2D Plot Group 1.

Click > Settings – Contour – Color and Style – Color table.

Select > GrayScale from the Pull-down menu.

Click > Plot.

See Figure 5.39.

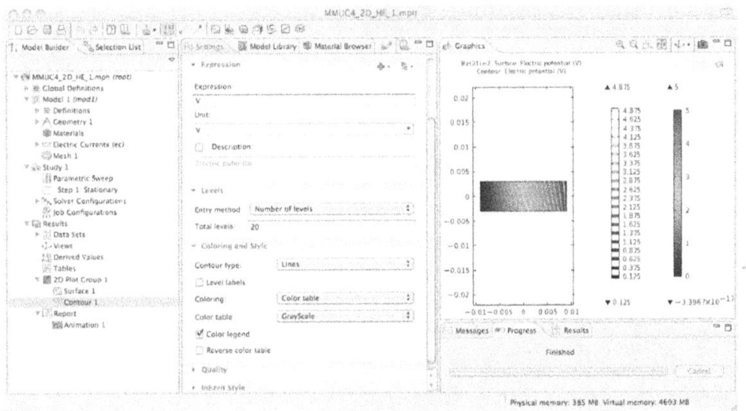

FIGURE 5.39 Hall Effect Model Computed Results, with Contour Plot.

Figure 5.39 shows the Hall Effect Model Computed results, with Contour Plot.

2D Hall Effect Model Animation

This animation shows how the Hall Voltage changes as the applied magnetic field changes.

In Model Builder,

Right-Click > Results – Report.

Select > Animation.

In Settings – Scene,

Verify > 2D Plot Group 1 is the Subject shown on the Pull-down menu.

In Settings – Animation – Output,

Select > Output type Movie from the Pull-down menu.

Select > Format GIF from the Pull-down menu.

NOTE *The detailed procedures for Animation in 4.x on the Macintosh and the PC are different. The Macintosh uses the GIF format. The PC uses the AVI format. If the modeler tries to use AVI on the Macintosh, an error results.*

In Settings – Animation – Frame Settings,

Click > Lock aspect ratio check box.

In Settings – Animation – Advanced,

Click > Antialiasing check box.

Click > Settings – Export.

NOTE *The movie will be exported to and run in the 4.x Graphics window, if there is no path designated in the Settings – Animation – Output File Name edit window.*

2D Hall Effect Model Movie

The modeler can save the movie as a unique file.

Click > Settings – Animation – Output Browse button.

Select the desired location for saving the movie.

Enter the desired File Name in the Save as edit window.

Click > Save.

Click > Settings – Export to run and save the model to the file.

Click > Save for the completed MMUC4_2D_HE_1.mph Hall Effect model.

2D Hall Effect Model Summary and Conclusions

The 2D Hall Effect model is a powerful tool that can be used to explore the Hall Effect in many different media (e.g. semiconductors, metals, semi-metals, etc.). As has been shown earlier in this chapter, the Hall Effect is easily and simply modeled with a 2D Electric Currents (ec) Interface.

FIRST PRINCIPLES AS APPLIED TO 2D MODEL DEFINITION

First Principles Analysis derives from the fundamental laws of nature. In the case of models using this Classical Physics Analysis approach, the laws of conservation in physics require that what goes in (as mass, energy, charge, etc.) must come out (as mass, energy, charge, etc.) or must accumulate within the boundaries of the model.

The careful modeler must be knowledgeable of the implicit assumptions and default specifications that are normally incorporated into the COMSOL Multiphysics software model when a model is built using the default settings.

Consider, for example, the two 2D models developed in this chapter. In these models, it is implicitly assumed there are no thermally related changes (mechanical, electrical, etc.). It is also assumed the materials are homogeneous and isotropic, except as specifically indicated in the Hall Effect, and there are no thin electrical contact barriers at the electrical junctions. None of these assumptions are typically true in the general case. However, by making such assumptions, it is possible to easily build a 2D First Approximation Model.

NOTE

A First Approximation Model is one that captures all the essential features of the problem that needs to be solved, without dwelling excessively on small details. A good First Approximation Model will yield an answer that enables the modeler to determine if he needs to invest the time and the resources required to build a more highly detailed model.

Also, the modeler needs to remember to name model parameters carefully as pointed out in Chapter 1.

REFERENCES

5.1 COMSOL Multiphysics Users Guide, Version 4.1, pp. 288–292

5.2 http://en.wikipedia.org/wiki/Electrochemical

5.3 http://en.wikipedia.org/wiki/Electropolishing

5.4 http://en.wikipedia.org/wiki/William_Gilbert_(astronomer)

5.5 http://en.wikipedia.org/wiki/Charles-Augustin_de_Coulomb

5.6 http://en.wikipedia.org/wiki/Joseph_Priestley

5.7 http://en.wikipedia.org/wiki/Georg_Ohm

5.8 http://en.wikipedia.org/wiki/Michael_Faraday

5.9 COMSOL Multiphysics Users Guide, Version 4.1, pp. 640–659

5.10 COMSOL Multiphysics Users Guide, Version 4.1, pp. 378–390

5.11 COMSOL Multiphysics Users Guide, Version 4.1, pp. 640–642

5.12 COMSOL Multiphysics Users Guide, Version 4.1, pp. 578–579

5.13 http://en.wikipedia.org/wiki/Georg_Ohm

5.14 Kittel, Charles, "Introduction to Solid State Physics", ISBN 0-471-87474-4

5.15 Sze, S.M., "Semiconductor Devices, Physics and Technology", ISBN 0-471- 87424-8

5.16 Ziman, J.M., "Principles of the Theory of Solids", First Edition, p. 185

5.17 http://en.wikipedia.org/wiki/Hall_effect

5.18 http://en.wikipedia.org/wiki/Edwin_Hall

5.19 http://en.wikipedia.org/wiki/Lorentz_Force

5.20 http://en.wikipedia.org/wiki/Hall_Voltage

5.21 COMSOL Multiphysics Users Guide, Version 4.1, p. 289

5.22 COMSOL Multiphysics Users Guide, Version 4.1, p. 373

Suggested Modeling Exercises

1. Build, mesh, and solve the 2D Electropolishing Model as presented earlier in this chapter.
2. Build, mesh, and solve the 2D Hall Effect Model as presented earlier in this chapter.
3. Change the values of the materials parameters and then build, mesh, and solve the 2D Electropolishing Model as an example problem.

4. Change the values of the materials parameters and then build, mesh, and solve the 2D Hall Effect Model as an example problem.
5. Change the value of the proportionality constant and then build, mesh, and solve the 2D Electropolishing Model as an example problem.
6. Change the value of the Hall coefficient and then build, mesh, and solve the 2D Hall Effect Model as an example problem.

CHAPTER 6

2D AXISYMMETRIC MODELING USING COMSOL MULTIPHYSICS 4.X

In This Chapter

GUIDELINES FOR 2D AXISYMMETRIC MODELING IN 4.X

NOTE

In 4.x, there are fundamentally two types of 2D geometries: 2D and 2D Axisymmetric. Chapter 5 introduced basic 2D models (x, y). 2D models are assumed to be homogeneous and isotropic in the third (non-planar) dimension. In this chapter, the modeler is introduced to the development and analysis of 2D Axisymmetric models. 2D Axisymmetric models have two geometric dimensions (r, z). 2D Axisymmetric models are planar representations of cylindrical 3D objects that are rotationally homogeneous and isotropic.

In this chapter, introductory 2D Axisymmetric models will be presented. Such 2D Axisymmetric models are more geometrically complex than 1D or 2D models. 2D Axisymmetric models have proven to be very valuable to the science and engineering communities, both in the past and currently, as first-cut evaluations of potential physical behavior under the influence of external stimuli. The 2D Axisymmetric model responses and other such ancillary information are gathered and screened early in a project for initial evaluation and potentially for later use in building higher-dimensionality (3D) field-based (electrical, magnetic, etc.) models.

Since the models in this and subsequent chapters are more complex to build and more difficult to solve than the models presented thus far, it is important that the modeler have available the tools necessary to most easily utilize the powerful capabilities of the 4.x software. In order to do that, the modeler should go to the main 4.x toolbar, Click > Options – Preferences – Model builder. When the Preferences – Model builder edit window is shown, Select > Show equation view checkbox and Show more options checkbox. Click > Apply {6.1}.

2D Axismmetric Modeling Considerations

2D Axisymmetric Modeling can be less difficult in some cases than 1D or 2D modeling, having fewer implicit assumptions, and yet potentially, it can still be a very challenging type of model to build, depending on the underlying physics involved. The least difficult aspect of 2D Axisymmetric model creation arises from the fact that the geometry is relatively simple. (In a 2D Axisymmetric model, the modeler has only a single plane as the modeling space.) However, the physics in a 2D Axisymmetric model can range from relatively easy to extremely complex.

In compliance with the laws of physics, a 2D Axisymmetric model implicitly assumes that energy flow, materials properties, environment, and all other conditions and variables that are of interest are homogeneous, isotropic, and/or constant, unless otherwise specified, throughout the entire domain of interest both within the model and through the boundary conditions and in the environs of the model.

The modeler needs to bear the above stated conditions in mind and insure that all of the modeling conditions and associated parameters (default settings) in each model created are properly considered, defined, verified, and/or set to the appropriate values.

It is always mandatory that the modeler be able to accurately anticipate the expected results of the model and accurately specify the manner in which those results will be presented. Never assume that any of the default values that are present when the model is created necessarily satisfy the needs or conditions of a particular model.

NOTE *Always verify that any parameters employed in the model are the correct value needed for that model. Calculated solutions that significantly deviate from the anticipated solution or from a comparison of values measured in an experimentally derived realistic model are probably indicative of one or more modeling errors either in the original model design, in the earlier model analysis, in the understanding of the underlying physics, or are simply due to human error.*

2D Axisymmetric Coordinate System

In 2D Axisymmetric models, there are two geometric coordinates, space (r) and space (z), and the temporal coordinate, time (t). In a steady-state solution to a 2D Axisymmetric model, parameters can only vary as a function of position in space (r) and space (z) coordinates. Such a 2D Axisymmetric model represents the parametric condition of the model in a time-independent mode (quasi-static). In a transient solution model, parameters can vary both by position in space (r) and/or space (z) and in time (t).

The space coordinates (r) and (z) typically represent distance coordinates throughout which the model is to calculate the change of the specified observables (i.e. temperature, heat flow, pressure, voltage, current, etc.) over the range of coordinate values ($r_{min} <= r <= r_{max}$) and ($z_{min} <= z <= z_{max}$). The time coordinate (t) represents the range of temporal values ($t_{min} <= t <= t_{max}$) from the beginning of observation period (t_{min}) to the end of observation period (t_{max}).

Heat Transfer Theory

Two substantially different approaches to modeling heat transfer are presented in this chapter. These different approaches demonstrate the power of the COMSOL 2D Axisymmetric modeling software. The modeling examples in this chapter demonstrate heat transfer using a stationary technique in the first model and using a transient technique in the second model.

Heat transfer is an extremely important design consideration. It is one of the most widely needed and applied technologies employed in applied physics and engineering. Most modern products or processes require an understanding of heat transfer either during development or during the use of the product or process (automobiles, plate glass fabrication, plastic extrusion, plastic products, houses, ice cream, etc.).

Heat transfer concerns have existed since early in the beginnings of humanity. At this point, however, it is understood that the science of thermodynamics and consequently today's understanding of heat transfer began with the work of Nicolas Leonard Sadi Carnot, as published in his 1824 paper {6.2} "Reflections on the Motive Power of Fire." The actual term "thermodynamics" is attributed to William Thomson (Lord Kelvin) {6.3}. Later contributions to the understanding of heat, heat transfer, and thermodynamics, in general, were made by James Prescott Joule {6.4}, Ludwig Boltzmann {6.5}, James Clerk Maxwell {6.6}, Max Planck {6.7}, and many others. The physical understanding and engineering use of thermodynamics play a very important role in the technological aspects of machine and process design in modern applied science, engineering, and medicine.

The first example presented herein, Cylinder Conduction, explores the 2D Axisymmetric Stationary modeling of heat transfer and temperature profiling for a thermally conductive material, implemented through use of the COMSOL Heat Transfer Interface.

The second 2D Axisymmetric Transient Heat Transfer Model example in this chapter explores the modeling of heat transfer over time (transient thermal response).

Heat Conduction Theory

Heat conduction is a naturally occurring process in all materials. It is readily observed in all aspects of modern life (refrigerators, freezers, microwave ovens, thermal ovens, engines, etc.). The heat transfer process allows both linear and rotational work to be done in the generation of electricity and the movement of vehicles. The initial understanding of transient heat transfer was developed by Newton {6.8} and started with Newton's Law of Cooling {6.9}:

$$\frac{dQ}{dt} = h * A * (T_S - T_E) \tag{6.1}$$

Where: $\frac{dQ}{dt}$ is the incremental energy lost in Joules per unit time (J/s).

A is the energy transmission surface area (m^2).

h is the heat transfer coefficient ($W/(m^{2}*K)$).

T_S is the surface temperature of the object losing heat (K).

T_E is the temperature of the environment gaining heat (K).

Subsequent work by Jean Baptiste Joseph Fourier {6.10}, based on Newton's Law of Cooling, developed the law for steady-state heat conduction (known as Fourier's Law {6.11}). Fourier's Law is expressed here in differential form as:

$$q = -k\nabla T \tag{6.2}$$

Where: q is the heat flux in Watts per square meter (W/m^2).

k is the thermal conductivity of the material ($W/(m*K)$).

∇T is the temperature gradient (K/m).

2D Axisymmetric Heat Conduction Modeling

The following numerical solution model (2D Axisymmetric Heat Conduction in a Cylinder Model) was originally developed by COMSOL as a tutorial model based on an example from the NAFEMS collection {6.12}. It was developed for distribution with the Multiphysics software as part of the Model Library. This model introduces two important basic concepts that apply to both applied physics and applied modeling: Axisymmetric Geometry (cylindrical) modeling and heat transfer modeling.

Heat transfer modeling is important in physical design and applied engineering problems. Typically, the modeler desires to understand heat generation during a process and either add heat or remove heat in order to achieve or maintain a desired temperature. Figure 6.1 shows a 3D rendition of the 2D Axisymmetric Cylinder Conduction geometry, as will be modeled herein. The dashed-line ellipses in Figure 6.1 indicate the 3D rotation that would need to occur to generate the 3D solid object from the 2D cross-section shown.

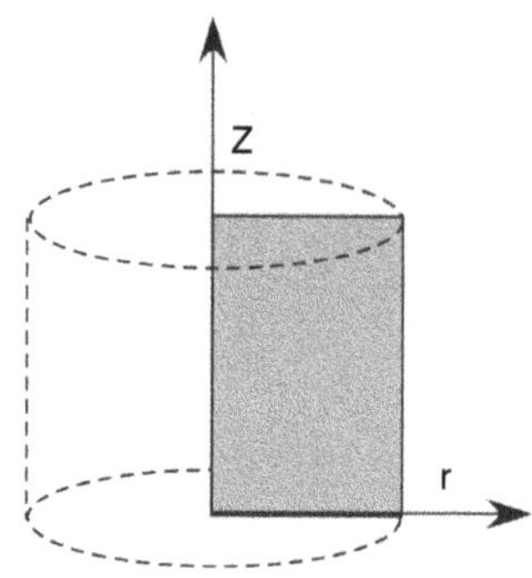

FIGURE 6.1 3D Rendition of the 2D Axisymmetric Cylinder Conduction Model.

2D Axisymmetric Heat Conduction in a Cylinder Model

NOTE

This model is derived from the COMSOL Cylinder Conduction Model. In this model, however, the selected thermally conductive solid is Niobium (Nb) {6.13, 6.14}. Niobium has a variety of uses, as an alloying element in steels, as an alloying element in titanium turbine blades, in superconductors, as an anti-corrosion coating, as an optical coating, and as an alloy in coinage.

2D AXISYMMETRIC BASIC MODELS

2D Axisymmetric Cylinder Conduction Model

Building the 2D Axisymmetric Cylinder Conduction Model

Startup 4.x.

Select > 2D axisymmetric.

Click > Next.

Click > Twistie for Heat Transfer.

Click > Heat Transfer in Solids *(ht)*.

Click > Add Selected.

Click > Next.

Select > Stationary in the Select Study Type window.

Click > Finish (Flag).

Click > Save As.

Enter MMUC4_2DA_HTC_1.mph.

See Figure 6.2.

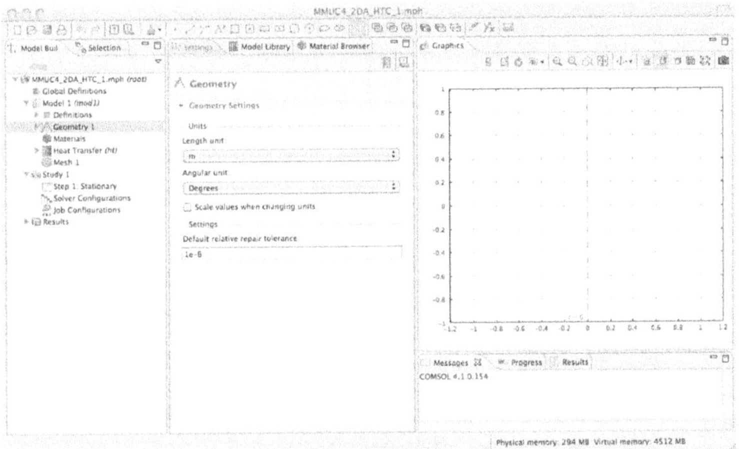

FIGURE 6.2 Desktop Display for the MMUC4_2DA_HTC_1.mph Model.

Figure 6.2 shows the Desktop Display for the MMUC4_2DA_HTC_1.mph model.

NOTE

The axis of rotation (symmetry) is indicated in the Desktop Display – Graphics window by a non-solid (dots and dashes) line at r = 0.

Right-Click > Global Definitions.

Select > Parameters from the Pop-up menu.

In the Settings – Parameters – Parameters edit window,

Enter the parameters shown in Table 6.1.

TABLE 6.1 Parameters Edit Window

Name	Expression	Description
k_Nb	52.335[W/(m*K)]	Thermal conductivity Nb
rho_Nb	8.57e3[kg/m^3]	Density Nb
Cp_Nb	2.7e2[J/(kg*K)]	Heat capacity Nb
T_0	2.7315e2[K]	Boundary temperature
q_0	5e5[W/m^2]	Heat flux

See Figure 6.3.

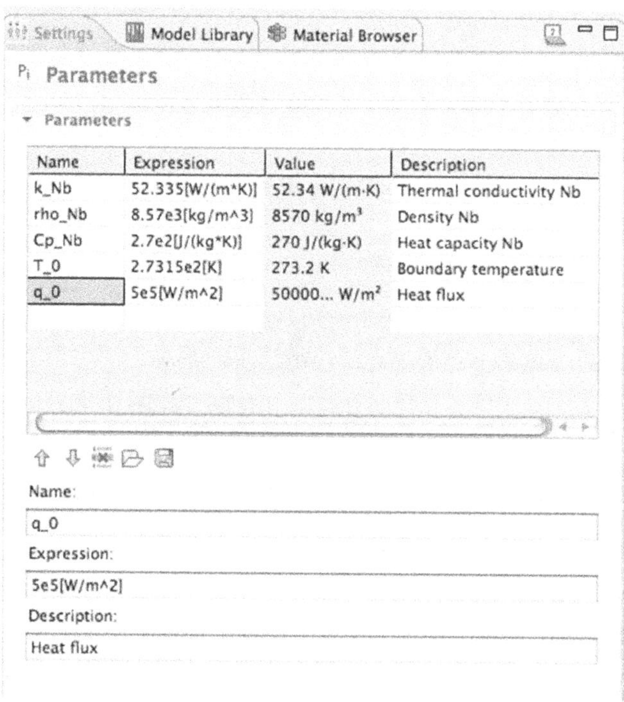

FIGURE 6.3 Settings - Parameters - Parameters Edit Window Filled.

Figure 6.3 shows the Settings – Parameters – Parameters edit window filled.

Right-Click > Model Builder – Geometry 1.

Select > Rectangle from the Pop-up menu.

Enter > 8e-2 in the Settings – Rectangle – Size – Width entry window.

Enter > 1.4e-1 in the Settings – Rectangle – Size – Height entry window.

Enter > 2e-2 in the Settings – Rectangle – Position – r entry window.

Click > Build All.

See Figure 6.4.

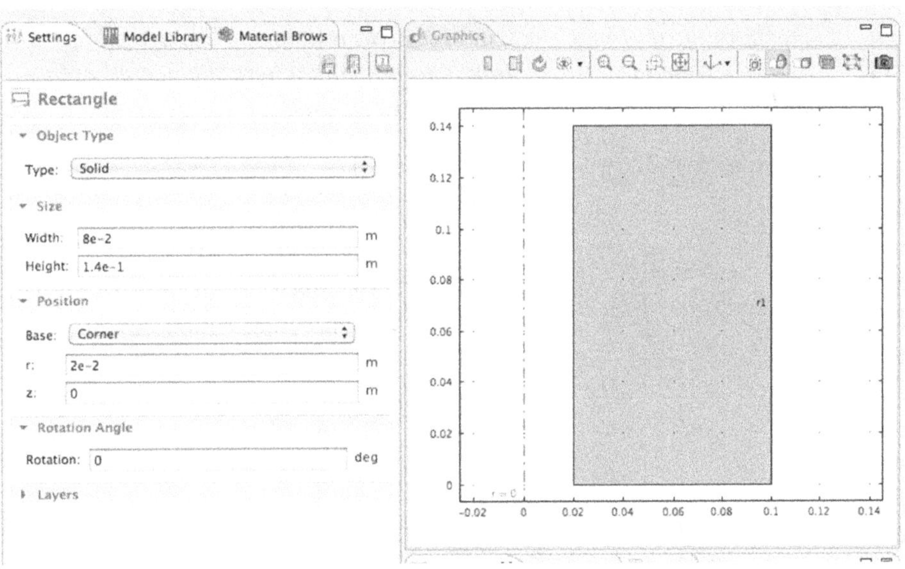

FIGURE 6.4 Settings – Rectangle Entry Windows and Graphics Window.

Figure 6.4 shows the Settings – Rectangle entry windows and Graphics window.

Right-Click > Model Builder – Geometry 1.

Select > Point from the Pop-up menu.

Enter > coordinates for point 1 using Table 6.2.

Click > Build Selected.

Right-Click > Model Builder – Geometry 1.

Select > Point from the Pop-up menu.

Enter > coordinates for point 2 using Table 6.2.

Click > Build Selected.

Click > Build All.

TABLE 6.2 Point Edit Windows

Name	Location r	Location z
point 1	2e-2	4e-2
point 2	2e-2	1e-1

NOTE *The boundary points 1 & 2 are constraint points. They are added to the perimeter of the rectangle to indicate where the heated area ends and the thermally insulated area begins.*

See Figure 6.5.

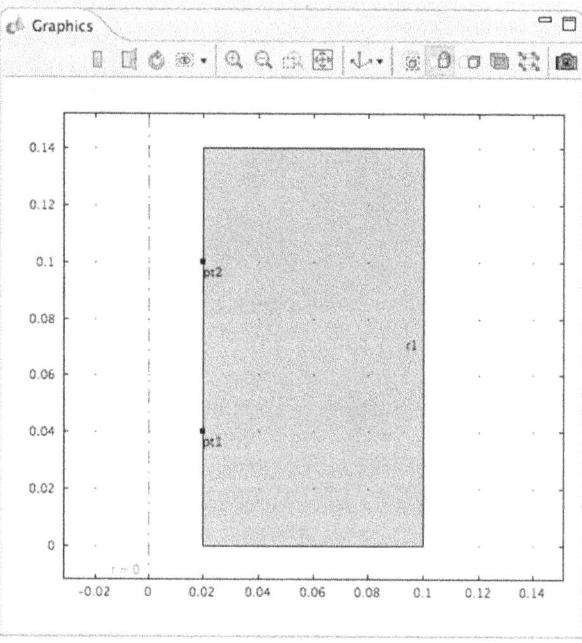

FIGURE 6.5 Model Geometry in the Graphics Window.

Figure 6.5 shows the Model Geometry in the Graphics window.

Heat Transfer (ht) Interface

Click > Model Builder – Model 1 – Twistie for the Heat Transfer *(ht)*.

Click > Model Builder – Model 1 – Heat Transfer *(ht)* – Heat Transfer in Solids 1.

Verify that Domain 1 is Selected in Settings – Heat Transfer in Solids – Domains – Selection.

Click > Settings – Heat Transfer in Solids – Heat Conduction – Thermal conductivity.

Select > User defined from the Pull-down menu.

Enter > k_Nb in the Thermal conductivity edit window.

Click > Settings – Heat Transfer in Solids – Thermodynamics – Density.

Select > User defined from the Pull-down menu.

Enter > rho_Nb in the Density edit window.

Click > Settings – Heat Transfer in Solids – Thermodynamics – Heat capacity at

constant pressure.

Select > User defined from the Pull-down menu.

Enter > Cp_Nb in the Heat capacity at constant pressure edit window.

See Figure 6.6.
Figure 6.6 shows the Settings – Heat Transfer in Solids edit windows.

Right-Click > Model Builder – Model 1 – Heat Transfer *(ht)*.

Select > Temperature from the Pop-up menu.

Shift-Click > Boundaries 2, 5, 6 in the Graphics window.

Click > Add to Selection in the Settings – Temperature – Boundaries panel.

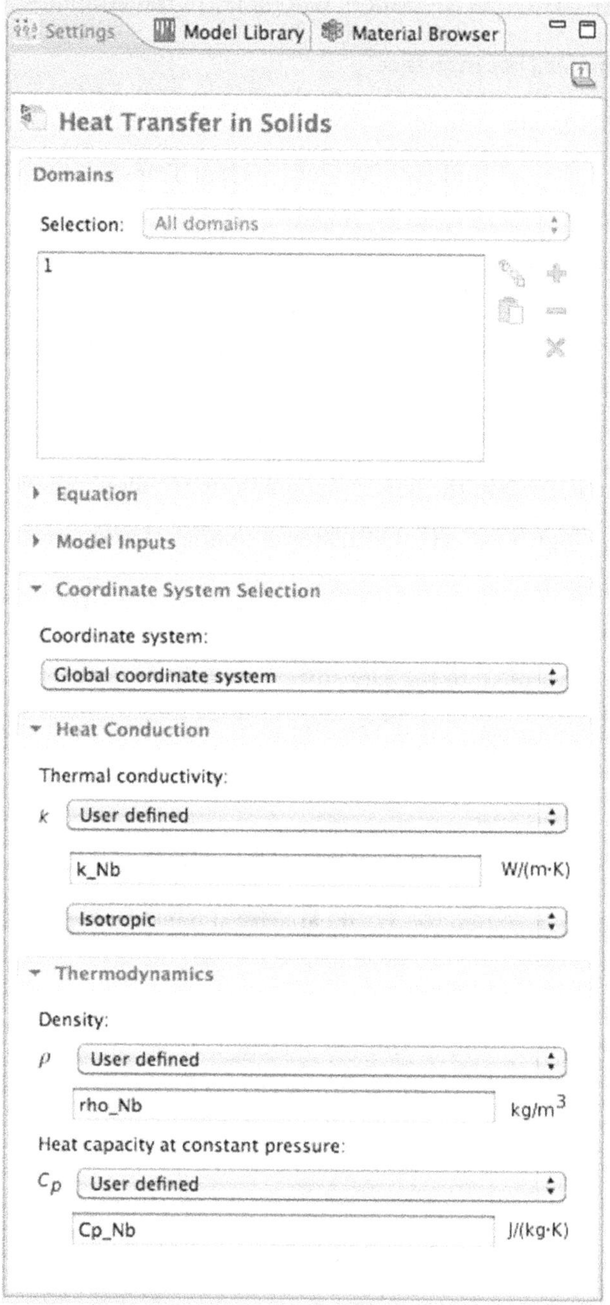

Settings | Model Library | Material Browser

Heat Transfer in Solids

Domains

Selection: All domains

1

▸ Equation

▸ Model Inputs

▾ Coordinate System Selection

Coordinate system:

Global coordinate system

▾ Heat Conduction

Thermal conductivity:

k User defined

k_Nb W/(m·K)

Isotropic

▾ Thermodynamics

Density:

ρ User defined

rho_Nb kg/m^3

Heat capacity at constant pressure:

C_p User defined

Cp_Nb J/(kg·K)

FIGURE 6.6 Settings – Heat Transfer in Solids Edit Windows.

Enter > T_0 in the Settings – Temperature – Temperature edit window.

See Figure 6.7.

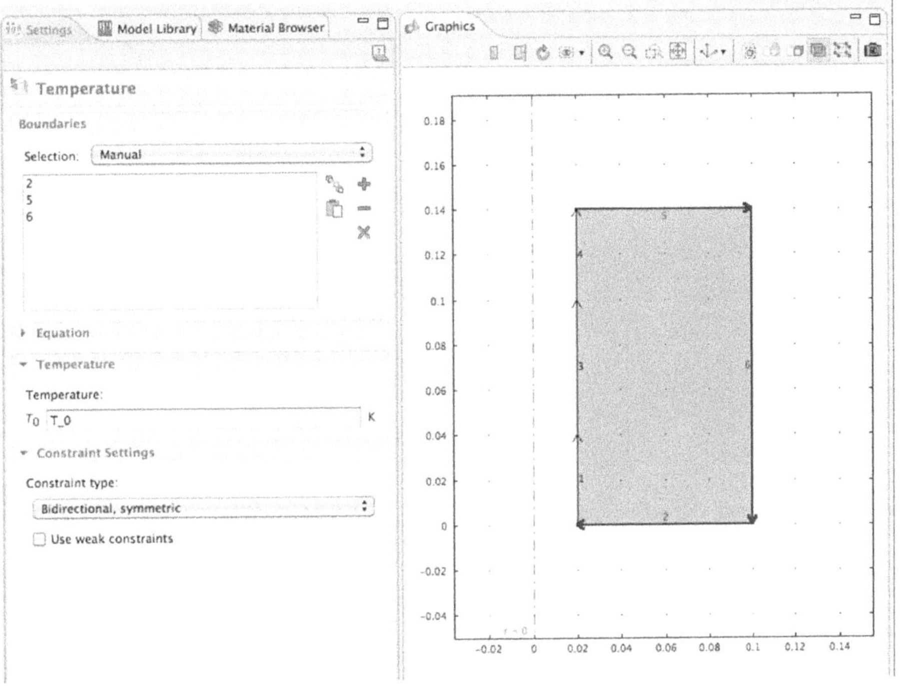

FIGURE 6.7 Settings – Temperature Edit Windows.

Figure 6.7 shows the Settings – Temperature edit windows.

Right-Click > Model Builder – Model 1 – Heat Transfer *(ht)*.

Select > Heat Flux from the Pop-up menu.

Click > Boundary 3 in the Graphics window.

Click > Add to Selection in the Settings – Heat Flux – Boundaries panel.

Enter > q_0 in the Settings – Heat Flux – General inward heat flux edit window.

See Figure 6.8.

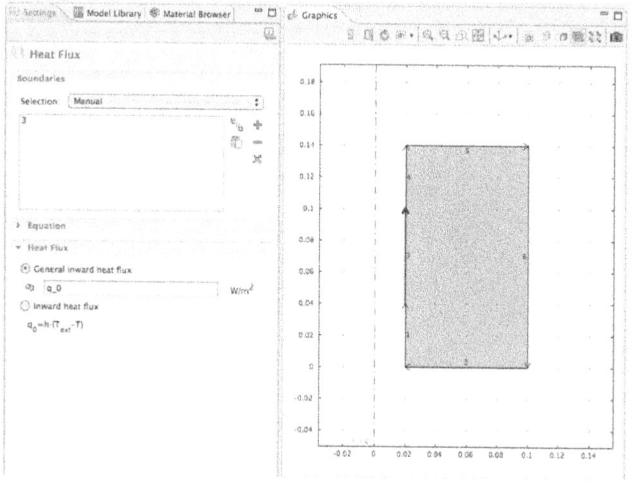

FIGURE 6.8 Settings – Heat Flux Edit Windows.

Figure 6.8 shows the Settings – Heat Flux edit windows.

Mesh 1

Right-Click > Model Builder – Model 1 – Mesh 1.

Select > Free Triangular from the pop-up menu.

Click > Settings – Build All button.

After meshing, the modeler should see a message in the message window about the number of elements (362 elements) in the mesh.

NOTE *The Free Triangular Mesh Type is chosen because it provides an adequate number of mesh elements and is relatively easy to solve.*

See Figure 6.9.

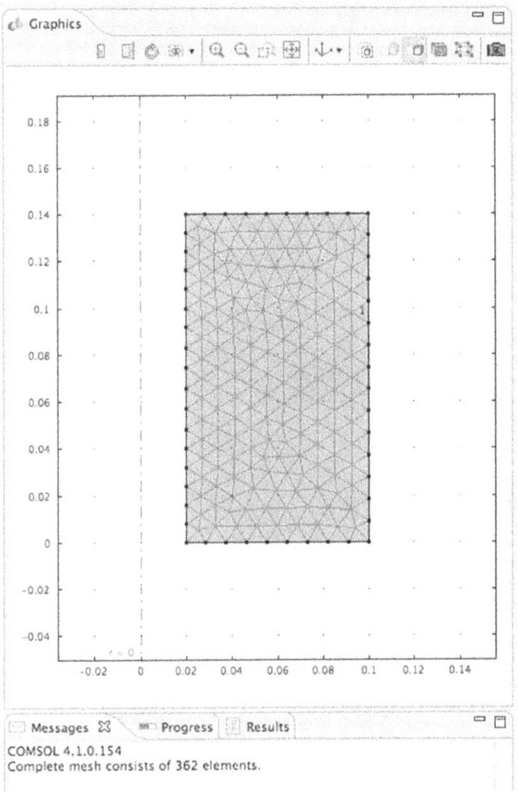

FIGURE 6.9 Desktop Display - Graphics - Meshed Domain.

Figure 6.9 shows the Desktop Display – Graphics – Meshed Domain.

Study 1

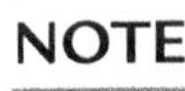

The default settings for the Stationary Study Type are appropriate for the solution of this problem and thus do not need to be modified.

In Model Builder,

Right-Click > Study 1 >Select > Compute.

Computed results, using the default display settings, are shown in Figure 6.10.

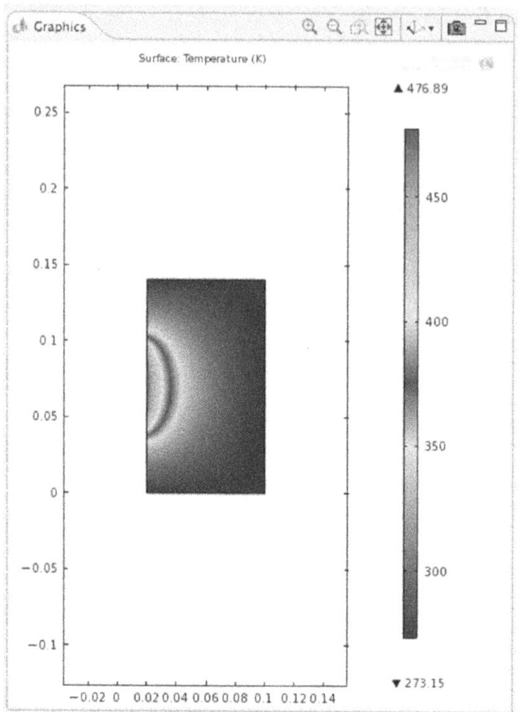

FIGURE 6.10 Computed Results, Using the Default Display Settings.

Figure 6.10 shows the computed results, using the default display settings.

Results

In Model Builder,

Click > Results > 2D Plot Group 1 twistie.

Right-Click > Model Builder – Results – 2D Plot Group 1.

Select > Contour from the Pull-down menu.

Click > Contour 1 in Model Builder – Results – 2D Plot Group 1.

In Settings – Contour – Coloring and Style,

Click > Color table Pull-down menu.

Select > GrayScale from the Color table Pull-down menu.

Click > Graphics – Zoom Extents.

Click > Plot.

See Figure 6.11.

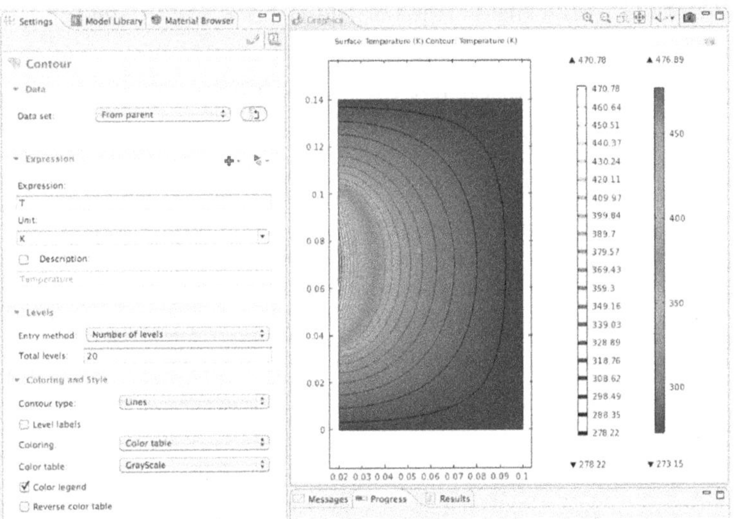

FIGURE 6.11 Desktop Display - Computed Solution.

Figure 6.11 shows the Desktop Display – Computed Solution.

2D Axisymmetric Heat Conduction in a Cylinder Model Summary and Conclusions

The 2D Axisymmetric Heat Conduction in a Cylinder Model is a powerful tool that can be used to explore heat transfer techniques in many different conductive media (e.g. metals, semimetals, alloys, graphene films, graphite films, etc.). Addition of the contour plot to the surface plot easily shows the non-linearity of the temperature distribution. As has been shown earlier in this chapter, the 2D Axisymmetric Heat Conduction in a Cylinder Model is easily and simply modeled with the Heat Transfer Interface.

2D Axisymmetric Transient Heat Transfer Model Considerations

In 1855, Adolf Eugen Fick {6.15} published his now fundamental and famous laws {6.16} for the diffusion (mass transport) of a first item (e.g. a gas, a liquid, etc.) through a second item (e.g. another gas, liquid, etc.).

Fick's First Law is written:

$$J = -D\nabla\Phi \tag{6.3}$$

Where: J = Diffusion Flux in moles per square meter per second [mol/(m^2*s)].

D = Diffusion Coefficient (diffusivity) in square meters per second [m^2/s].

Φ = Concentration in moles per cubic meter [mol/m^3].

∇ = Gradient operator.

NOTE *It should be noted that the minus sign in front of the diffusivity indicates that the item that is flowing (gas, heat, money, etc.) flows from the higher concentration to the lower concentration.*

See Figure 6.12

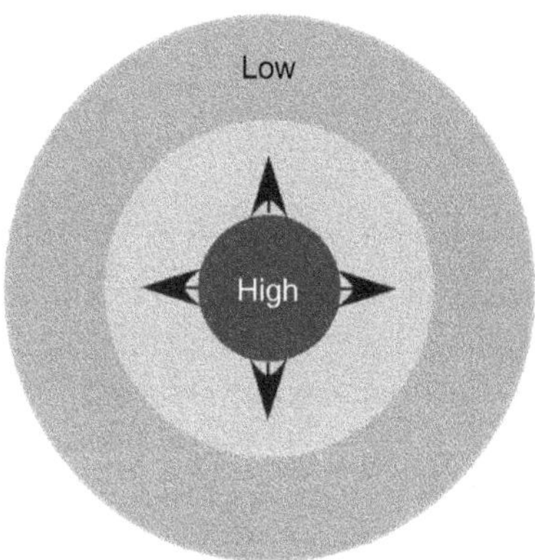

FIGURE 6.12 Graphical Rendition of Fick's First Law.

Fick's Second Law calculates the details of how diffusion (mass transport) causes the diffusing item's concentration to change as a function of time (under transient conditions).

Fick's Second Law is written:

$$\frac{\partial \Phi}{\partial t} = D\nabla^2 \Phi \tag{6.4}$$

Where: D = Diffusion Coefficient (diffusivity) in square meters per second [m^2/s].

Φ = Concentration vector in moles per cubic meter [mol/m^3].

∇^2 = Laplace operator.

If modified slightly, where:

$$D = a = \frac{k}{\rho C_p} \tag{6.5}$$

Where: α = Thermal Diffusion Coefficient (thermal diffusivity) in square meters per second [m^2/s].

k = Thermal conductivity in Watts per meter-degree Kelvin [W/(m*K)].

ρ = Material density in kilograms per cubic meter [kg/m^3].

C_p = Specific heat in Joules per kilograms-degree Kelvin [J/(kg*K)].

Then, Fick's Second Law becomes the Heat Equation:

$$\frac{\partial u}{\partial t} = a\nabla^2 u \tag{6.6}$$

Where: u = u(x, y, z, t), a function of the three spatial parameters (x, y, z) and time (t).

α = Thermal Diffusion Coefficient (thermal diffusivity) in square meters per second [m^2/s].

∇^2 = Laplace operator.

The heat equation has been broadly used to solve many different diffusion problems {6.17}. In the case of thermal diffusion, the spatial parameters are the geometric coordinates. Examples of other uses of the heat equation are easily found in physics, finance, chemistry, and many other application areas.

2D Axisymmetric Transient Heat Transfer Model

This is the model of a sphere of Niobium. The sphere of Niobium has an initial temperature of 273.15 K (0 degC) and is suspended (immersed) in a medium that applies a constant temperature (1000 degC) to the exterior surface of that sphere, starting at time zero.

Building the 2D Axisymmetric Transient Heat Transfer Model

Startup 4.x.

Select > 2D axisymmetric.

Click > Next.

Click > Twistie for Heat Transfer in the Add Physics window.

Click > Heat Transfer – Heat Transfer in Solids *(ht)*.

Click > Add Selected.

Click > Next.

Select > Time Dependent in the Select Study Type window.

Click > Finish (Flag).

Click > Save As.

Enter MMUC4_2DA_HTTD_1.mph.

See Figure 6.13.

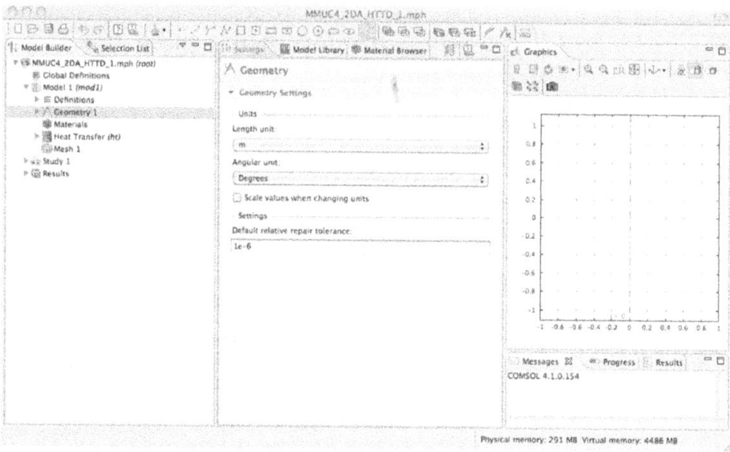

FIGURE 6.13 Desktop Display for the MMUC4_2DA_HTTD_1.mph Model.

Figure 6.13 shows the Desktop Display for the MMUC4_2DA_HTTD_1.mph model.

Right-Click > Global Definitions.
Select > Parameters from the Pop-up menu.
In the Settings – Parameters – Parameters edit window,
Enter the parameters shown in Table 6.3.

TABLE 6.3 Parameters Edit Window

Name	Expression	Description
k_Nb	52.335[W/(m*K)]	Thermal conductivity Nb
rho_Nb	8.57e3[kg/m^3]	Density Nb
Cp_Nb	2.7e2[J/(kg*K)]	Heat capacity Nb
T_0	2.7315e2[K]	Initial temperature
T_appl	1.27315e3[K]	Boundary temperature

See Figure 6.14.

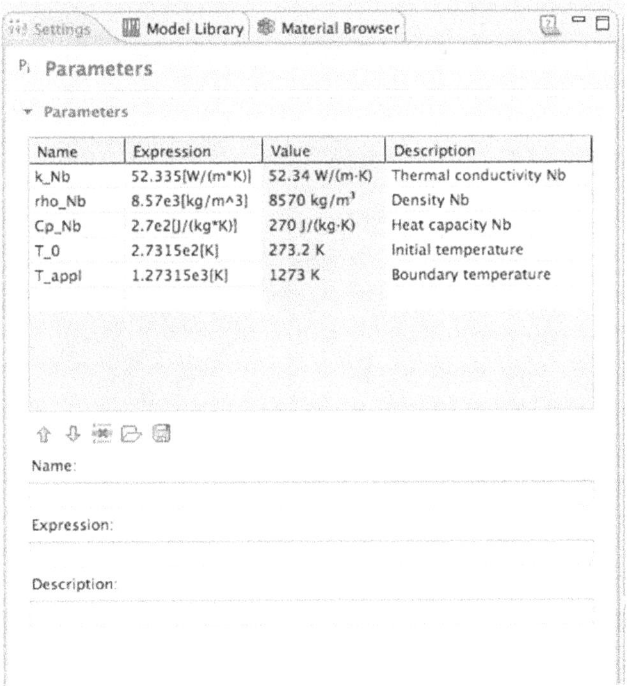

FIGURE 6.14 Settings - Parameters - Parameters Edit Window Filled.

Figure 6.14 shows the Settings – Parameters – Parameters edit window filled.

Create 2D Axisymmetric Sphere

NOTE *In the next few steps, a circle will be created and then overlaid by a rectangle. Once the Boolean difference is taken between the two objects, the geometric remainder is then 2D Axisymmetric cross-section of a sphere.*

In Model Builder – Model 1 (mod1),

Right-Click > Model Builder – Geometry 1.

Select > Circle from the Pop-up menu.

Enter > 0.4 in the Settings – Circle – Size – Radius entry window.

Click > Build All.

See Figure 6.15.

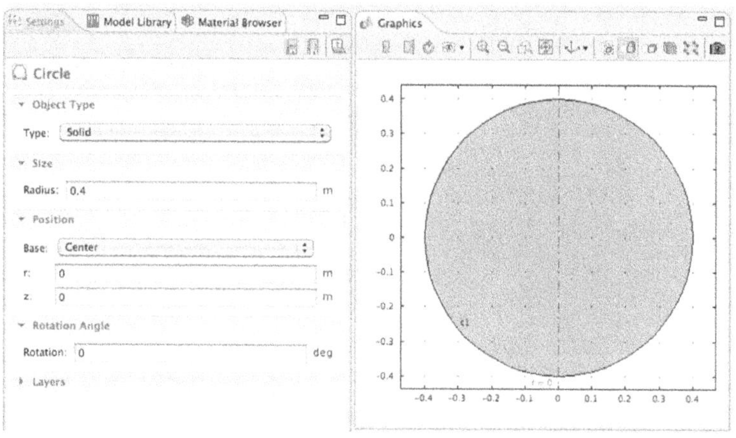

FIGURE 6.15 Settings - Circle Entry Windows.

Figure 6.15 shows the Settings – Circle entry windows.

Right-Click > Model Builder – Geometry 1.

Select > Rectangle from the Pop-up menu.

Enter > 0.4 in the Settings – Rectangle – Size – Width entry window.

Enter > 0.8 in the Settings – Rectangle – Size – Height entry window.

Click > Settings – Rectangle – Position – Base.

Select > Corner from the Pop-up menu.

Enter > -0.4 in the Settings – Rectangle – Position – r entry window.

Enter > -0.4 in the Settings – Rectangle – Position – z entry window.

Click > Build All.

See Figure 6.16.

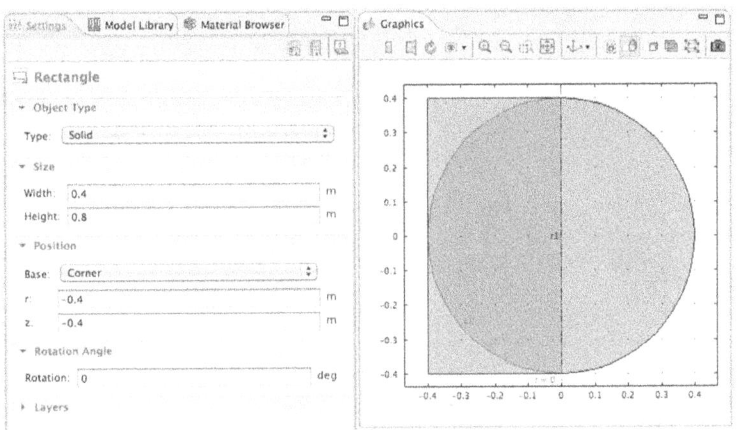

FIGURE 6.16 Settings - Rectangle Entry Windows.

Figure 6.16 shows the Settings – Rectangle entry windows.

Right-Click > Model Builder – Geometry 1.

Select > Boolean Operations – Difference from the Pop-up menu.

Click > c1 in the Graphics window.

Click > Add to Selection for the Settings – Difference – Objects to add window (top window).

Click > Activate Selection for the Settings – Difference – Objects to subtract window (bottom window).

Click > r1 in the Graphics window.

Click > Add to Selection for the Settings – Difference – Objects to subtract window (bottom window).

See Figure 6.17.

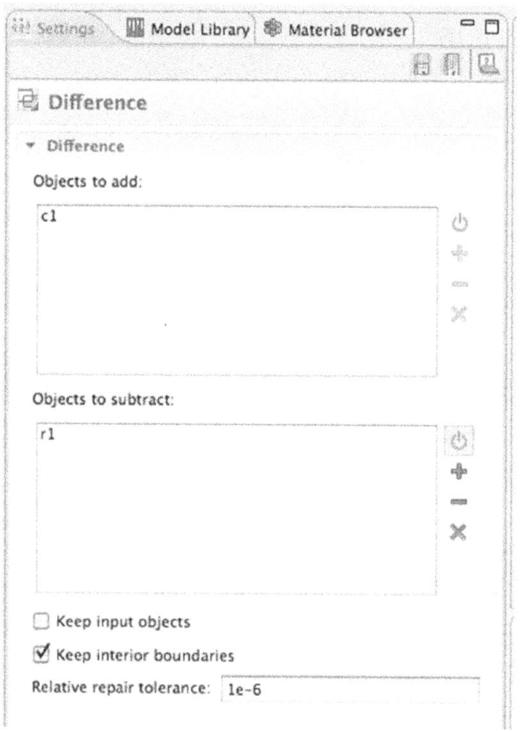

FIGURE 6.17 Settings – Difference Entry Windows.

Figure 6.17 shows the Settings – Difference entry windows.

Click > Build All.

See Figure 6.18.

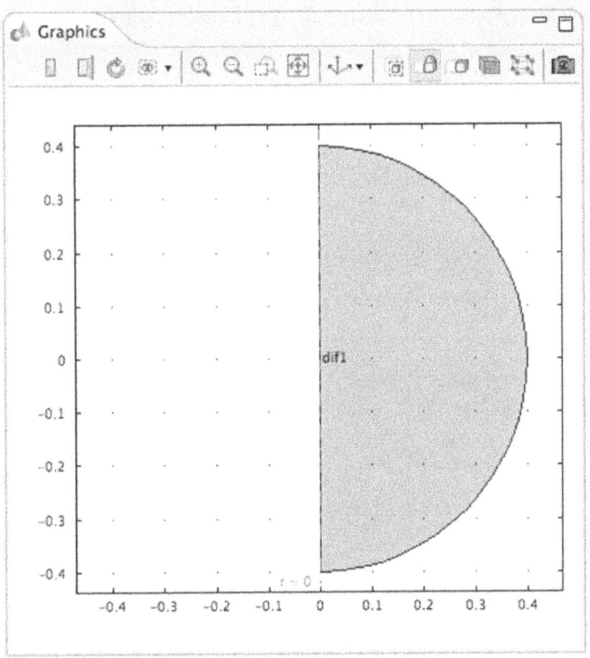

FIGURE 6.18 Graphics Window – Axisymmetric Sphere.

Figure 6.18 shows the Graphics Window – Axisymmetric Sphere.

Heat Transfer (ht) Interface

Click> Twistie for the Model Builder – Model 1 – Heat Transfer (ht).

Click> Model Builder – Model 1 – Heat Transfer (ht) – Heat Transfer in Solids 1.

Click > Settings – Heat Transfer in Solids 1 – Heat Conduction – Thermal conductivity k Pull-down menu.

Select > User defined.

Enter > k_Nb in the Thermal conductivity edit window.

Click > Settings – Heat Transfer in Solids 1 – Thermodynamics – Density ρ Pull-down menu.

Select > User defined.

Enter > rho_Nb in the Density edit window.

Click > Settings – Heat Transfer in Solids 1 – Thermodynamics – Heat capacity at constant pressure C_p Pull-down menu.

Select > User defined.

Enter > Cp_Nb in the Density edit window.

See Figure 6.19.

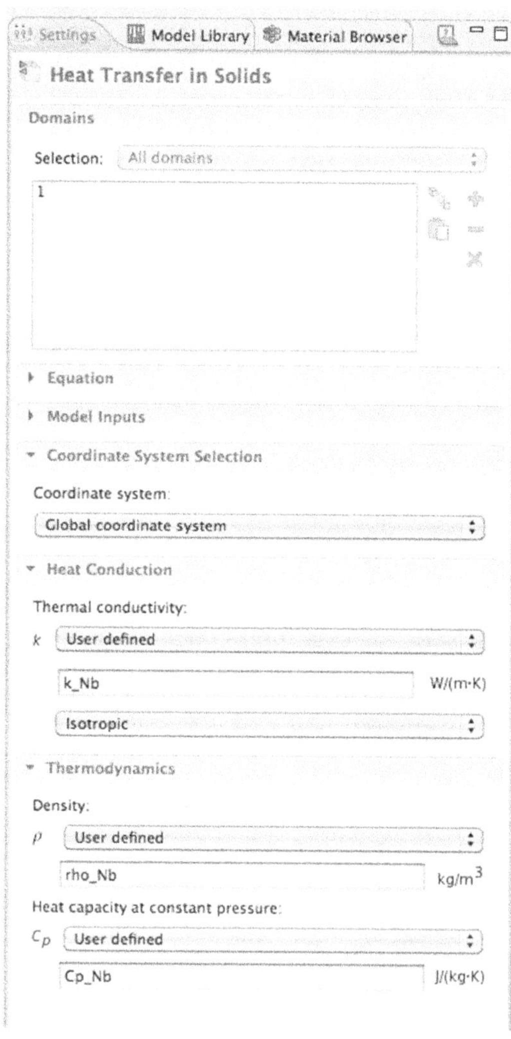

FIGURE 6.19 Settings – Heat Transfer in Solids 1 – Heat Conduction Edit Windows.

Figure 6.19 shows the Settings – Heat Transfer in Solids 1 – Heat Conduction edit windows.

Click > Model Builder – Model 1 – Heat Transfer (ht) – Initial Values 1.

Enter > T_0 in Settings – Initial Values – Initial Values – Temperature edit window.

See Figure 6.20.

FIGURE 6.20 Settings - Initial Values - Initial Values - Temperature Edit Window.

Figure 6.20 shows the Settings – Initial Values – Initial Values – Temperature edit window.

Right-Click > Model Builder – Model 1 – Heat Transfer (ht).

Select > Temperature from the Pop-up menu.

Click > Model Builder – Model 1 – Heat Transfer (ht) – Temperature 1.

Shift-Click > Boundaries 2 & 3 in the Graphics window.

Click > Add to Selection in the Settings – Temperature – Boundaries edit window.

Enter > T_appl in the Settings – Temperature – Temperature – Temperature edit window.

See Figure 6.21.

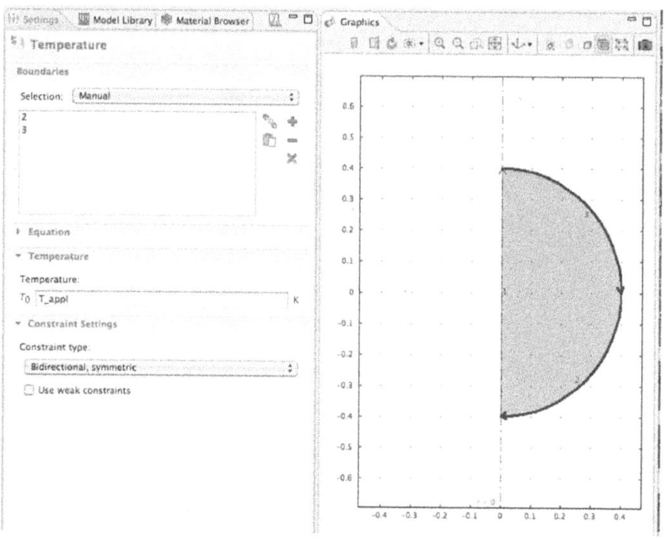

FIGURE 6.21 Settings – Temperature and Graphics Windows.

Figure 6.21 shows the Settings – Temperature and Graphics windows.

Mesh 1

NOTE *The Free Triangular Mesh is easy to use and converges well and so it is chosen to use with this model.*

Right-Click > Model Builder – Model 1 – Mesh 1.

Select > Free Triangular.

Click > Build All button.

After meshing, the modeler should see a message in the message window about the number of elements (267 elements) in the mesh.

See Figure 6.22.

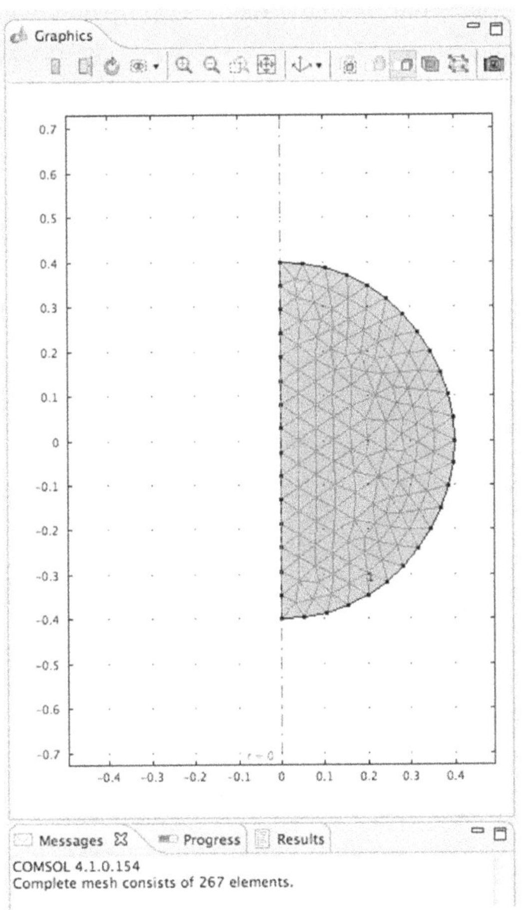

FIGURE 6.22 Desktop Display - Graphics - Meshed Domain.

Figure 6.22 shows the Desktop Display – Graphics – Meshed Domain.

Study 1

Click > Model Builder – Study 1 – Twistie.

Click > Study 1 – Time Dependent.

Click > Range button in Settings – Time Dependent – Study Settings – Times.

Enter > Start = 0, Stop = 300, Step = 2 in the Range Pop-up edit window.

Click > Replace button in the Range Pop-up edit window.

See Figure 6.23.

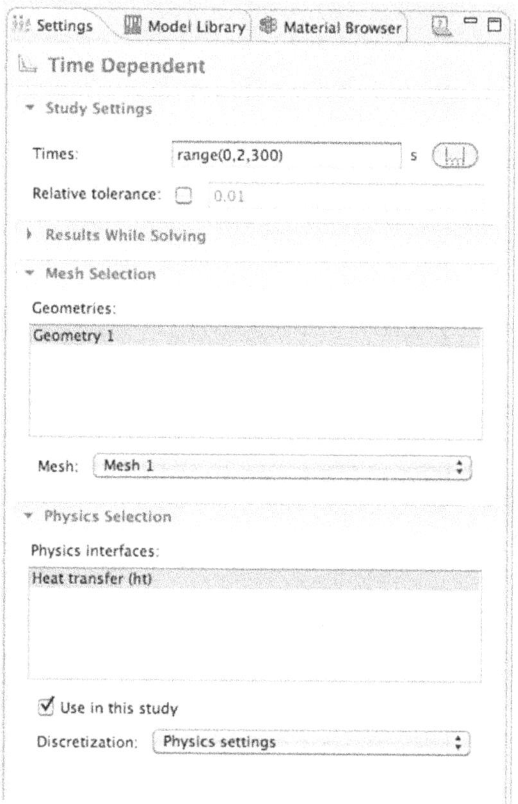

FIGURE 6.23 Settings – Time Dependent – Study Settings.

Figure 6.23 shows the Settings – Time Dependent – Study Settings.

In Model Builder,

Right-Click Study 1 > Select > Compute.

Click > Zoom Extents.

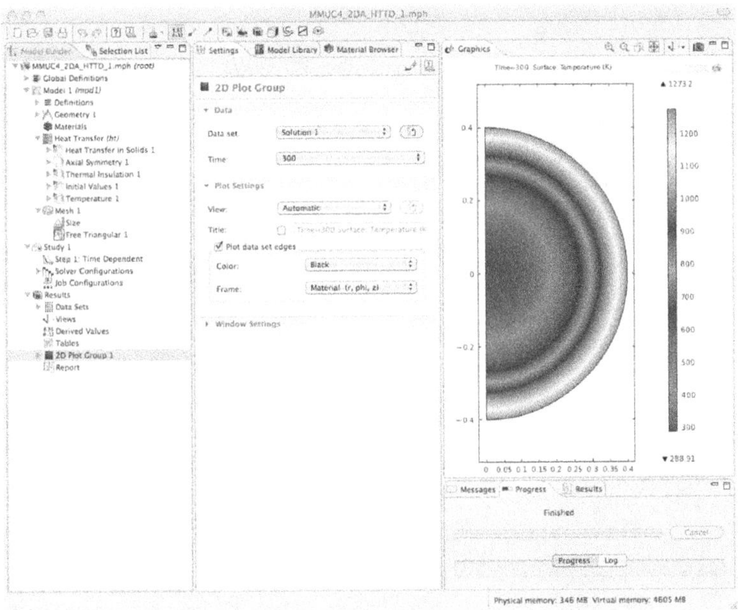

FIGURE 6.24 2D Axisymmetric Transient Heat Transfer Model Results, Using the Default Display Settings, Plus Zoom Extents.

Computed results, using the default display settings, plus Zoom Extents, are shown in Figure 6.24.

Figure 6.24 shows the 2D Axisymmetric Transient Heat Transfer Model results, using the default display settings, plus Zoom Extents.

2D Results

NOTE *This step converts the units of the plot from Kelvin to Celsius.*

In Model Builder,

Click > Results – 2D Plot Group 1 twistie.

Click > Results – 2D Plot Group 1 – Surface 1.

Click > Settings – Surface – Expression – Unit.

Select > degC from the Pull-down menu.

Click > Plot.

See Figure 6.25.

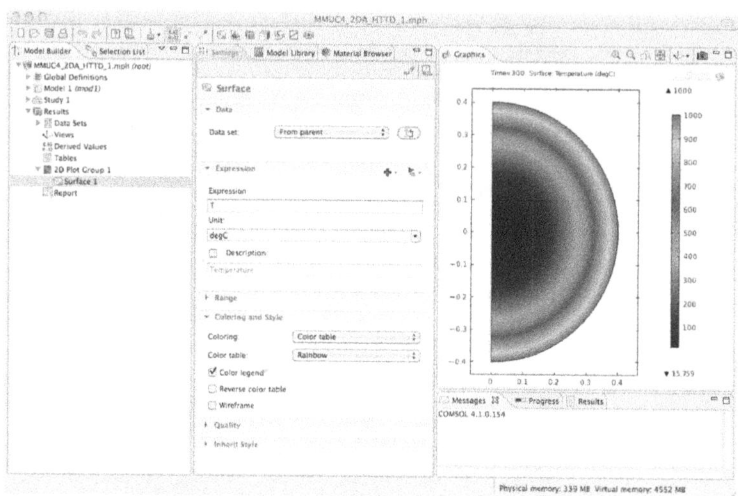

FIGURE 6.25 2D Axisymmetric Transient Heat Transfer Model Results in Degrees Celsius.

Figure 6.25 shows the 2D Axisymmetric Transient Heat Transfer Model results in degrees Celsius.

NOTE *Since this model is 2D Axisymmetric, a 3D model can be generated from the 2D solution by creating a revolution data set {6.18}. This saves all the effort that would be required to build and calculate the 3D version of this model. In order to generate the 3D version, a new data set needs to be generated from the 2D data set.*

3D Results

3D Data Set

In Model Builder,

Right-Click > Results – Data Sets.

Select > Revolution 2D from the Pop-up menu.

Enter > 0.4 in the Settings – Revolution 2D – Axis Data z Point 2 edit window.

Click > Settings – Revolution 2D – Revolution Layers twistie.

Enter > 180 in the Settings – Revolution 2D – Revolution Layers – Revolution angle edit window.

See Figure 6.26.

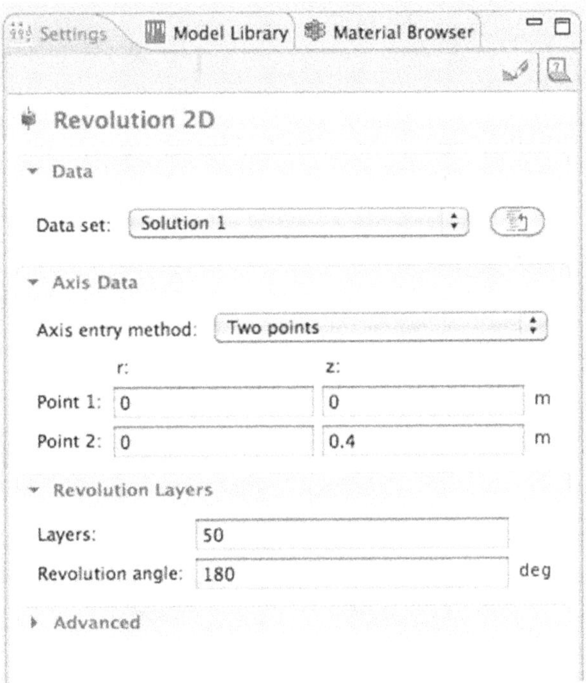

FIGURE 6.26 Settings - Revolution 2D Edit Windows.

Figure 6.26 shows the Settings – Revolution 2D edit windows.

3D Plot

In Model Builder,

Right-Click > Results.

Select > 3D Plot Group from the Pull-down menu.

Right-Click > Results – 3D Plot Group 2.

Select > Surface from the Pull-down menu.

Click > Plot.

Click > Results – 3D Plot Group 2 – Surface 1.

Click > Settings – Surface – Data – Data Set.

Select > Revolution 2D 1 from the Pull-down menu.

Click > Settings – Surface – Expression – Unit.

Select > degC from the Pull-down menu.

Click > Plot.

See Figure 6.27.

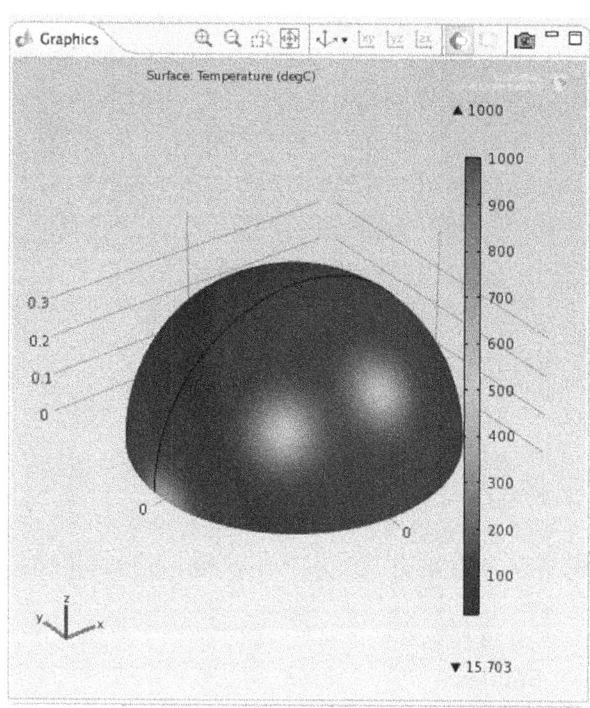

FIGURE 6.27 Initial Revolution 2D Axisymmetric Transient Heat Transfer Model (3D) Results in Degrees Celsius.

Figure 6.27 shows the Initial Revolution 2D Axisymmetric Transient Heat Transfer Model (3D) results in degrees Celsius.

Click > Go to XY View two times.

See Figure 6.28.

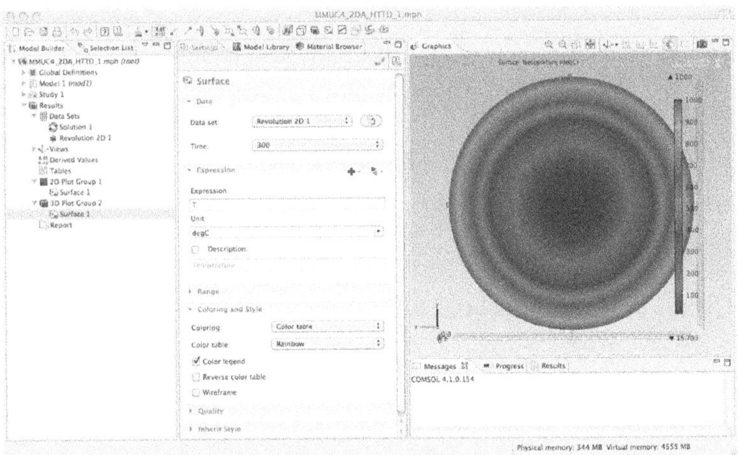

FIGURE 6.28 Revolution 2D Axisymmetric Transient Heat Transfer Model (3D) Results in Degrees Celsius.

Figure 6.28 shows the Revolution 2D Axisymmetric Transient Heat Transfer Model (3D) results in degrees Celsius.

Revolution 2D Axisymmetric Transient Heat Transfer Model (3D) Animation

This animation shows how the temperature changes as the time changes.

In Model Builder,

Right-Click > Results – Report.

Select > Animation.

In Settings – Scene,

Select > 3D Plot Group 2 from the Pull-down menu.

Click > Refresh.

In Settings – Animation – Output,

Select > Output type Movie from the Pull-down menu.

Select > Format GIF from the Pull-down menu.

Enter > 30 in the Frames per second edit window.

NOTE *The detailed procedures for Animation in 4.x on the Macintosh and the PC are different. The Macintosh uses the GIF format. The PC uses the AVI format. If the modeler tries to use AVI on the Macintosh, an error results.*

In Settings – Animation – Frame Settings,

Click > Lock aspect ratio check box.

In Settings – Animation – Advanced,

Click > Antialiasing check box.

Click > Settings – Export.

NOTE *The movie will be exported to and run in the 4.x Graphics window, if there is no path designated in the Settings – Animation – Output Filename edit window.*

Revolution 2D Axisymmetric Transient Heat Transfer Model (3D) Movie

The modeler can save the movie as a unique file.

Click > Settings – Animation – Output Browse button.

Select the desired location for saving the movie.

Enter the desired File Name in the Save-as edit window.

Click > Save.

Click > Settings – Export to run and save the model to the file.

Click > Save the completed MMUC4_2DA_HTTD_1.mph Revolution 2D Axisymmetric Transient Heat Transfer Model (3D) model.

Revolution 2D Axisymmetric Transient Heat Transfer Model (3D) Model Summary and Conclusions

The 2D Axisymmetric Transient Heat Transfer Model is a powerful tool that can be used to explore the heat transfer in many different media (e.g. semiconductors, metals, semimetals, insulators, etc.). It can also easily be converted to a 3D as has been shown earlier in this chapter.

FIRST PRINCIPLES AS APPLIED TO 2D AXISYMMETRIC MODEL DEFINITION

First Principles Analysis derives from the fundamental laws of nature. In the case of models using this Classical Physics Analysis approach, the laws of conservation in physics require that what goes in (as mass, energy, charge, etc.) must come out (as mass, energy, charge, etc.) or must accumulate within the boundaries of the model.

The careful modeler must be knowledgeable of the implicit assumptions and default specifications that are normally incorporated into the COMSOL Multiphysics software model when a model is built using the default settings.

Consider, for example, the two 2D models developed in this chapter. In these models, it is implicitly assumed that there are no thermally related changes (mechanical, electrical, etc.). It is also assumed that the materials are homogeneous and isotropic, except as specifically indicated and that there are no thin insulating contact barriers at the thermal junctions. None of these assumptions are typically true in the general case. However, by making such assumptions, it is possible to easily build a 2D Axisymmetric and/or a 3D (2D Revolution) First Approximation Model.

NOTE *A First Approximation Model is one that captures all the essential features of the problem that needs to be solved, without dwelling excessively on all of the small details. A good First Approximation Model will yield an answer that enables the modeler to determine if he needs to invest the time and the resources required to build a more highly detailed model.*

Also, the modeler needs to remember to name model parameters carefully as pointed out in Chapter 1.

REFERENCES

6.1 COMSOL Multiphysics Users Guide, Version 4.1, pp. 288-292

6.2 http://en.wikipedia.org/wiki/Nicolas_Léonard_Sadi_Carnot

6.3 http://en.wikipedia.org/wiki/William_Thomson,_1st_Baron_Kelvin

6.4 http://en.wikipedia.org/wiki/James_Prescott_Joule

6.5 http://en.wikipedia.org/wiki/Boltzmann

6.6 http://en.wikipedia.org/wiki/James_Clerk_Maxwell

6.7 http://en.wikipedia.org/wiki/Planck

6.8 http://en.wikipedia.org/wiki/Isaac_Newton

6.9 http://en.wikipedia.org/wiki/Newton%27s_law_of_cooling

6.10 http://en.wikipedia.org/wiki/Joseph_Fourier

6.11 http://en.wikipedia.org/wiki/Conduction_(heat)

6.12 A.D. Cameron, et al., *NAFEMS Benchmark Tests for Thermal Analysis (Summary)*, NAFEMS Ltd., 1986

6.13 http://en.wikipedia.org/wiki/Niobium

6.14 Kittel, Charles, "Introduction to Solid State Physics", ISBN 0-471-87474-4, pp. 324

6.15 http://en.wikipedia.org/wiki/Adolf_Eugen_Fick

6.16 http://en.wikipedia.org/wiki/Fick%27s_law_of_diffusion

6.17 http://en.wikipedia.org/wiki/Heat_equation

6.18 COMSOL Multiphysics Users Guide, Version 4.1, pp. 684

Suggested Modeling Exercises

1. Build, mesh, and solve the 2D Axisymmetric Cylinder Conduction Model as presented earlier in this chapter.
2. Build, mesh, and solve the 2D Axisymmetric Transient Heat Transfer Model as presented earlier in this chapter.
3. Change the values of the materials parameters and then build, mesh, and solve the 2D Axisymmetric Cylinder Conduction Model as an example problem.
4. Change the values of the materials parameters and then build, mesh, and solve the 2D Axisymmetric Transient Heat Transfer Model as an example problem.
5. Change the value of the heat flux and then build, mesh, and solve the 2D Axisymmetric Cylinder Conduction Model as an example problem.
6. Change the value of the radius to reduce the size of the sphere and then build, mesh, and solve the 2D Axisymmetric Transient Heat Transfer Model as an example problem.

CHAPTER 7

2D SIMPLE MIXED MODE MODELING USING COMSOL MULTIPHYSICS 4.X

In This Chapter

- Guidelines for 2D Simple Mixed Mode Modeling in 4.x
    - 2D Simple Mixed Mode Modeling Considerations
- 2D Simple Mixed Mode Models
    - 2D Electrical Impedance Sensor Model
    - 2D Metal Layer on a Dielectric Block Model
- First Principles as Applied to 2D Simple Mixed Mode Model Definition
- References
- Suggested Modeling Exercises

GUIDELINES FOR 2D SIMPLE MIXED MODE MODELING IN 4.X

NOTE

In this chapter, simple mixed mode 2D models will be presented. Such 2D models are typically more conceptually complex than the models that were presented in earlier chapters of this text. 2D simple mixed mode models have proven to be very valuable to the science and engineering communities, both in the past and currently, as first-cut evaluations of potential systemic physical behavior under the influence of mixed external stimuli. The 2D mixed mode model responses and other such ancillary information can be gathered and screened early in a project for a first-cut evaluation. That initial information can potentially be used later as guidance in building higher-dimensionality (3D) field-based (electrical, magnetic, etc.) models.

Since the models in this and subsequent chapters are more complex and more difficult to solve than the models presented thus far, it is important that the modeler have available the tools necessary to most easily utilize the powerful capabilities of the 4.x software. In order to do that, if you have not done this previously, the modeler should go to the main 4.x toolbar, Click > Options – Preferences – Model builder. When the Preferences – Model builder edit window is shown, Select > Show equation view checkbox and Show more options checkbox. Click > Apply {7.1}.

2D Simple Mixed Mode Modeling Considerations

2D Models can in some cases be less difficult than some 1D or 3D models, having fewer implicit assumptions, and yet potentially, 2D models can still be a very challenging type of model to build, depending on the underlying physics. The least difficult aspect of 2D model creation arises from the fact that the geometry is relatively simple. However, the physics in a 2D simple mixed mode model can range from relatively simple to extremely complex.

In compliance with the laws of physics, a 2D model implicitly assumes that energy flow, materials properties, environment, and all other conditions and variables that are of interest are homogeneous, isotropic and/or constant, unless otherwise specified, throughout the entire domain of interest both within the model and through the boundary conditions and in the environs of the model.

The modeler needs to bear the above stated conditions in mind and insure that all of the modeling conditions and associated parameters (default settings) in each model created are properly considered, defined, verified and/or set to the appropriate values.

NOTE

It is always mandatory that the modeler be able to accurately anticipate the expected results of the model and accurately specify the manner in which those results will be presented. Never assume that any of the default values that are present when the model is created necessarily satisfy the needs or conditions of a particular model.

Always verify that any parameters employed in the model are the correct value needed for that model. Calculated solutions that significantly deviate from the anticipated solution or from a comparison of values measured in an experimentally derived realistic model are probably indicative of one or more modeling errors either in the original model design, in the earlier model analysis, in the understanding of the underlying physics or are simply due to human error.

2D Coordinate System

In a 2D model, if parameters can only vary as a function of the position (x) and (y) coordinates, then such a 2D model represents the parametric condition of the model in a time independent mode (stationary). In a time dependent study or frequency domain study model, parameters can vary both with position in (x) and/or (y) and with time (t).

See Figure 7.1.

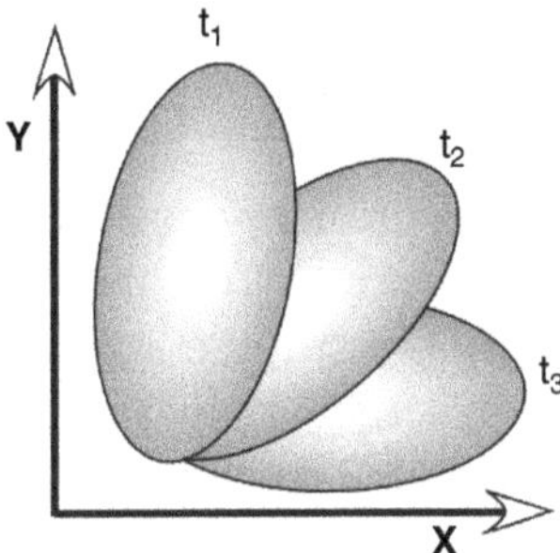

FIGURE 7.1 2D Coordinate System, Plus Time.

Figure 7.1 shows the 2D Coordinate System, plus Time.

NOTE

In all 2D models, there is a specified third geometric dimension with a default value of 1[m]. The properties of the model being constructed are isotropically mapped throughout that depth. That depth is known as the Out-of-Plane Thickness. That Out-of-Plane Thickness should be changed as needed by the modeler to ensure that the resulting calculations reflect the parameters of the physical system being modeled. The Out-of-Plane Thickness edit window can be found by Clicking on the 2D Physics Interface being employed.

See Figure 7.2.

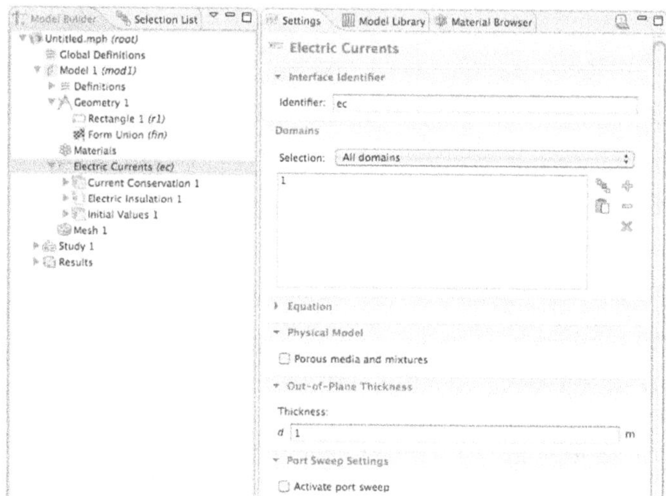

FIGURE 7.2 Out-of-Plane Thickness Edit Window.

Figure 7.2 shows the Out-of-Plane Thickness edit window.

In this chapter, two new primary concepts are introduced to the modeler: the Frequency Domain study {7.2} and the Highly Conductive Layer {7.3}. The Frequency Domain study corresponds to a frequency sweep. That is, the model solution is essentially a collection of (quasi-static) solutions, where the input signals are sinusoidal waves, except that one or more of the dependent variables (f(x, y, t)) changes with time. The space coordinates (x) and (y) typically represent a distance coordinate throughout which the model is to calculate the change of the specified observables (i.e. temperature, heat flow, pressure, voltage, current, etc.) over the range of values ($x_{min} <= x <= x_{max}$) and ($y_{min} <= y <= y_{max}$). The time coordinate (t) represents the range of values ($t_{min} <= t <= t_{max}$) from the beginning of the observation period (t_{min}) to the end of the observation period (t_{max}).

Electrical Impedance Theory

The concept of electrical impedance {7.4}, as used in Alternating Current (AC) theory {7.5}, is an expansion on the basic concept of resistance as illustrated by Ohm's Law {7.6}, in Direct Current theory.

Ohm's Law was discovered by Georg Ohm and as published in 1827, is:

$$I = \frac{V}{R} \tag{7.1}$$

Where: *I = Current in Amperes [A].*

V = Voltage (Electromotive Force) in Volts [V].

R = Resistance in Ohms [ohm].

In AC theory, both voltage (V) and current (I) alternate periodically as a function of time. Typically, the alternating behavior (frequency (f)) of the voltage and current are separately represented as either a single sinusoidal wave or as a sum of several sinusoidal waves.

NOTE *The analysis of complex waveforms is typically handled by Fourier Analysis {7.7}. That topic will not be presented herein. However, the reader is encouraged to expand his modeling horizons by exploring waveform analysis further.*

In this case, however, for clarity, the exploration of the concept of impedance will be confined to single frequency analysis. The concept of impedance was developed and named by Oliver Heaviside {7.8} in 1886. Arthur E. Kennelly {7.9} reformulated impedance in the currently used complex number formulation in 1893.

The first factor that needs to be considered, when expanding modeling calculations from the DC realm (frequency equals zero (f = 0)) to the AC realm (frequency greater than zero (f > 0)), is that the resistance (R) maps into the impedance (Z), as follows {7.10}:

$$Z = R + j\left(\omega L - \frac{1}{\omega C}\right) = R + jX = \left(R^2 + X^2\right)^{1/2} e^{j\tan^{-1}(X/R)} \tag{7.2}$$

Where: Z = Complex Impedance [ohm].

R = Resistance in Ohms [ohm].

$j = (-1)^{1/2}$ (imaginary unit).

$\omega = 2\pi f$ = angular frequency {7.11}.

X = Reactance [ohm] {7.12}.

L = Inductance [henry] [H].

C = Capacitance [farad] [F].

The relative vector phase relationship of an AC voltage applied to a simple series circuit containing resistance, inductance, and capacitance is shown in Figure 7.3.

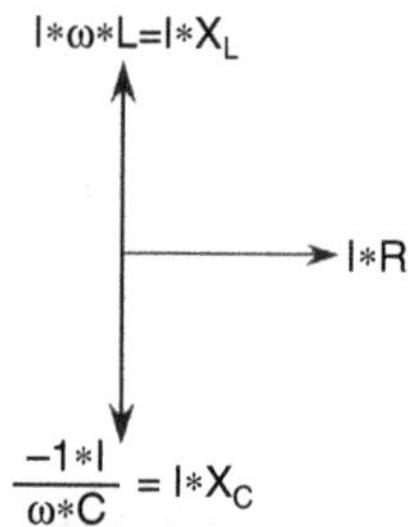

FIGURE 7.3 AC Voltage Resistive/Reactive Vector Phase Diagram.

Figure 7.3 shows the AC voltage resistive/reactive vector phase diagram.

A second factor that needs to be considered by the modeler, when modeling in the AC realm, is the skin depth (d) {7.13}. In any material, as a function of the complex permittivity {7.14}, electromagnetic waves (AC) will be attenuated (dissipated, turned into heat, etc.) and shifted in phase as a function of the distance (depth) traveled in that material.

Consider, for example, for a transverse electromagnetic wave propagating in the Z direction, the voltage relationship would be expressed as follows:

$$E_x = E_0 * e^{-kz} = E_0 * e^{-az*} e^{-j*\beta z} \tag{7.3}$$

Where: E_x = Transverse Electromagnetic wave propagating in the Z Direction.

E_0 = Scalar Voltage Amplitude.

k = Complex Propagation constant.

j = $(-1)^{1/2}$.

e = Base of Natural Logarithms.

α = Attenuation Constant.

β = Wave Solution Constant.

And where α is:

$$a = \omega * \left(\frac{\mu\varepsilon}{2} \left(1 + \left(1 + \left(\frac{\sigma}{\omega\varepsilon} \right)^2 \right)^{\frac{1}{2}} \right) \right)^{\frac{1}{2}} \tag{7.4}$$

And where:

ε = permittivity.

μ = permeability.

ω = angular frequency.

σ = conductivity.

For a good conductor, where $1 << \sigma/\omega\varepsilon$, the 1s in the above equation can be ignored and then α becomes:

$$a = \sqrt{\frac{\omega\mu\sigma}{2}} \tag{7.5}$$

The skin depth (δ) is the point at which the amplitude of the signal of interest decreases to $E_0{}^* e^{-1}$. Therefore δ is:

$$\delta = \frac{1}{a} \tag{7.6}$$

The first model presented in this chapter, 2D Electric Impedance Sensor Model (MMUC4_2D_EIS_1.mph), explores the sensing of two (2) small volume differential conductivity regions in a block of material that has a bulk conductivity of $1e^{-3}$ S/m and a relative permittivity of 12. The model is implemented using the AC/DC Electric Currents Physics Interface and is solved using a Frequency Domain Study.

NOTE

In the process of building and solving the MMUC4_2D_EIS_1.mph model, it will become obvious to the modeler that the power of the Electric Impedance Sensing technique lies in the non-invasive nature of this tool. In order to use this tool, the modeler does not need to know in advance what is inside the "black box" and does not need to destroy (cut open) the "black box" to determine the presence of areas of differential electronic properties and their physical locations.

2D SIMPLE MIXED MODE MODELS

2D Electric Impedance Sensor Model

Building the 2D Electric Impedance Sensor Model

Startup 4.x.

Select > 2D.

Click > Next.

Click > Twistie for AC/DC.

Click > Electric Currents (ec).

Click > Add Selected.

Click > Next.

Select > Frequency Domain in the Select Study Type window.

Click > Finish (Flag).

Click > Save As.

Enter MMUC4_2D_EIS_1.mph.

Click > Save.

See Figure 7.4.

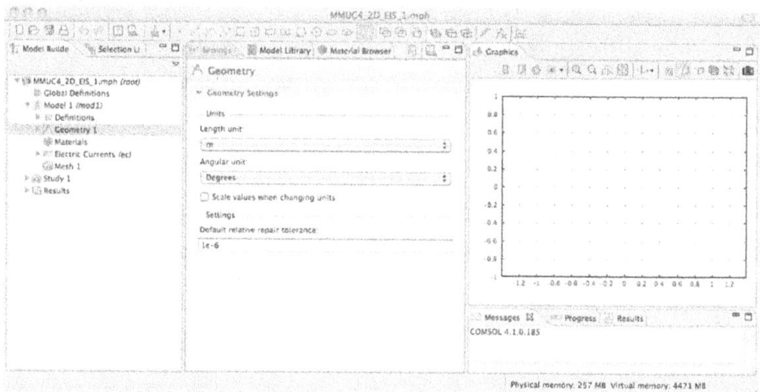

FIGURE 7.4 Desktop Display for the MMUC4_2D_EIS_1.mph Model.

Figure 7.4 shows the Desktop Display for the MMUC4_2D_EIS_1.mph model.

Right-Click > Global Definitions.

Select > Parameters from the Pop-up menu.

In the Settings – Parameters – Parameters edit window,

Enter the parameters shown in Table 7.1.

TABLE 7.1 Parameters Edit Window

Name	Expression	Description
sig_bulk	1[mS/m]	Conductivity bulk
eps_r_bulk	12	Permittivity relative bulk
y_0	-0.3[m]	Cavity 0 center y
x_0	0.3[m]	Cavity 0 center x
r_0	0.053[m]	Cavity 0 radius
y_1	-0.3[m]	Cavity 1 center y
x_1	-0.3[m]	Cavity 1 center x
r_1	0.047[m]	Cavity 1 radius
freq_0	1[MHz]	Frequency measurement

See Figure 7.5.

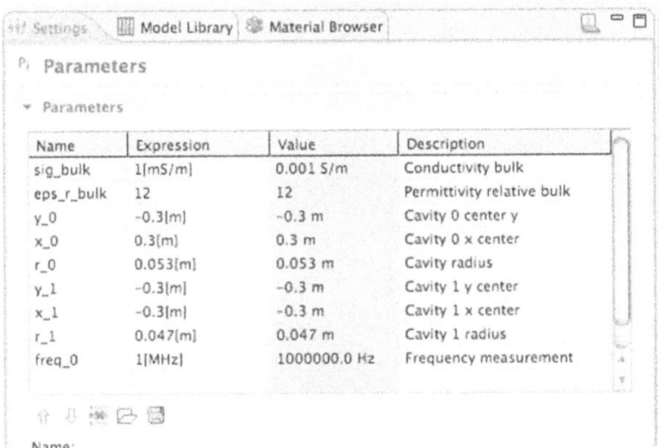

FIGURE 7.5 Settings - Parameters - Parameters Edit Window Filled.

Figure 7.5 shows the Settings – Parameters – Parameters edit window filled.

Right-Click > Global Definitions.

Select > Variables from the Pop-up menu.

In the Settings – Variables – Variables edit window,

Enter the variables shown in Table 7.2.

TABLE 7.2 Parameters Edit Window

Name	Expression	Description
sigma_1	sig_bulk*(((x-x_0-x0)^2+(y-y_0)^2)>r_0^2)	Conductivity cavity 0 local
eps_r_1	1+(eps_r_bulk-1)*(((x-x_0-x0)^2+(y-y_0)^2)>r_0^2)	Permittivity cavity 0 local
sigma_2	sig_bulk*(((x-x_1-x0)^2+(y-y_1)^2)>r_1^2)	Conductivity cavity 1 local
eps_r_2	1+(eps_r_bulk-1)*(((x-x_1-x0)^2+(y-y_1)^2)>r_1^2)	Permittivity cavity 1 local
sigma_local	(sigma_1+sigma_2)/2	Conductivity total local
eps_r_local	(eps_r_1+eps_r_2)/2	Permittivity total local

See Figure 7.6.

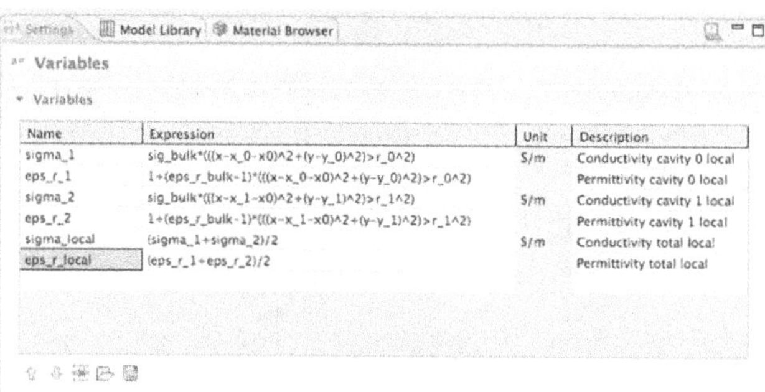

FIGURE 7.6 Settings – Variables – Variables Edit Window Filled.

Figure 7.6 shows the Settings – Variables – Variables edit window filled.

Geometry

Right-Click > Model Builder > Model 1 > Geometry 1.

Select > Rectangle from the Pop-up menu.

Enter > 1 in the Settings – Rectangle – Size – Width entry window.

Enter > 0.5 in the Settings – Rectangle – Size – Height entry window.

Enter > -0.5 in the Settings – Rectangle – Position – x entry window.

Enter > -0.5 in the Settings – Rectangle – Position – y entry window.

Click > Build All.

Right-Click > Model Builder – Geometry 1.

Select > Point from the Pop-up menu.

Enter > coordinates for point 1 using Table 7.3.

Click > Build Selected.

Select > Point from the Pop-up menu.

Enter > coordinates for point 2 using Table 7.3.

Click > Build Selected.

TABLE 7.3 Point Edit Windows

Name	Location x	Location y
point 1	-0.01	0
point 2	0.01	0

Click > Build All.

NOTE

The boundary points 1 & 2 are constraint points. They are added to the perimeter of the rectangle to indicate where the two sides of the electrode area end and where each of the two electrically insulated areas begin.

See Figure 7.7.

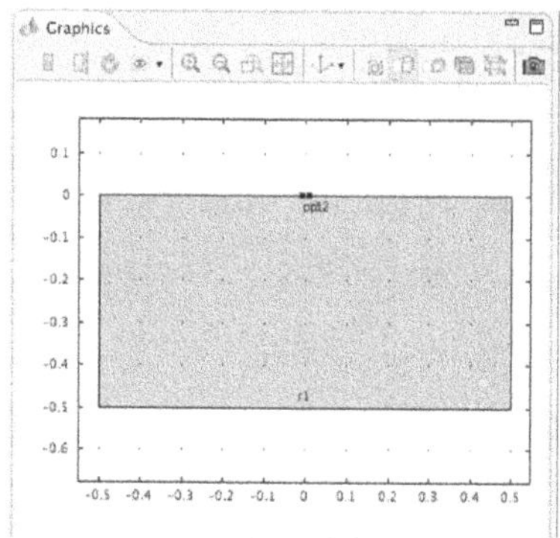

FIGURE 7.7 Model Geometry in the Graphics Window.

Figure 7.7 shows the Model Geometry in the Graphics window.

Electric Currents (ec) Interface

Click > Model Builder – Model 1 – Twistie for the Electric Currents (ec).

Click > Model Builder – Model 1 – Electric Currents (ec).

Verify that Domain 1 is Selected in Settings – Electric Currents – Domains – Selection.

Click > Settings – Electric Currents – Twistie for Equation.

Select > Frequency Domain from the Equation form Pull-down menu.

Select > User defined from the Frequency Pull-down menu.

Enter > freq_0 in the Frequency edit window.

NOTE *The modeler should note that the default setting of the Out-of-Plane Thickness (1[m]) is correct for this model and should not be changed.*

Click> Model Builder – Model 1 – Electric Currents (ec) – Current Conservation 1.

Click > Settings – Current Conservation – Conduction Current – Electrical conductivity.

Select > User defined from the Electrical conductivity Pull-down menu.

Enter > sigma_local in the Electrical conductivity edit window.

Click > Settings – Current Conservation – Electric Field – Relative Permittivity.

Select > User defined from the Relative permittivity Pull-down menu.

Enter > eps_r_local in the Electrical conductivity edit window.

See Figure 7.8.

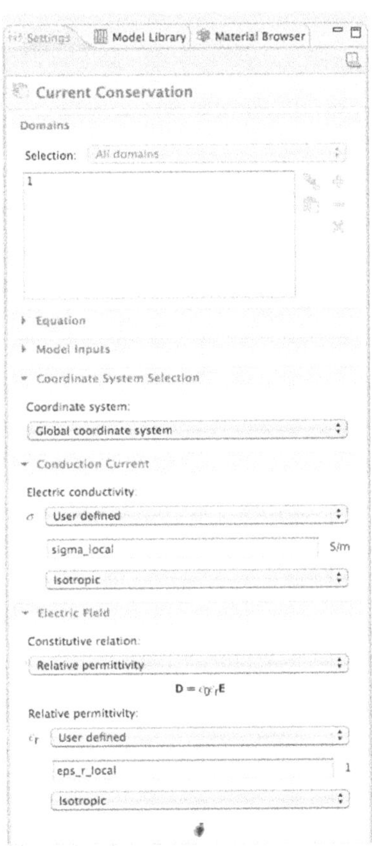

FIGURE 7.8 Settings - Current Conservation Edit Windows.

Figure 7.8 shows the Settings – Current Conservation edit windows.

Right-Click > Model Builder – Model 1 – Electric Currents (ec).

Select > Terminal from the Pop-up menu.

Click > Boundary 4 only in the Graphics window.

Click > Add to Selection in the Settings – Terminal – Boundaries panel.

Enter > 1 in the Settings – Terminal – Current edit window.

See Figure 7.9.

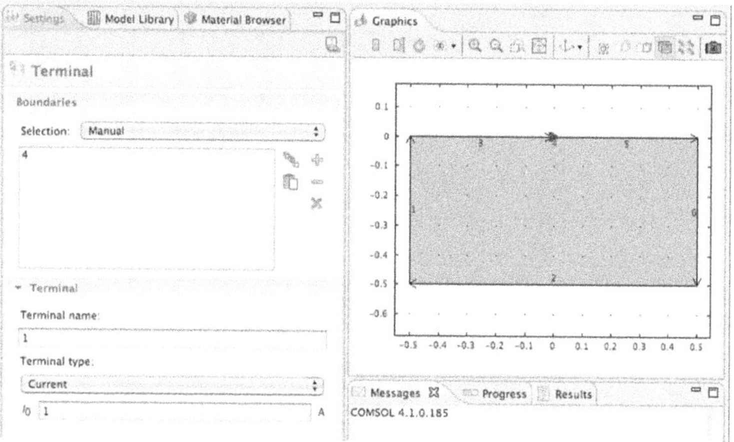

FIGURE 7.9 Settings - Terminal Edit Windows.

Figure 7.9 shows the Settings – Terminal edit windows.

Right-Click > Model Builder – Model 1 – Electric Currents (ec).

Select > Ground from the Pop-up menu.

Shift-Click > Boundaries 1, 2, 6 in the Graphics window.

Click > Add to Selection in the Settings – Ground – Boundaries panel.

See Figure 7.10.

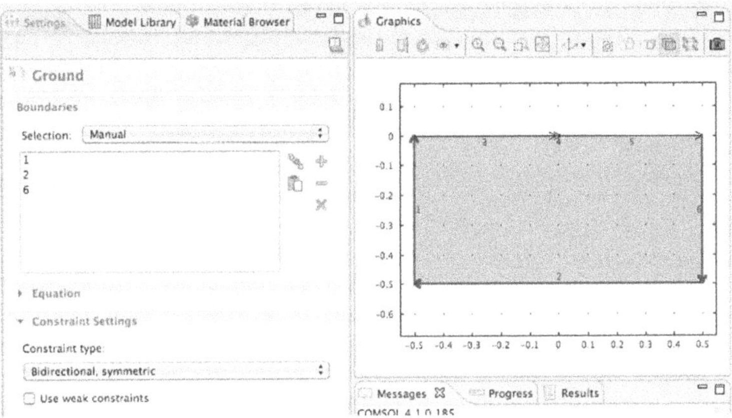

FIGURE 7.10 Settings - Ground Edit Windows.

Figure 7.10 shows the Settings – Ground edit windows.

Mesh 1

Right-Click > Model Builder – Model 1 – Mesh 1.

Select > Free Triangular from the Pop-up menu.

NOTE *The Free Triangular Mesh Type is chosen because, once adjusted, it will provide an adequate number of mesh elements and is relatively easy to solve.*

Click > Size.

Click > Settings – Size – Twistie for Element Size Parameters.

Enter > 0.01 in the Maximum element size edit window.

Click > Build All.

After meshing, the modeler should see a message in the message window about the number of elements (12474 elements) in the mesh.

See Figure 7.11.

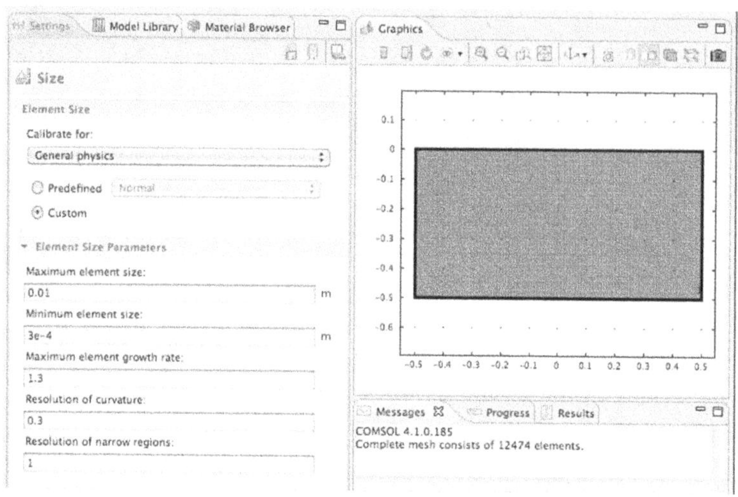

FIGURE 7.11 Settings - Size - Graphics - Meshed Domain.

Figure 7.11 shows the Settings – Size – Graphics – Meshed Domain.

Study 1

NOTE *The default settings for the Frequency Domain Study Type are not appropriate for the solution of this problem and thus need to be modified.*

Right-Click > Model Builder – Study 1.

Select > Show Default Solver.

Click > Model Builder – Study 1 twistie.

Click > Model Builder – Study 1– Solver Configurations twistie.

Click > Model Builder – Study 1– Solver Configurations – Solver 1 twistie.

Click > Model Builder – Study 1– Solver Configurations – Solver 1– Stationary Solver 1 twistie.

Click > Model Builder – Study 1– Solver Configurations – Solver 1– Stationary Solver 1 – Parametric 1.

Click > Settings – Parametric – General – Defined by study step.

Select > User defined in the Pull-down menu.

Enter > x0 in the Settings – Parametric – General – Parameter names edit window.

Click > Range button to enter the parameter values.

Enter > Start = –0.5, Stop = 0.5, and Step = 0.01 in the appropriate Range edit windows.

Click > Replace button.

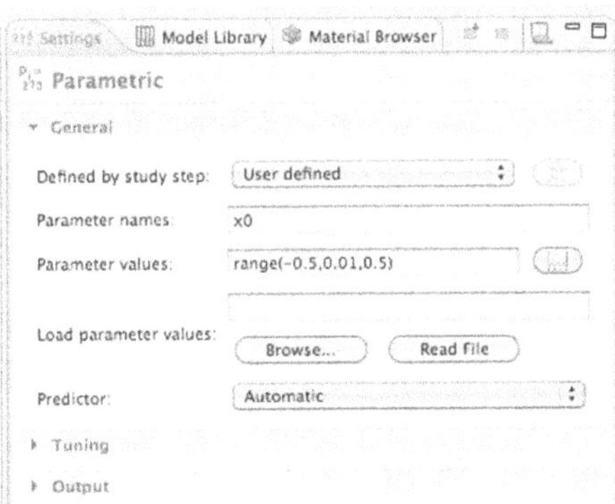

FIGURE 7.12 Settings - Parametric - General - Parameter Names Edit Windows.

See Figure 7.12.

Figure 7.12 Settings – Parametric – General – Parameter names edit windows.

In Model Builder, Right-Click Study 1 > Select > Compute.

Computed results, using the default display settings, are shown in Figure 7.13.

Figure 7.13 shows the computed results, using the default display settings.

NOTE

The modeler should note that the computed results, using the default display settings, do not readily show the location and size of the calculated voids. In order to easily observe those results, it is necessary to adjust the display settings parameters to increase the solution contrast in the display plot.

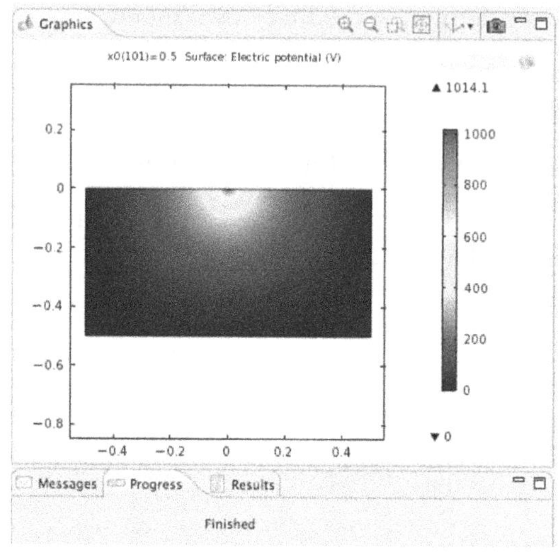

FIGURE 7.13 Computed Results, Using the Default Display Settings.

Results

In Model Builder,

Click > Results > 2D Plot Group 1.

Click > Settings – 2D Plot Group – Parameter value (x0).

Select > 0 from the Pull-down list.

Click > Model Builder – Results – 2D Plot Group 1 twistie.

Click > Model Builder – Results – 2D Plot Group 1 – Surface 1.

Click > Zoom Extents.

Enter > 20*log10(root.mod1.ec.normJ) in the Settings – Surface – Expression edit window.

NOTE

*The equation 20*log10 (variable) transforms the variable data from a linear plot into a logarithmic plot. The variable data is converted into decibels {7.15}. The net effect is to improve the plot contrast so that the calculated voids are displayed properly.*

Similarly, the minimum, maximum, and color table parameters are chosen to improve the plot display. The modeler should try different values and observe quality and contrast of the resulting plots.

The variable name {7.16} beginning with (root.) establishes the exact path within the COMSOL Multiphysics model. In this case, the path goes as follows: root (Model Builder) > mod1 (Model 1) > ec (Electric Currents) > normJ (Normal Current).

Click > Plot.

Click > Settings – Surface – Range twistie.

Select > Settings – Surface – Range – Manual color range check box.

Enter > Minimum = -35 and Maximum = 35.

Click > Settings – Surface – Coloring and Style – Color table.

Select > Wave from the Pull-down menu.

Click > Plot (if needed).

See Figure 7.14.

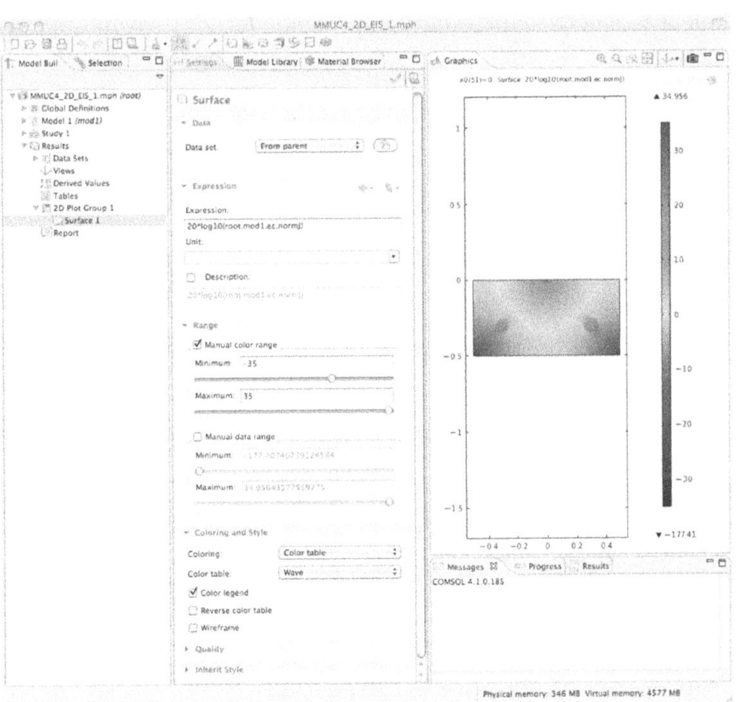

FIGURE 7.14 Desktop Display - Computed Solution with Modified Display Parameters.

Figure 7.14 shows the Desktop Display – Computed Solution with modified display parameters.

NOTE *The modeler should note that the calculated voids are now displayed clearly.*

More Results

The computed solution can also be displayed as a function of the impedance.

Right-Click > Model Builder – Results.

Select > 1D Plot Group from the Pop-up menu.

Right-Click > Model Builder – Results – 1D Plot Group 2.

Select > Global from the Pop-up menu.

Click > Settings – Global – Expressions – Replace Expression button.

Select > Electric Currents – Impedance (ec.Z11) from the Pop-up menu.

Click > Plot.

See Figure 7.15.

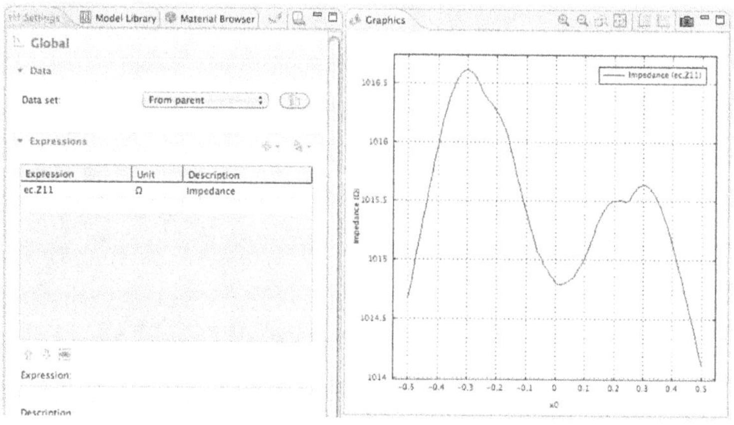

FIGURE 7.15 1D Impedance Plot for the Two Void 2D EIS Solution.

Figure 7.15 shows the 1D Impedance Plot for the two void 2D EIS Solution.

The phase angle of the impedance can also be displayed.

Right-Click > Model Builder – Results.

Select > 1D Plot Group from the Pop-up menu.

Right-Click > Model Builder – Results – 1D Plot Group 3.

Select > Global from the Pop-up menu.

Click > Settings – Global – Expressions – Replace Expression.

Select > Impedance (ec.Z11) from the Pop-up menu.

Enter > 180*arg(ec.Z11)/pi in the Expression edit window.

NOTE *The function arg {7.17} yields the phase angle of the function enclosed within the parentheses.*

Click > Plot.

See Figure 7.16.

Figure 7.16 shows the Phase Angle of the 1D Impedance Plot for the two void 2D EIS Solution.

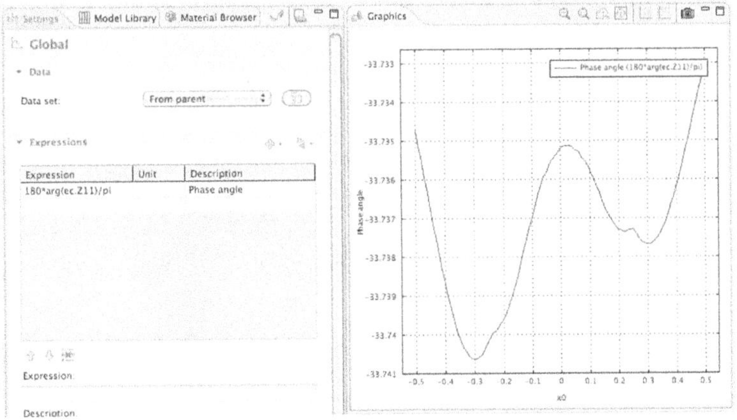

FIGURE 7.16 Phase Angle of the 1D Impedance Plot for the Two Void 2D EIS Solution.

2D Electric Impedance Sensor Model Summary and Conclusions

The 2D Electric Impedance Sensor Model is a powerful tool that can be used to develop the design of an Electric Impedance Sensor Machine to nondestructively explore materials and search for inclusions, voids, or any other differences in the electrical properties of imbedded regions of a block of material. With this model, the modeler can easily vary all of the machines parameters and optimize the design before the first prototype is physically built. This technique works best when the contrast in the properties of the materials is largest. The technique can be used either statically or dynamically and has broad application support in industry.

2D Metal Layer on a Dielectric Block Model Considerations

The modeler will recall from Chapter 6, that in 1855, Adolf Eugen Fick {7.18} published his now fundamental and famous laws {7.19} for the diffusion (mass transport) of a first item (e.g. a gas, a liquid, etc.) through a second item (e.g. another gas, liquid, etc.).

Fick's First Law is written:

$$J = -D\nabla\Phi \tag{7.7}$$

Where: J = Diffusion Flux in moles per square meter per second [$mol/(m^2*s)$].

D = Diffusion Coefficient (diffusivity) in square meters per second [m^2/s].

Φ = Concentration in moles per cubic meter [mol/m^3].

∇ = Gradient operator.

NOTE *It should be noted that the minus sign in front of the diffusivity indicates that the item that is flowing (gas, heat, money, etc.). It flows from the higher concentration to the lower concentration.*

See Figure 7.17

Figure 7.17 shows a graphical rendition of Fick's First Law.

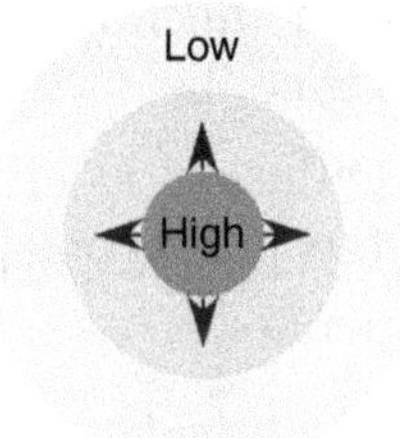

FIGURE 7.17 Graphical Rendition of Fick's First Law.

Fick's Second Law calculates the details of how diffusion (mass transport) causes the diffusing item's concentration to change as a function of time (under transient conditions).

Fick's Second Law is written:

$$\frac{\partial \Phi}{\partial t} = D\nabla^2\Phi \tag{7.8}$$

Where: D = Diffusion Coefficient (diffusivity) in square meters per second [m^2/s].

Φ = Concentration vector in moles per cubic meter [mol/m^3].

∇^2 = Laplace operator.

If modified slightly, where:

$$D = a = \frac{k}{\rho C_p} \tag{7.9}$$

Where: a = Thermal Diffusion Coefficient (thermal diffusivity) in square meters per second [m^2/s].

k = Thermal conductivity in Watts per meter-degree Kelvin [W/(m*K)].

ρ = Material density in kilograms per cubic meter [kg/m^3].

C_p = Specific heat in Joules per kilograms-degree Kelvin [J/(kg*K)].

Then, Fick's Second Law becomes the Heat Equation:

$$\frac{\partial u}{\partial t} = a\nabla^2 u \tag{7.10}$$

Where: u = u(x, y, z, t), a function of the three spatial parameters (x, y, z) and time (t).

α = Thermal Diffusion Coefficient (thermal diffusivity) in square meters per second [m^2/s].

∇^2 = Laplace operator.

NOTE

The heat equation has been broadly used to solve many different diffusion problems {7.20}. In the case of thermal diffusion, the spatial parameters are the geometric coordinates. Examples of other uses of the heat equation are easily found in physics, finance, chemistry, and many other application areas.

2D Metal Layer on a Dielectric Block Model

NOTE

This is a 2D model of heat conduction through a silica glass block that has a physically thin (0.0001m), high thermal conductivity metallic layer (copper) on the top of the block. The silica glass block has an initial temperature of 300K. At time zero, the left side of the block is fixed at a temperature of 300K (T_L) and the right side of the block is fixed at 600K (T_H). The model calculates the temperature distribution within the block as time progresses and heat flows from right to left in the block.

See Figure 7.18

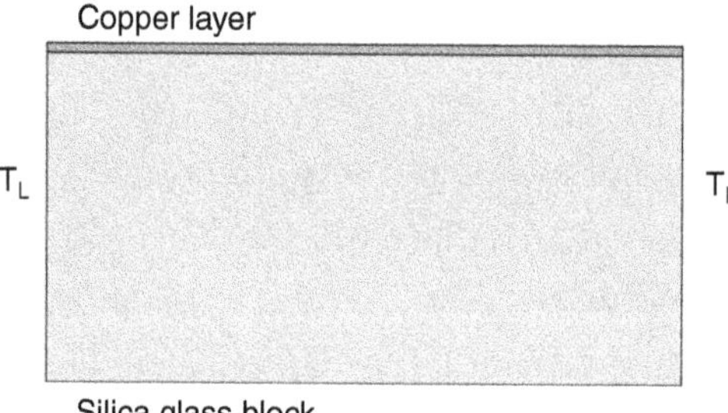

FIGURE 7.18 Silica Glass Block with a Copper Layer.

Figure 7.18 shows a Silica Glass Block with a Copper Layer.

Building the 2D Metal Layer on a Dielectric Block Model

Startup 4.x.

Select > 2D.

Click > Next.

Click > Twistie for Heat Transfer in the Add Physics window.

Click > Heat Transfer – Heat Transfer in Solids (ht).

Click > Add Selected.

Click > Next.

Select > Time Dependent in the Select Study Type window.

Click > Finish (Flag).

Click > Save As.

Enter MMUC4_2D_CpSG_1.mph.

See Figure 7.19.

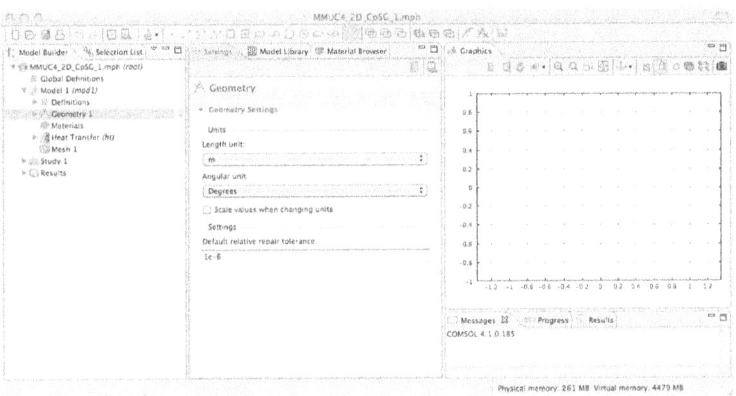

FIGURE 7.19 Desktop Display for the MMUC4_2D_CpSG_1.mph Model.

Figure 7.19 shows the Desktop Display for the MMUC4_2D_CpSG_1. mph model.

Geometry 1

NOTE *In the first study implemented in this model, the copper layer will be approximated by the Highly Conductive Layer function in the Heat Transfer in Solids Interface. In the second study implemented in this model, the copper layer will be a physical copper layer. The modeler can then directly compare the results of the calculations done for both studies, on nominally the same problem, and then he can determine the trade-offs for modeling by either route.*

In Model Builder – Model 1 (mod1),

Right-Click > Model Builder – Geometry 1.

Select > Rectangle from the Pop-up menu.

Enter > 0.02[m] in the Settings – Rectangle – Width entry window.

Enter > 0.01[m] in the Settings – Rectangle – Height entry window.

Click > Build All.

See Figure 7.20.

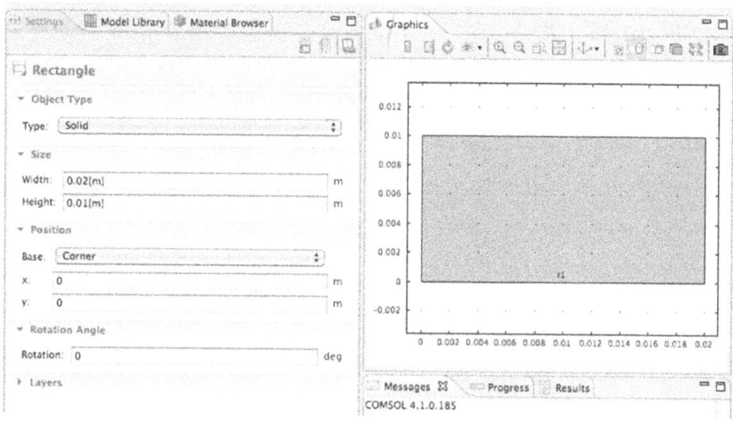

FIGURE 7.20 Settings - Rectangle Entry Windows.

Figure 7.20 shows the Settings – Rectangle entry windows.

Materials

Right-Click> Model Builder – Model 1 – Materials,

Select > Open Material Browser from the Pop-up menu.

Click > Settings – Materials Browser – Twistie of the Built-in materials library.

Select > Built-in Silica glass.

Right-Click > Silica glass.

Select > Add Material to Model.

Right-Click> Model Builder – Model 1 – Materials.

Select > Open Material Browser from the Pop-up menu.

Click > Settings – Materials Browser – Twistie of the Built-in materials library.

Select > Built-in Copper.

Right-Click > Copper.

Select > Add Material to Model.

Click > Copper.

Click > Settings – Material – Geometric Scope – Geometric entity level.

Select > Boundary from the Pull-down menu.

Click > Boundary 3 in the Graphics window.

Click > Add to Selection in the Settings – Material – Geometric Scope – Boundary edit window.

See Figure 7.21.

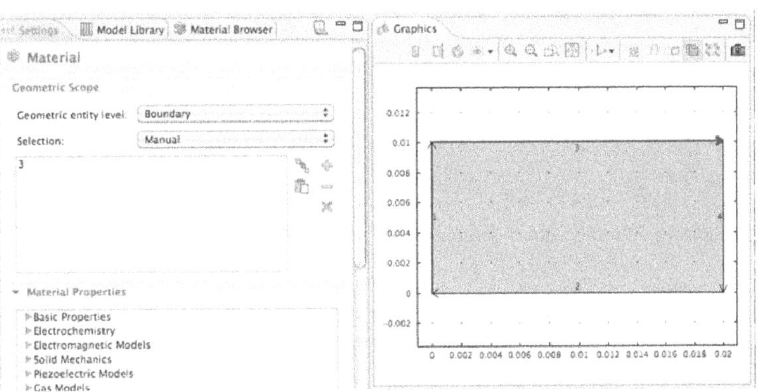

FIGURE 7.21 Settings - Material - Geometric Scope - Boundary Edit Window.

Figure 7.21 shows the Settings – Material – Geometric Scope – Boundary edit window.

Heat Transfer (ht) Interface

Right-Click> Model Builder – Model 1 – Heat Transfer (ht).

Select > Temperature from the Pop-up menu.

Click > Model Builder – Model 1 – Heat Transfer (ht) – Temperature 1.

Click > Boundary 1 in the Graphics window.

Click > Add to Selection in the Settings – Temperature – Boundaries edit window.

Enter > 300[K] in the Settings – Temperature – Temperature – Temperature edit window.

See Figure 7.22.

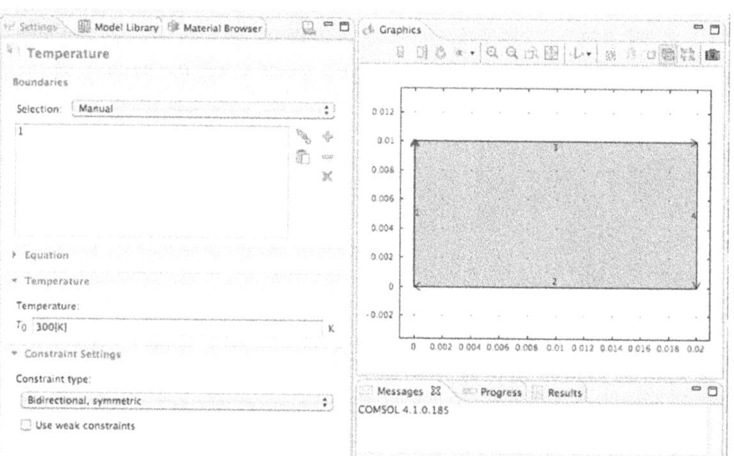

FIGURE 7.22 Settings - Temperature - Temperature - Temperature Edit Window.

Figure 7.22 shows the Settings – Temperature – Temperature – Temperature edit window.

Right-Click> Model Builder – Model 1 – Heat Transfer (ht).

Select > Temperature from the Pop-up menu.

Click > Model Builder – Model 1 – Heat Transfer (ht) – Temperature 2.

Click > Boundary 4 in the Graphics window.

Click > Add to Selection in the Settings – Temperature – Boundaries edit window.

Enter > 600[K] in the Settings – Temperature – Temperature – Temperature edit window.

See Figure 7.23.

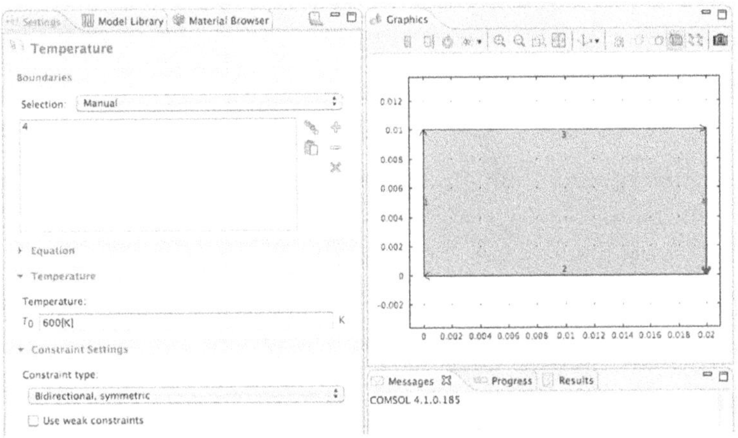

FIGURE 7.23 Settings - Temperature - Temperature - Temperature Edit Window.

Figure 7.23 shows the Settings – Temperature – Temperature – Temperature edit window.

Right-Click> Model Builder – Model 1 – Heat Transfer (ht).

Select > Highly Conductive Layer from the Pop-up menu.

Click > Model Builder – Model 1 – Heat Transfer (ht) – Highly Conductive Layer 1.

Click > Boundary 3 in the Graphics window.

Click > Add to Selection in the Settings – Temperature – Boundaries edit window.

Enter > 1e-4[m] in the Settings – Highly Conductive Layer – Model Inputs – Layer thickness edit window.

See Figure 7.24.

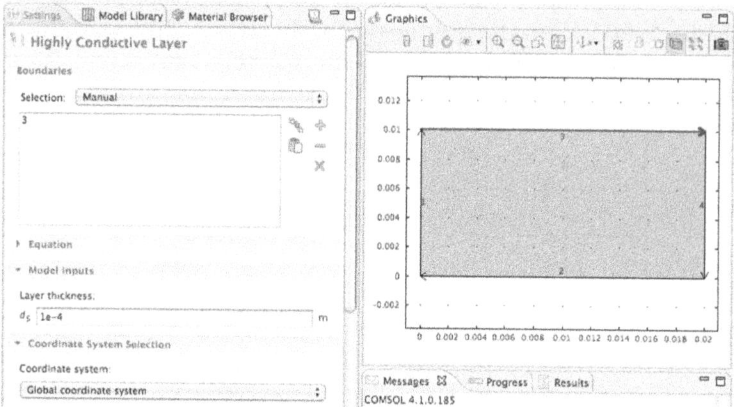

FIGURE 7.24 Settings - Highly Conductive Layer - Model Inputs - Layer Thickness Edit Window.

Figure 7.24 shows the Settings – Highly Conductive Layer – Model Inputs – Layer thickness edit window.

Click > Model Builder – Model 1 – Heat Transfer (ht) – Initial Values 1.

Enter > 300[K] in the Settings – Initial Values – Initial Values – Temperature edit window.

See Figure 7.25.

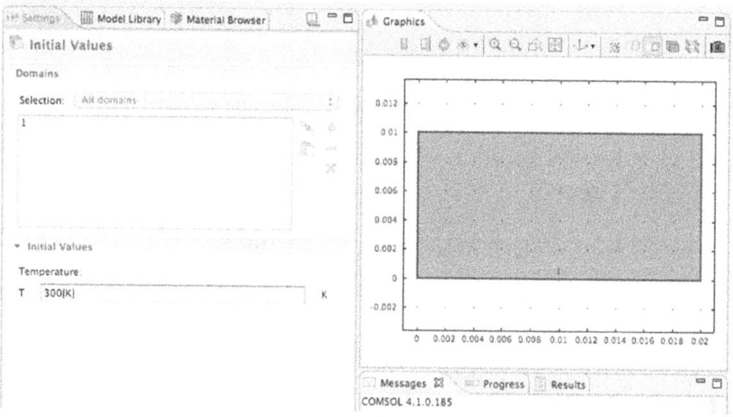

FIGURE 7.25 Settings - Initial Values - Initial Values - Temperature Edit Window.

Figure 7.25 shows the Settings – Initial Values – Initial Values – Temperature edit window.

Mesh 1

NOTE *The Free Triangular Mesh is easy to use and converges well and so it is chosen to use with this model.*

Right-Click > Model Builder – Model 1 – Mesh 1.

Select > Free Triangular.

Click > Build All button.

After meshing, the modeler should see a message in the message window about the number of elements (316 elements) in the mesh.

See Figure 7.26.

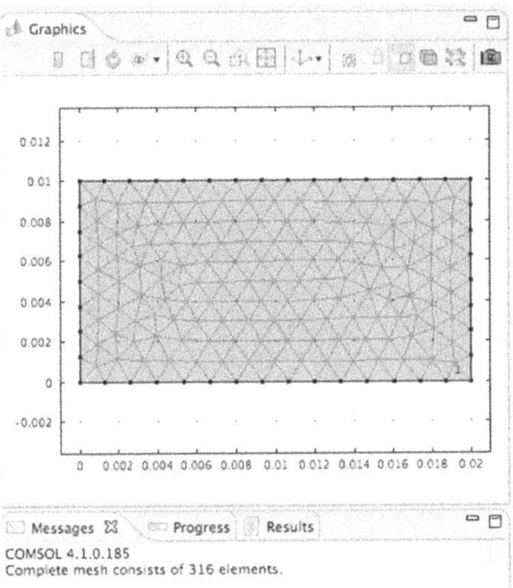

FIGURE 7.26 Desktop Display - Graphics - Meshed Domain.

Figure 7.26 shows the Desktop Display – Graphics – Meshed Domain.

Study 1

Click > Model Builder – Study 1– Twistie.

Click > Study 1 – Step 1: Time Dependent.

Click > Range button in Settings – Time Dependent – Study Settings – Times.

Enter > Start = 0, Stop = 60, Step = 5 in the Range Pop-up edit window.

Click > Replace button in the Range Pop-up edit window.

See Figure 7.27.

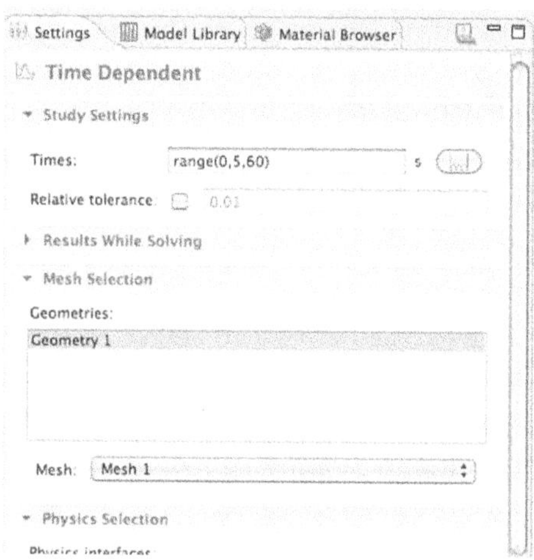

FIGURE 7.27 Settings – Time Dependent – Study Settings.

Figure 7.27 shows the Settings – Time Dependent – Study Settings.

In Model Builder,

Right-Click > Study 1 >Select > Compute.

Results

Click > Zoom Extents.

Computed results, using the default display settings, plus Zoom Extents, are shown in Figure 7.28.

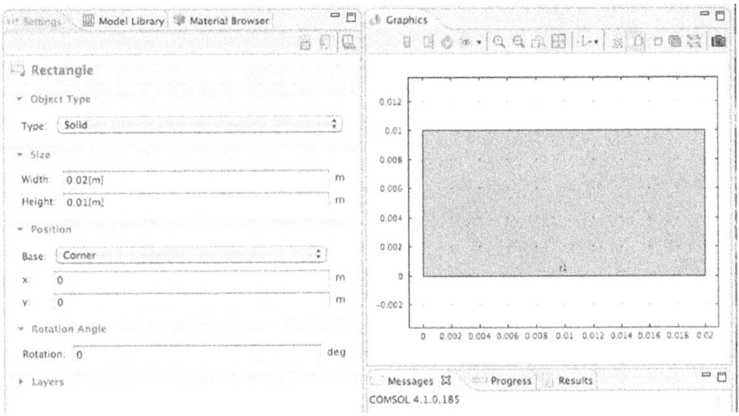

FIGURE 7.28 2D Metal Layer on a Dielectric Block Model Results, Using the Default Display Settings, Plus Zoom Extents.

Figure 7.28 shows the 2D Metal Layer on a Dielectric Block Model results, using the default display settings, plus Zoom Extents.

Building the 2nd 2D Metal Layer on a Dielectric Block Model

NOTE

In the first study implemented in this model, the copper layer was approximated by the Highly Conductive Layer function in the Heat Transfer in Solids Interface. In this, the second study implemented in this model, the copper layer will be a geometrical copper layer. The modeler will now be able to directly compare the results of both the first model and the second model calculations, on nominally the same problem. The modeler can now determine the relative trade-offs required when he chooses to model by one approach or the other.

Right-Click > Model Builder – MMUC4_2D_CpSG_1.mph (root).

Select > Add Model.

Select > 2D.

Click > Next.

Click > Twistie for Heat Transfer in the Add Physics window.

Click > Heat Transfer – Heat Transfer in Solids (ht).

Click > Add Selected.

Click > Next.

Select > Time Dependent in the Select Study Type window.

Click > Finish (Flag).

Geometry 2

In Model Builder – Model 2 (mod2),

Right-Click > Model Builder – Geometry 2.

Select > Rectangle from the Pop-up menu.

Enter > 0.02[m] in the Settings – Rectangle – Width entry window.

Enter > 0.01[m] in the Settings – Rectangle – Height entry window.

Click > Build All.

See Figure 7.29.

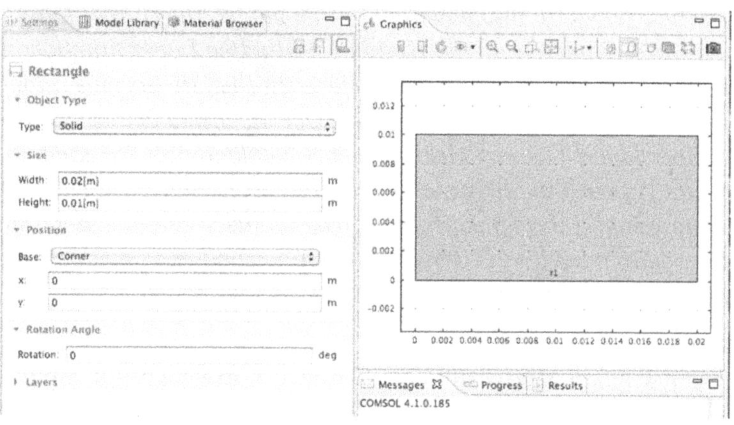

FIGURE 7.29 Settings - Rectangle Entry Windows.

Figure 7.29 shows the Settings – Rectangle Entry Windows.

In Model Builder – Model 2 (mod2),

Right-Click > Model Builder – Geometry 2.

Select > Rectangle from the Pop-up menu.

Enter > 0.02[m] in the Settings – Rectangle – Width entry window.

Enter > 1e-4[m] in the Settings – Rectangle – Height entry window.

Enter > 0.01[m] in the Settings – Rectangle – Position y entry window.

Click > Build All.

See Figure 7.30.

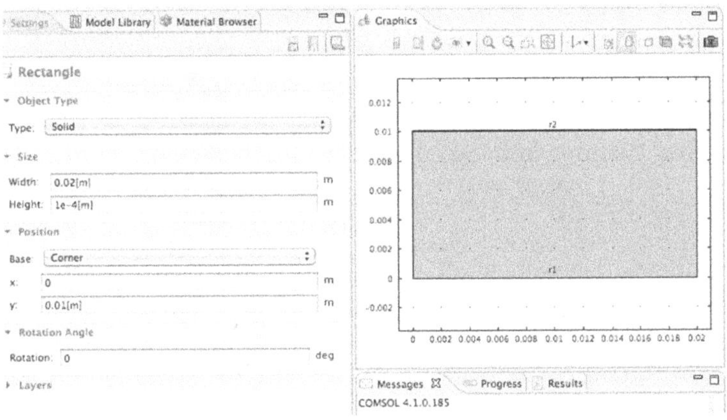

FIGURE 7.30 Settings - Rectangle Entry Windows.

Figure 7.30 shows the Settings – Rectangle Entry Windows.

Materials

Right-Click > Model Builder – Model 2 – Materials.

Select > Open Material Browser from the Pop-up menu.

Click > Settings – Materials Browser – Twistie of the Built-in materials library.

Select > Built-in Silica glass.

Right-Click > Silica glass.

Select > Add Material to "geom2".

Right-Click > Model Builder – Model 1 – Materials.

Select > Open Material Browser from the Pop-up menu.

Click > Settings – Materials Browser – Twistie of the Built-in materials library.

Select > Built-in Copper.

Right-Click > Copper.

Select > Add Material to “geom2”.

Click > Model Builder – Model 2 – Materials – Silica Glass.

Click > Settings – Material – Geometric Scope – Domain Selection edit window.

Select > 2.

Click > Remove from Selection.

See Figure 7.31.

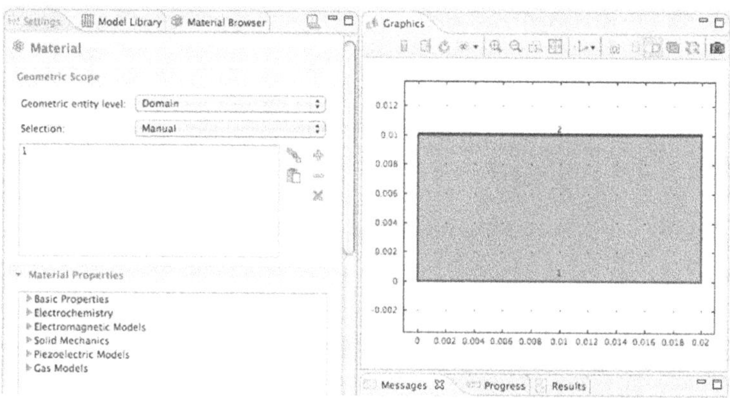

FIGURE 7.31 Settings - Material - Geometric Scope - Domain Selection Edit Window.

Figure 7.31 shows the Settings – Material – Geometric Scope – Domain Selection edit window.

Click > Model Builder – Model 2 – Materials – Copper.

Expand > Domain 2 edge.

NOTE *Click on the Zoom Box tool. Apply the Zoom Box tool to the upper left hand corner of the geometrical figure displayed in the Graphics window.*

See Figure 7.32.

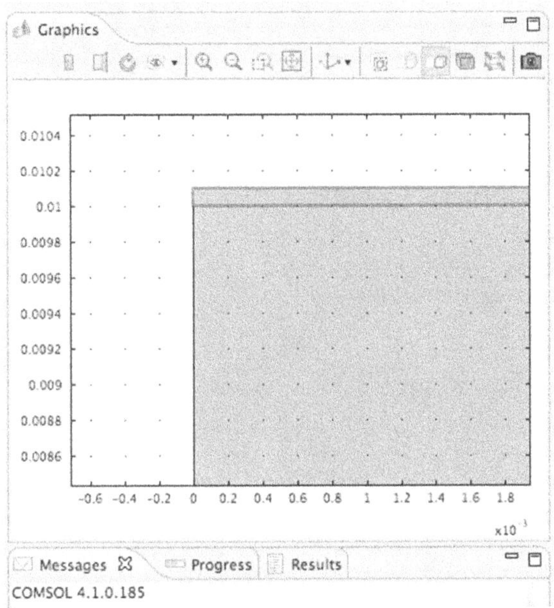

FIGURE 7.32 Expanded Geometry in Graphics Window.

Figure 7.32 shows the Expanded Geometry in Graphics window.

Select > Domain 2 in the Graphics window.

Click > Add to Selection.

See Figure 7.33.

Figure 7.33 shows the Settings – Material – Geometric Scope – Domain Selection edit window.

Click > Zoom Extents.

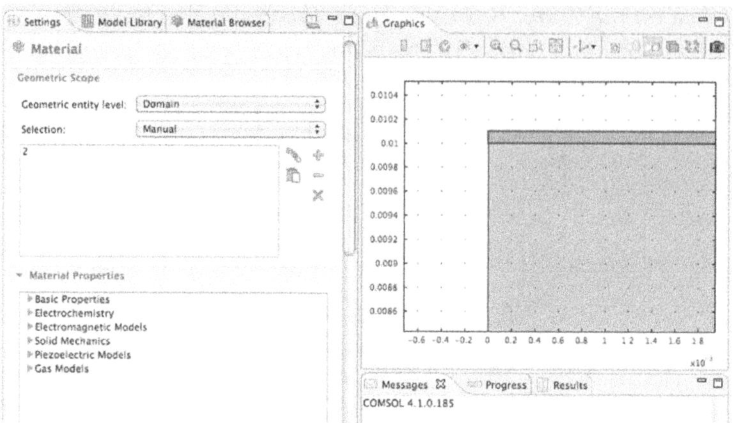

FIGURE 7.33 Settings - Material - Geometric Scope - Domain Selection Edit Window.

See the full size Domain 2 in Figure 7.34.

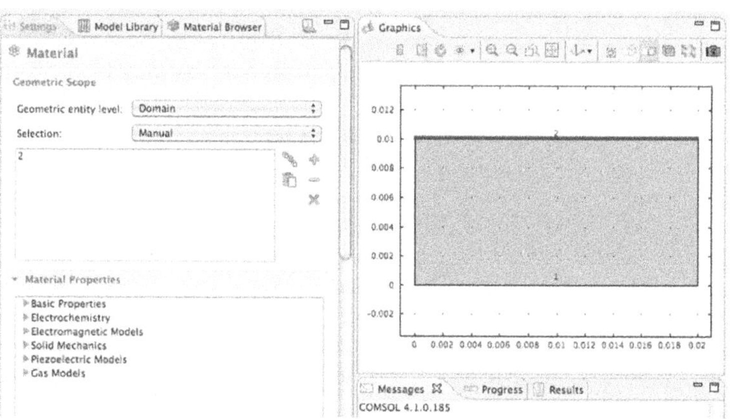

FIGURE 7.34 Settings - Material - Geometric Scope - Domain Selection Edit Window.

Figure 7.34 shows the Settings – Material – Geometric Scope – Domain Selection edit window.

Heat Transfer 2 (ht2) Interface

Right-Click> Model Builder – Model 2 – Heat Transfer 2 (ht2).

Select > Temperature from the Pop-up menu.

Click > Model Builder – Model – Heat Transfer (ht) – Temperature 1.

Expand > Domain 1 & 2 top left corner in Graphics window.

Shift-Click > Boundaries 1 & 3.

Click > Add to Selection in the Settings – Temperature – Boundaries edit window.

Enter > 300[K] in the Settings – Temperature – Temperature – Temperature edit window.

See Figure 7.35.

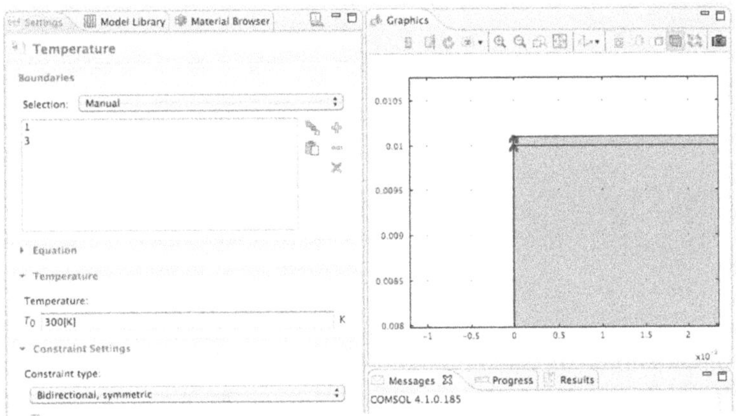

FIGURE 7.35 Settings - Temperature - Temperature - Temperature Edit Window.

Figure 7.35 shows the Settings – Temperature – Temperature – Temperature edit window.

Click > Zoom Extents.

Right-Click > Model Builder – Model 2 – Heat Transfer 2 (ht2).

Select > Temperature from the Pop-up menu.

Click > Model Builder – Model 2 – Heat Transfer 2 (ht2) –Temperature 2.

Expand > Domain 1 & 2 top right corner in Graphics window.

Shift-Click > Boundaries 6 & 7.

Click > Add to Selection in the Settings – Temperature – Boundaries edit window.

Enter > 600[K] in the Settings – Temperature – Temperature – Temperature edit window.

See Figure 7.36.

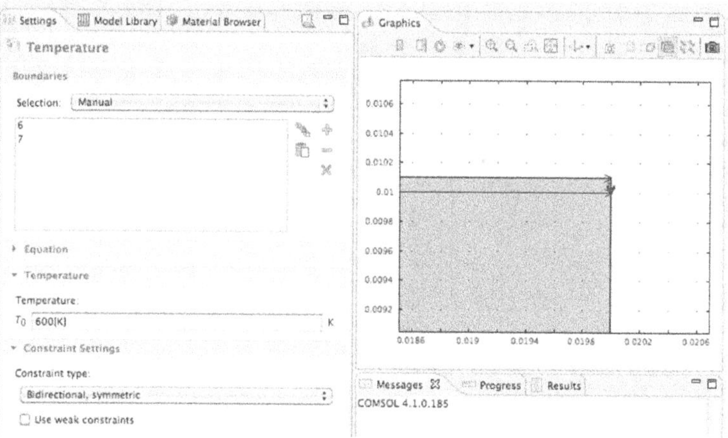

FIGURE 7.36 Settings - Temperature - Temperature - Temperature Edit Window.

Figure 7.36 shows the Settings – Temperature – Temperature – Temperature edit window.

Click > Zoom Extents.

Click > Model Builder – Model 2 – Heat Transfer 2 (ht2) – Initial Values 1.

Enter > 300[K] in the Settings – Initial Values – Initial Values – Temperature edit window.

See Figure 7.37.

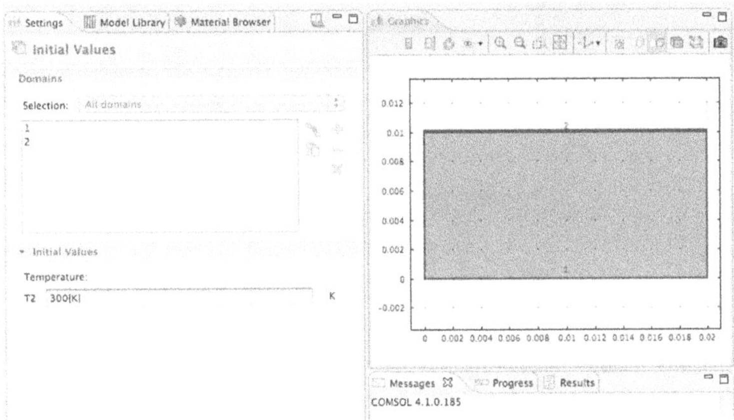

FIGURE 7.37 Settings – Initial Values – Initial Values – Temperature Edit Window.

Figure 7.37 shows the Settings – Initial Values – Initial Values – Temperature edit window.

Mesh 2

NOTE *The Free Triangular Mesh is easy to use and converges well and so it is chosen to use with this model.*

Right-Click > Model Builder – Model 1 – Mesh 2.

Select > Free Triangular.

Click > Build All button.

After meshing, the modeler should see a message in the message window about the number of elements (2083 elements) in the mesh.

See Figure 7.38.

Figure 7.38 shows the Desktop Display – Graphics – Meshed Domain.

Study 2

Click > Model Builder – Study 2 – Twistie.

Click > Study 2 – Step 1: Time Dependent.

Click > Range button in Settings – Time Dependent – Study Settings – Times.

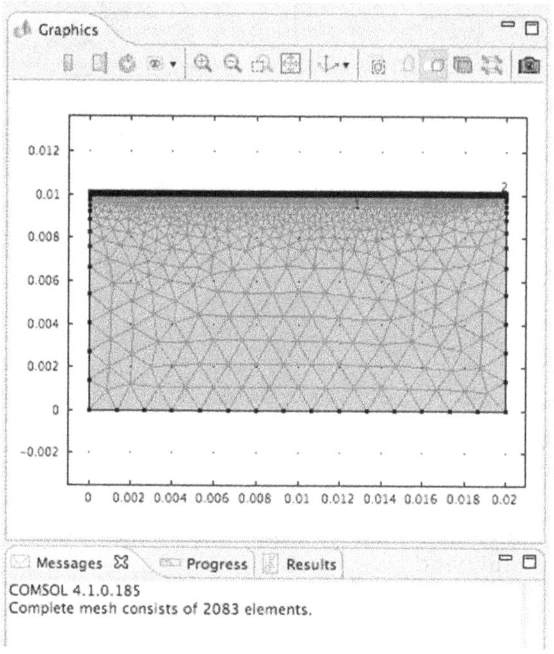

FIGURE 7.38 Desktop Display - Graphics - Meshed Domain.

Enter > Start = 0, Stop = 60, Step = 5 in the Range Pop-up edit window.

Click > Replace button in the Range Pop-up edit window.

See Figure 7.39.

Figure 7.39 shows the Settings – Time Dependent – Study Settings.

In Model Builder,

Right-Click Study 2 > Select > Compute.

Results

Click > Zoom Extents.

Computed results, using the default display settings, plus Zoom Extents, are shown in Figure 7.40.

Figure 7.40 shows the 2D Metal Layer on a Dielectric Block Model results, using the default display settings, plus Zoom Extents.

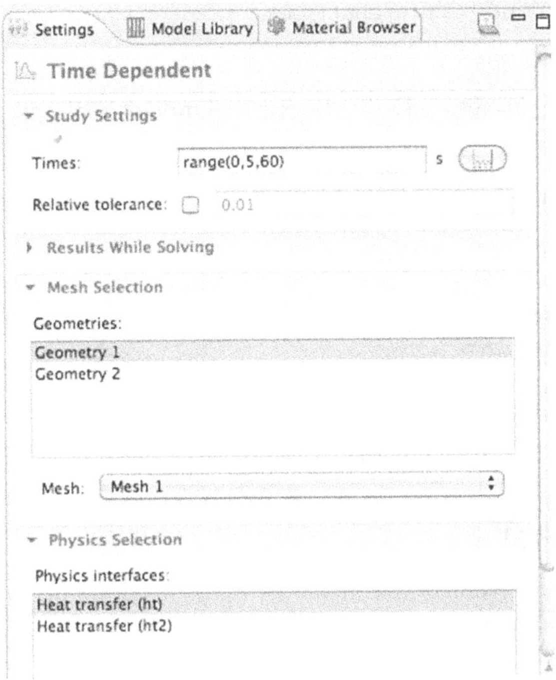

FIGURE 7.39 Settings – Time Dependent – Study Settings.

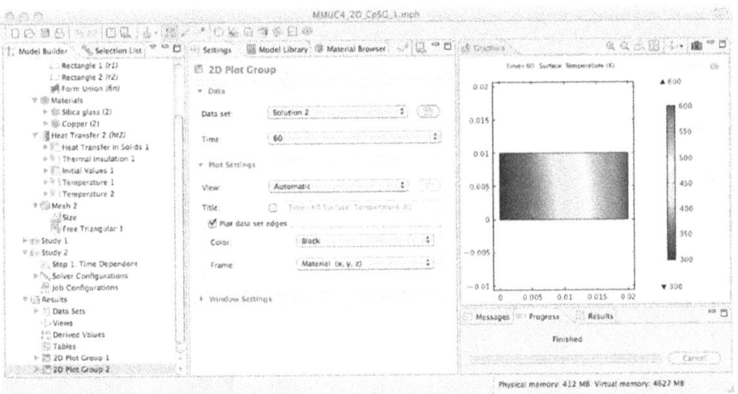

FIGURE 7.40 2D Metal Layer on a Dielectric Block Model Results, Using the Default Display Settings, Plus Zoom Extents.

Results Comparison of Study 1 and Study 2

Click > 2D Plot Group 2.

Click > Settings – 2D Plot Group – Windows Settings Twistie.

Click > Windows title check box.

Enter > Copper Layer Plot.

Click > Settings – 2D Plot Group – Windows Settings – Plot window.

Select > New Window from the Plot window Pull-down menu.

Click > Plot.

Click > Zoom Extents.

NOTE *To see the Plots from both models at the same time, use the following instructions.*

Right-Click > Copper Layer Plot tab.

Select > Move – View from the Pop-up menu.

Drag the upper edge of the Plot Outline to the boundary between the Graphics and Messages windows and then Click.

Drag the border between the two plot windows up until the two windows are the same size.

Click > Zoom Extents for each plot window.

See Figure 7.41.

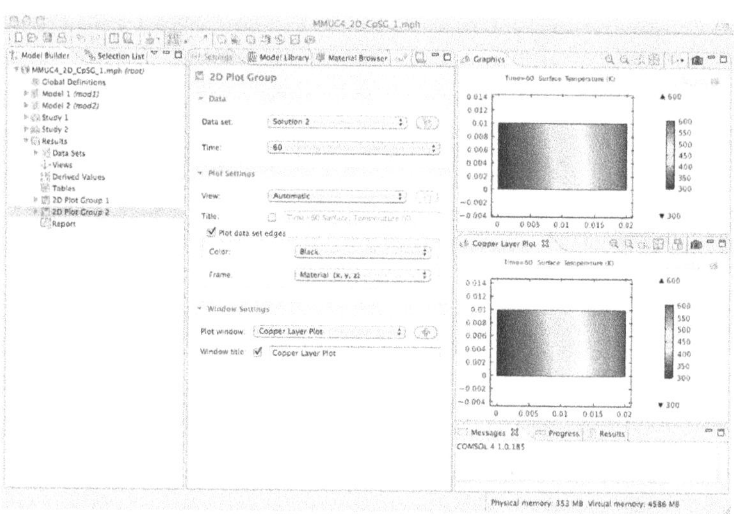

FIGURE 7.41 Simultaneous Comparison of the Highly Conductive Layer Model and the Copper Layer Model Solutions.

Figure 7.41 shows the Simultaneous Comparison of the Highly Conductive Layer Model and the Copper Layer Model Solutions.

NOTE *The modeler can readily observe that the Highly Conductive Layer Model and the Copper Layer Model result in exactly the same solution.*

2D Metal Layer on a Dielectric Block Models Summary and Conclusions

The 2D Metal Layer on a Dielectric Block Model is a powerful tool that can be used to develop the design of numerous different models for heat transfer problems. With this model, the modeler can easily vary all of the machine's parameters and optimize the design to obtain the best heat transfer before the first prototype is physically built. This technique works best when the contrast in the properties of the materials is largest. The technique can be used either statically or dynamically and has broad application support in industry.

FIRST PRINCIPLES AS APPLIED TO 2D SIMPLE MIXED MODE MODEL DEFINITION

First Principles Analysis derives from the fundamental laws of nature. In the case of models using this Classical Physics Analysis approach, the laws of conservation in physics require that what goes in (as mass, energy, charge, etc.) must come out (as mass, energy, charge, etc.) or must accumulate within the boundaries of the model.

The careful modeler must be knowledgeable of the implicit assumptions and default specifications that are normally incorporated into the COMSOL Multiphysics software model when a model is built using the default settings.

Consider, for example, the two 2D models developed in this chapter. In these models, it is implicitly assumed that there are no thermally related changes (mechanical, electrical, etc.). It is also assumed that the materials are homogeneous and isotropic, except as specifically indicated and that there are no thin insulating contact barriers at the thermal junctions. None of these assumptions are typically true in the general case. However, by making such assumptions, it is possible to easily build a 2D Simple Mixed Mode First Approximation Model.

NOTE

A First Approximation Model is one that captures all the essential features of the problem that needs to be solved, without dwelling excessively on all of the small details. A good First Approximation Model will yield an answer that enables the modeler to determine if he needs to invest the time and the resources required to build a more highly detailed model.

Also, the modeler needs to remember to name model parameters carefully as pointed out in Chapter 1.

REFERENCES

7.1 COMSOL Multiphysics Users Guide, Version 4.1, pp. 288-292

7.2 COMSOL Multiphysics Users Guide, Version 4.1, pp. 661

7.3 Heat Transfer Module Users Guide, Version 4.1, pp. 39

7.4 http://en.wikipedia.org/wiki/Electrical_impedance

7.5 http://en.wikipedia.org/wiki/Alternating_current

7.6 http://en.wikipedia.org/wiki/Ohm%27s_Law

7.7 http://en.wikipedia.org/wiki/Fourier_Analysis

7.8 http://en.wikipedia.org/wiki/Oliver_Heaviside

7.9 http://en.wikipedia.org/wiki/Arthur_E._Kennelly

7.10 Scott, W.T., The Physics of Electricity and Magnetism, John Wiley and Sons, Second Edition, 1966, Chapter 9.2-9.4, pp 461-485

7.11 http://en.wikipedia.org/wiki/Angular_frequency

7.12 http://en.wikipedia.org/wiki/Reactance_(electronics)

7.13 http://en.wikipedia.org/wiki/Skin_depth

7.14 http://en.wikipedia.org/wiki/Permittivity

7.15 http://en.wikipedia.org/wiki/Decibels

7.16 COMSOL Multiphysics Users Guide, Version 4.1, pp. 93

7.17 COMSOL Multiphysics Users Guide, Version 4.1, pp. 86

7.18 http://en.wikipedia.org/wiki/Adolf_Eugen_Fick

7.19 http://en.wikipedia.org/wiki/Fick%27s_law_of_diffusion

7.20 http://en.wikipedia.org/wiki/Heat_equation

Suggested Modeling Exercises

1. Build, mesh, and solve the 2D Electric Impedance Sensor Model as presented earlier in this chapter.
2. Build, mesh, and solve the 2D Metal Layer on a Dielectric Block Model as presented earlier in this chapter.
3. Change the values of the materials parameters and then build, mesh, and solve the 2D Electric Impedance Sensor Model as an example problem.
4. Change the values of the materials parameters and then build, mesh, and solve the 2D Metal Layer on a Dielectric Block Model as an example problem.
5. Change the value of the boundary temperatures and then build, mesh, and solve the 2D Metal Layer on a Dielectric Block Model as an example problem.
6. Change the value of the geometries of the layer and the block and then solve the 2D Metal Layer on a Dielectric Block Model as an example problem.

CHAPTER 8

2D COMPLEX MIXED MODE MODELING USING COMSOL MULTIPHYSICS 4.X

In This Chapter

- Guidelines for 2D Complex Mixed Mode Modeling in 4.x
  - 2D Complex Mixed Mode Modeling Considerations
- 2D Complex Mixed Mode Models
  - 2D Copper Electroplating Model
  - 2D Electrocoalescence Oil/Water Separation Model
- First Principles as Applied to 2D Complex Mixed Mode Model Definition
- References
- Suggested Modeling Exercises

GUIDELINES FOR 2D COMPLEX MIXED MODE MODELING IN 4.X

NOTE

In this chapter, complex mixed mode 2D models will be presented. Such 2D models are typically more conceptually complex than the models that were presented in earlier chapters of this text. 2D complex mixed mode models have proven to be very valuable to the science and engineering communities, both in the past and currently, as first-cut evaluations of potential systemic physical behavior under the influence of mixed external stimuli. The 2D mixed mode model responses and other such ancillary information can be gathered and screened early in a project for a first-cut evaluation. That initial information can potentially be used later as guidance in building higher-dimensionality (3D) field-based (electrical, magnetic, etc.) models.

Since the models in this and subsequent chapters are more complex and more difficult to solve than the models presented thus far, it is important that the modeler have available the tools necessary to most easily utilize the powerful capabilities of the 4.x software. In order to do that, if you have not done this previously, the modeler should go to the main 4.x toolbar, Click > Options – Preferences – Model builder. When the Preferences – Model builder edit window is shown, Select > Show equation view checkbox and Show more options checkbox. Click > Apply {8.1}.

2D Complex Mixed Mode Modeling Considerations

2D Models can in some cases be less difficult than some 1D or 3D models, having fewer implicit assumptions. Potentially, 2D models can still be a very challenging type of model to build, depending on the underlying physics. The least difficult aspect of 2D model creation arises from the fact that the geometry is relatively simple. However, the physics in a 2D complex mixed mode model can range from relatively simple to extremely complex. This is especially true of the differential mass transport in fluid media models featured in this chapter.

In compliance with the laws of physics, a 2D model implicitly assumes that energy flow, materials properties, environment, and all other conditions and variables that are of interest are homogeneous, isotropic, and/or constant, unless otherwise specified. This condition is assumed to be true throughout the entire domain of interest both within the model and through the boundary conditions and in the environs of the model.

The modeler needs to bear the above stated conditions in mind and carefully ensure that all of the modeling conditions and associated parameters (default settings) in each model created are properly considered, defined, verified, and/or set to the appropriate values.

NOTE

It is always mandatory that the modeler be able to accurately anticipate the expected results of the model and accurately specify the manner in which those results will be presented. Never assume that any of the default values that are present when the model is created necessarily satisfy the needs or conditions of a particular model.

Always verify that any parameters employed in the model are the correct value needed for that model. Calculated solutions that significantly deviate from the anticipated solution or from a comparison of values measured in an experimentally derived realistic model are probably indicative of one or more modeling errors either in the original model design, in the earlier model analysis, in the understanding of the underlying physics, or are simply due to human error.

2D Coordinate System

In a 2D model, if parameters can only vary as a function of the position (x) and (y) coordinates, then such a 2D model represents the parametric condition of the model in a time independent mode (stationary). In a time dependent study or frequency domain study model, parameters can vary both with position in (x) and/or (y) and with time (t).

See Figure 8.1.

Figure 8.1 shows the 2D coordinate system, plus time.

NOTE

In some 2D models in some physics interfaces, there may be a specified third geometric dimension with a default value of 1[m]. The properties of the model being constructed are isotropically mapped throughout that depth. That depth is known as the Out-of-Plane Thickness. That Out-of-Plane Thickness should be changed as needed by the modeler to ensure the resulting calculations reflect the parameters of the physical system being modeled. The Out-of-Plane Thickness edit window can be found by Clicking on the 2D Physics Interface being employed.

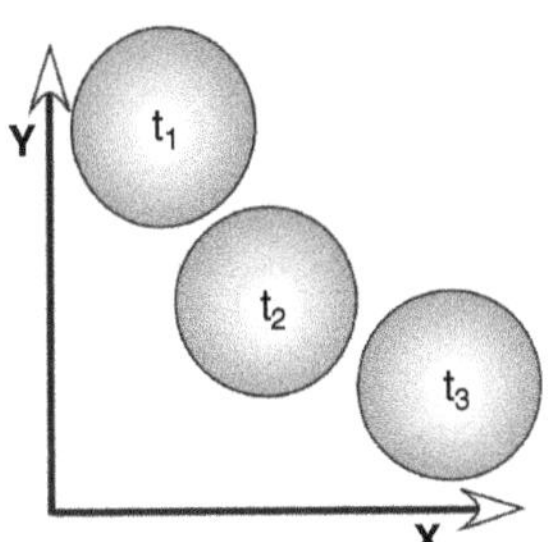

FIGURE 8.1 2D Coordinate System, Plus Time.

See Figure 8.2.

Figure 8.2 shows the Out-of-Plane Thickness edit window.

In this chapter, two new primary concepts are introduced to the modeler: mass transport of copper ions through a fluid medium resulting in the electrodeposition (electroplating) of a copper layer {8.2} and the mass transport and electrocoalescence of water droplets in an oil medium {8.3}. The first model in this chapter employs the Nernst-Planck Equation (chnp)

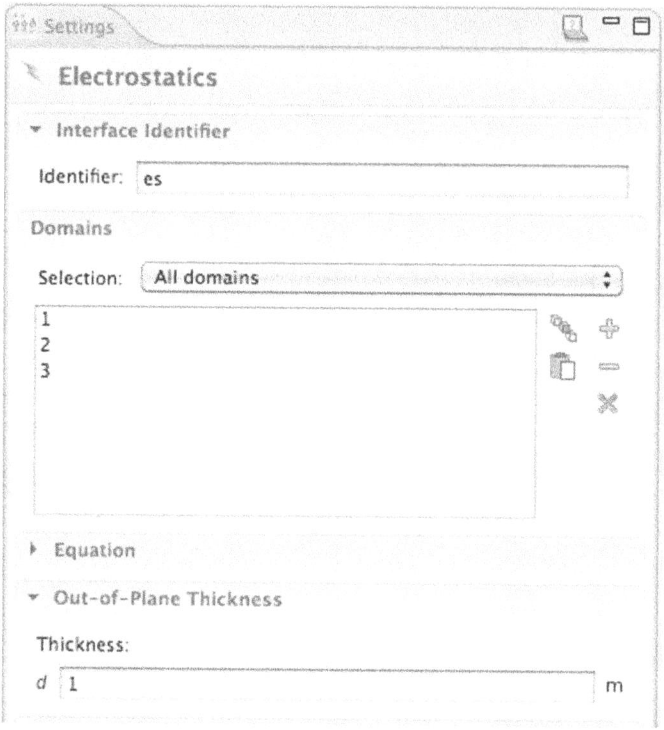

FIGURE 8.2 Out-of-Plane Thickness Edit Window.

{8.4} found in the Chemical Species Transport physics interface. The second model in this chapter employs an Electrostatic (es) field to induce the electrocoalescence of water droplets in an oil medium under the Laminar Two-Phase Flow, Phase Field (tpf) {8.5} physics interface.

Electroplating Theory

The electroplating processes that are utilized currently are an outgrowth of the discoveries made during the underlying scientific work done in the study of the fundamentals of electrochemistry {8.6} by John Frederic Daniell (1836){8.7} and Michael Faraday (1812) {8.8}.

NOTE *There is a very large literature of specific, detailed electroplating processes just for copper and copper alloys {8.9} in and of themselves. In fact, many books have been written that include specific solution process chemistry and time/power process details. Exploration of that literature is left to the modeler.*

The object of the model presented in this chapter is to demonstrate specifically, on a first approximation basis, the modeling of one example of the copper sulfate electroplating process in which the acidity is in the moderate (pH 4) range.

Many of the more common electroplating processes utilize consumable anodes (copper) as the source of copper ion replenishment for the electroplating bath.

See Figure 8.3.

Figure 8.3 shows a typical consumable anode copper electroplating bath.

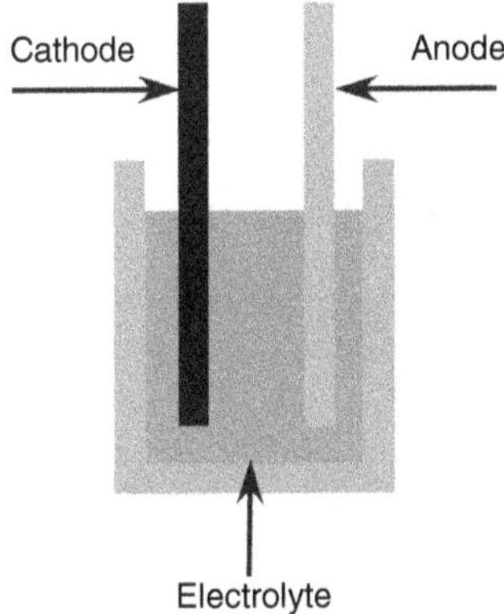

FIGURE 8.3 Typical Consumable Anode Copper Electroplating Bath.

Since in this case the model is for a first approximation to the copper electroplating process, trace additives that are typically included in the bath {8.10} to improve the second order properties of the deposited copper (brightness, porosity, adhesion, etc.) are not included within this model.

NOTE *The inclusion of additional trace additives and the related ions will increase the difficulty of solving the model without necessarily clarifying the fundamental understanding needed in the process of implementing the modeling of the electroplating process.*

The Nernst-Planck Equation {8.11} governs the ionic mass transport process that occurs in the fluid electrolyte. The Nernst-Planck Equation is a conservation of mass equation and is an extension of Fick's Law of diffusion {8.12} to include electrostatic forces.

The flux of each of the ions in the electrolyte is given by the Nernst-Planck equation as follows:

$$N_i = -D_i \nabla c_i - z_i u_i F c_i \nabla V \tag{8.1}$$

Where: $\mathbf{N}_i$ = Mass Transport Vector [mol/(m^2*s].

D_i = Diffusivity of the ith species in the electrolyte [m^2/s].

c_i = Concentration of the ith species in the electrolyte [mol/m^3].

z_i = Charge of the ith species in the electrolyte [1] (unitless).

u_i = Mobility of the ith species in the electrolyte [(mol*m^2)/(J*s)].

F = Faraday's constant [A*s/mol].

V = Potential in the fluid [V].

And the mobility u_i can be expressed as:

$$u_i = \frac{D_i}{RT} \tag{8.2}$$

Where: D_i = Diffusivity of the ith species in the electrolyte [m^2/s].

R = Universal gas constant 8.31447[J/mol*K].

T = Temperature [K].

The material balances for each species are expressed as:

$$\frac{\partial c_i}{\partial t} = -\nabla \bullet N_i \tag{8.3}$$

Where: c_i = Concentration of the ith species in the electrolyte [mol/m^3].

$\mathbf{N}_i$ = Mass Transport Vector [mol/(m^2*s].

t = time [t].

The electroneutrality condition is given as follows:

$$\sum_i -z_i c_i = 0 \tag{8.4}$$

Where: z_i = Charge of the ith species in the electrolyte [1] (unitless).

c_i = Concentration of the ith species in the electrolyte [mol/m^3].

The boundary conditions at the anode and the cathode are determined by the assumed electrochemical reaction and the Butler-Volmer equation {8.13}.

NOTE

At this point in building this model it is necessary to make an assumption about the exact nature of the process by which copper is deposited on the cathode. There are two reactions that occur. They are: $Cu^{2+}+e^- = Cu^+$ and $Cu^+ +e^- = Cu$. Typically, because not all things are equal and it is known that the Rate Determining Step (RDS) (slowest) is the $Cu^{2+}+e^- = Cu^+$, by about a factor of 1000 {8.14}, it is herein assumed that the $Cu^{2+}+e^- = Cu^+$ step is in equilibrium.

That being the case, then the cathode mass transport is:

$$N_{Cu^{2+}} \bullet n = \frac{i_0}{2F}\left(\exp\left(\frac{1.5F\eta_{cat}}{RT}\right) - \frac{c_{Cu^{2+}}}{c_{Cu^{2+},ref}}\exp\left(-\frac{0.5F\eta_{cat}}{RT}\right)\right) \tag{8.5}$$

Where: $\mathbf{N}_i$ = Mass Transport Vector [mol/(m^2*s].

$\mathbf{n}$ = normal vector.

i_0 = Exchange current density [A/m^2].

R = Universal gas constant [J/(mol*K)].

$c_{Cu^{2+}}$ = Concentration of the Cu^{2+} species in the electrolyte [mol/m^3].

$c_{Cu^{2+},ref}$ = Reference concentration of the Cu^{2+} species in the electrolyte [mol/m^3].

η_{cat} = Cathode overpotential [V].

F = Faraday's constant [A*s/mol].

T = Temperature [K].

It also follows from the above that the anode mass transport is:

$$N_{Cu^{2+}} \bullet n = \frac{i_0}{2F}\left(\exp\left(\frac{1.5F\eta_{an}}{RT}\right) - \frac{c_{Cu^{2+}}}{c_{Cu^{2+},ref}}\exp\left(-\frac{0.5F\eta_{an}}{RT}\right)\right) \tag{8.6}$$

Where: $\mathbf{N}_i$ = Mass Transport Vector [mol/(m^2*s].

$\mathbf{n}$ = normal vector.

i_0 = Exchange current density [A/m^2].

R = Universal gas constant [J/(mol*K)].

c_{Cu}^{2+} = Concentration of the Cu^{2+} species in the electrolyte [mol/m^3].

$c_{Cu}^{2+}{}_{,ref}$ = Reference concentration of the Cu^{2+} species in the electrolyte [mol/m^3].

η_{an} = Anode overpotential [V].

F = Faraday's constant [A*s/mol].

T = Temperature [K].

For the insulating boundaries, where the mass transport is zero,

$$N_{Cu^{2+}} \bullet n = 0 \tag{8.7}$$

Where: $\mathbf{N}_{Cu}^{2+}$ = Mass Transport Vector [mol/(m^2*s].

$\mathbf{n}$ = normal vector.

And for sulfate ions, the insulating applies everywhere, thus,

$$N_{SO_4^{2+}} \bullet n = 0 \tag{8.8}$$

Where: $\mathbf{N}_{SO4}^{2+}$ = Mass Transport Vector [mol/(m^2*s].

$\mathbf{n}$ = normal vector.

The first model presented in this chapter, 2D Copper Electroplating Model (MMUC4_2D_CE_1.mph), calculates the deposition of a copper layer on the cathode from an electroplating bath, as shown in Figure 8.4. This model demonstrates the use of the Nernst-Planck Equations (chnp) in the Chemical Species Transport Interface. The Nernst-Planck Equations are used in conjunction with the Moving Mesh (ale) in the Mathematics Interface.

See Figure 8.4.

Figure 8.4 shows a 2D Copper Electroplating model.

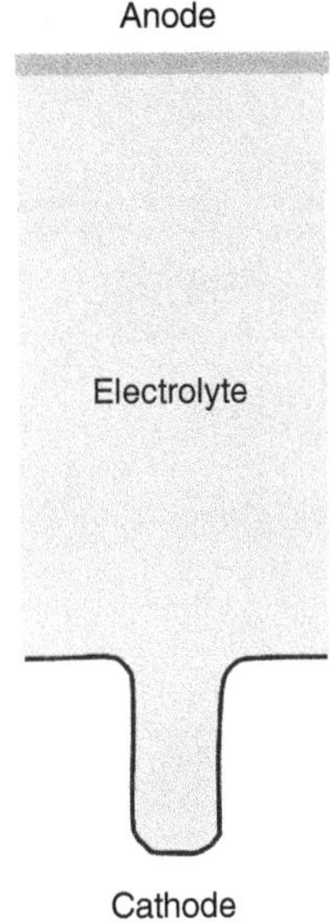

FIGURE 8.4 2D Copper Electroplating Model.

2D COMPLEX MIXED MODE MODELS

2D Copper Electroplating Model

Building the 2D Copper Electroplating Model

Startup 4.x.

Select > 2D.

Click > Next.

Click > Twistie for Chemical Species Transport Interface.

Click > Nernst-Planck Equations *(chnp)*.

Click > Add Selected.

Click > Twistie for Mathematics Interface.

Click > Twistie for Mathematics – Deformed Mesh.

Click > Moving Mesh *(ale)*.

Click > Add Selected.

Click > Next.

Select > Study Type – Time Dependent.

Click > Finish (Flag).

Click > Save As.

Enter MMUC4_2D_CE_1.mph.

Click > Save.

See Figure 8.5.

Figure 8.5 shows the Desktop Display for the MMUC4_2D_CE_1.mph model.

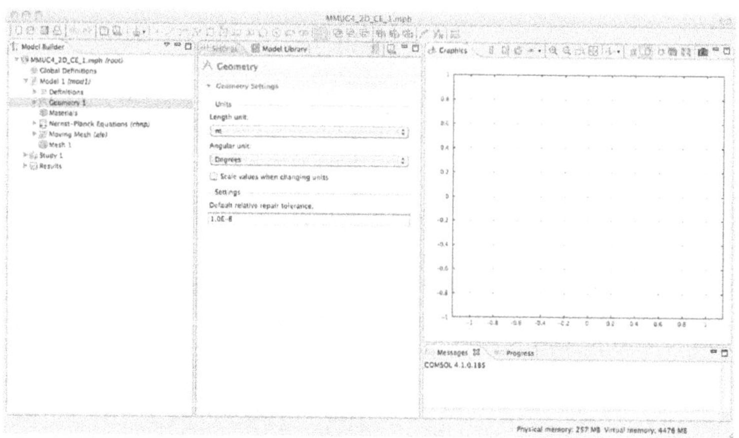

FIGURE 8.5 Desktop Display for the MMUC4_2D_CE_1.mph Model.

Right-Click > Model Builder – Global Definitions.

Select > Parameters from the Pop-up menu.

In the Settings – Parameters – Parameters edit window,

Enter the parameters shown in Table 8.1.

NOTE *Numerical parameter values are followed by [units] except when the numerical value is dimensionless. Then the numerical parameter is followed by [1].*

TABLE 8.1 Parameters

Name	Expression	Description
Cinit	500[mol/(m^3)]	Initial concentration
D_c1	2e-9[m^2/s]	Diffusivity
D_c2	D_c1	Diffusivity
T0	298[K]	System temperature
i0	150[A/m^2]	Exchange current density
phi_eq	0[V]	Relative equilibrium potential
alpha	0.75[1]	Symmetry factor
phi_s_anode	0.0859[V]	Anode potential
phi_s_cathode	–0.0859[V]	Cathode potential
z_net	2[1]	Net species charge
z_c1	z_net[1]	Charge, species c1
z_c2	-z_net[1]	Charge, species c2
um_c1	D_c1/R_const/T0	Mobility, species c1
um_c2	um_c1	Mobility, species c2
MCu	63.546e-3[kg/mol]	Cu molar mass
rhoCu	7.7264e3[kg/m^3]	Cu density
alpha1	0.5[1]	Symmetry factor
alpha2	1.5[1]	Symmetry factor

See Figures 8.6 and 8.7.

Figure 8.6 shows the Settings – Parameters – Parameters edit window first 10 parameters.

Settings | Model Library

Parameters

Parameters

Name	Expression	Value	Description
Cinit	500[mol/(m^3)]	500 mol/m³	Initial concentration
D_c1	2e-9[m^2/s]	2.0E-9 m²/s	Diffusivity
D_c2	D_c1	2.0E-9 m²/s	Diffusivity
T0	298[K]	298 K	System temperature
i0	150[A/m^2]	150 A/m²	Exchange current density
phi_eq	0[V]	0 V	Relative equilibrium potential
alpha	0.75[1]	0.75	Symmetry factor
phi_s_anode	0.0859[V]	0.0859 V	Anode potential
phi_s_cathode	-0.0859[V]	-0.0859 V	Cathode potential
z_net	2[1]	2	Net species charge

Name:

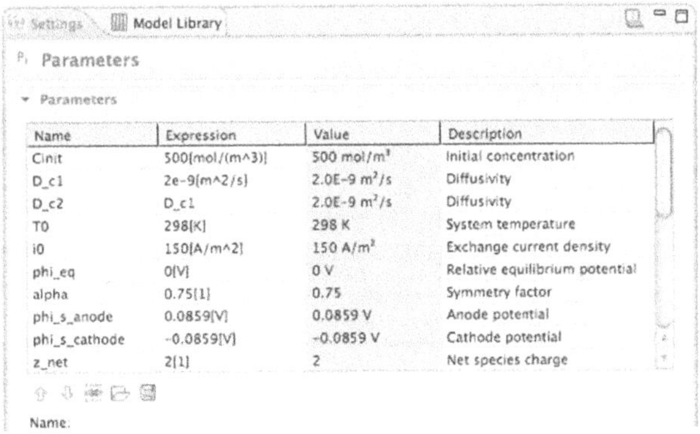

FIGURE 8.6 Settings – Parameters – Parameters Edit Window First 10 Parameters.

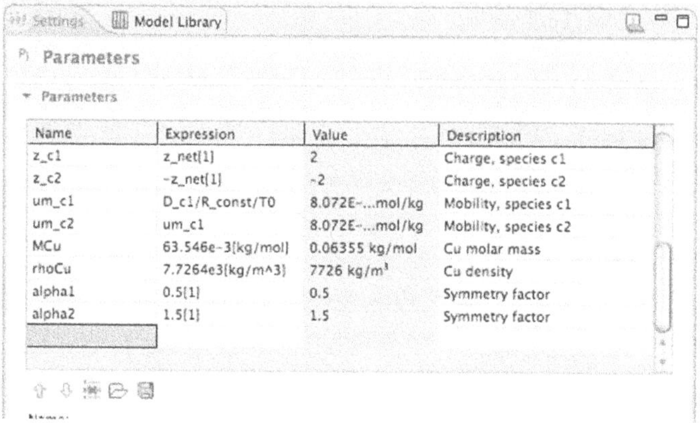

FIGURE 8.7 Settings - Parameters - Parameters Edit Window Second 8 Parameters.

Figure 8.7 shows the Settings – Parameters – Parameters edit window second 8 parameters.

Right-Click > Model Builder – Model 1 – Definitions.

Select > Variables from the Pop-up menu.

In the Settings – Variables – Variables edit window,

Enter the variables shown in Table 8.2.

See Figure 8.8.

TABLE 8.2 Variables

Name	**Expression**	**Description**
i_anode	i0*(exp(alpha*z_net*F_const/R_const/T0*(phi_s_anode-V-phi_eq))-c1/Cinit*exp(-alpha1*z_net*F_const/R_const/T0*(phi_s_anode-V-phi_eq)))	Anode current density
i_cathode	i0*(exp(alpha2*z_net*F_const/R_const/T0*(phi_s_cathode-V-phi_eq))-c1/Cinit*exp(-alpha1*z_net*F_const/R_const/T0*(phi_s_cathode-V-phi_eq)))	Cathode current density
growth	i_cathode*MCu/rhoCu/z_net/F_const	Cathode deposition rate
n_growth	i_anode*MCu/rhoCu/z_net/F_const	Anode deposition rate
displ_x	abs(x-X)	Absolute displacement in x direction

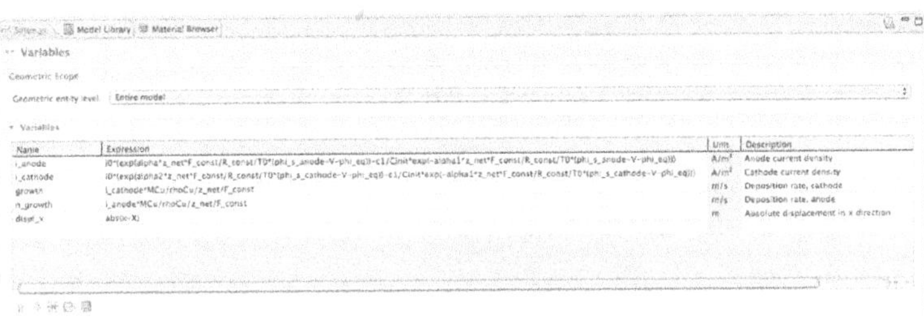

FIGURE 8.8 Settings - Variables - Variables Edit Window Filled.

Figure 8.8 shows the Settings –Variables –Variables edit window filled.

Geometry

Rectangle 1

Right-Click > Model Builder – Model 1 – Geometry 1.

Select > Rectangle from the Pop-up menu.

Enter > 1.6e-5 in the Settings – Rectangle – Size –Width entry window.

Enter > 3e-5 in the Settings – Rectangle – Size – Height entry window.

Enter > -0.8e-5 in the Settings – Rectangle – Position – x entry window.

Enter > 1e-5 in the Settings – Rectangle – Position – y entry window.

Click > Build All.

Rectangle 2

Right-Click > Model Builder – Model 1 – Geometry 1.

Select > Rectangle from the Pop-up menu.

Enter > 0.4e-5 in the Settings – Rectangle – Size – Width entry window.

Enter > 1e-5 in the Settings – Rectangle – Size – Height entry window.

Enter > -0.2e-5 in the Settings – Rectangle – Position – x entry window.

Enter > 0 in the Settings – Rectangle – Position – y entry window.

Click > Build All.

Click > Zoom Extents.

Union 1

Right-Click > Model Builder – Model 1 – Geometry 1,

Select > Boolean Operations – Union from the Pop-up menu.

Shift-Click > Rectangles r1 & r2 in the Graphics window.

Click > Add to Selection in Settings – Union – Union.

Uncheck > Keep Interior Boundaries.

Click > Build All.

See Figure 8.9.

Figure 8.9 shows the Initial Model Geometry in the Graphics window.

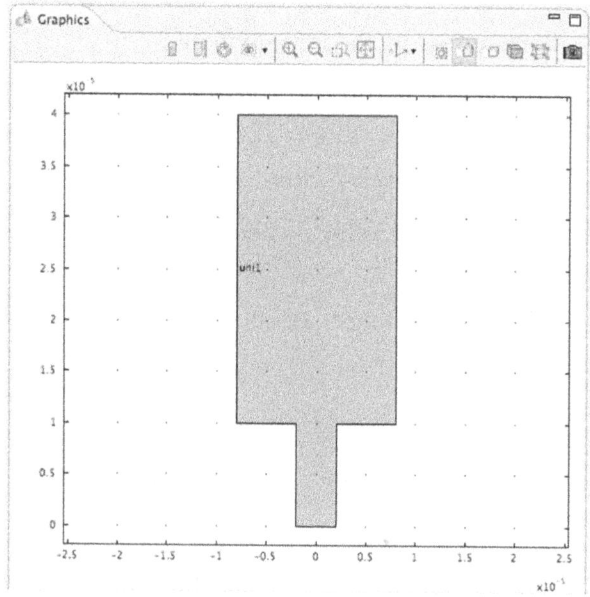

FIGURE 8.9 Initial Model Geometry in the Graphics Window.

Fillet 1

NOTE *Adding a Fillet to the cathode corners (rounding) reduces some of the potential calculational difficulties that might have arisen had the corners remained sharp.*

Right-Click > Model Builder – Model 1 – Geometry 1.

Select > Fillet from the Pop-up menu.

Shift-Click > Vertices 3, 4, 5, & 6 only in the Graphics window.

Click > Add to Selection in Settings – Fillet – Points.

Enter > 7.5e-7 in the Settings – Fillet – Radius edit window.

Click > Build Selected.

See Figure 8.10.

Figure 8.10 shows the Final Model Geometry in the Graphics window.

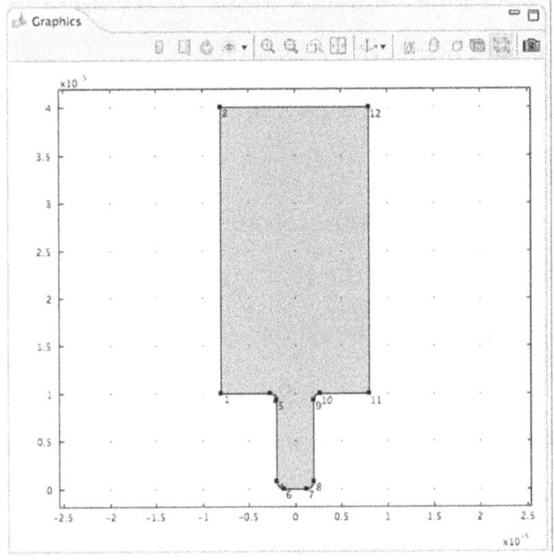

FIGURE 8.10 Final Model Geometry in the Graphics Window.

Selection 1

Right-Click > Model Builder – Model 1 – Definitions.

Select > Selection from the Pop-up menu.

Right-Click > Model Builder – Model 1 – Definitions – Selection 1.

Select > Rename from the Pop-up menu.

Enter anode in the Rename Selection edit window.

Click > OK.

Click > Settings – Selection – Geometric Scope – Geometric entity level.

Select > Boundary from the Pop-up menu.

Select > Boundary 3 in the Graphics window.

Click > Add to Selection in Settings – Selection – Geometric Scope.

Selection 2

Right-Click > Model Builder – Model 1 – Definitions.

Select > Selection from the Pop-up menu.

Right-Click > Model Builder – Model 1 – Definitions – Selection 2.

Select > Rename from the Pop-up menu.

Enter cathode in the Rename Selection edit window.

Click > OK.

Click > Settings – Selection – Geometric Scope – Geometric entity level.

Select > Boundary from the Pop-up menu.

Click > All Boundaries checkbox.

Shift-Click > Boundaries 1, 3, & 8 in the Graphics window.

Click > Remove from Selection in Settings – Selection – Geometric Scope.

See Figure 8.11.

Figure 8.11 shows the cathode boundary selection.

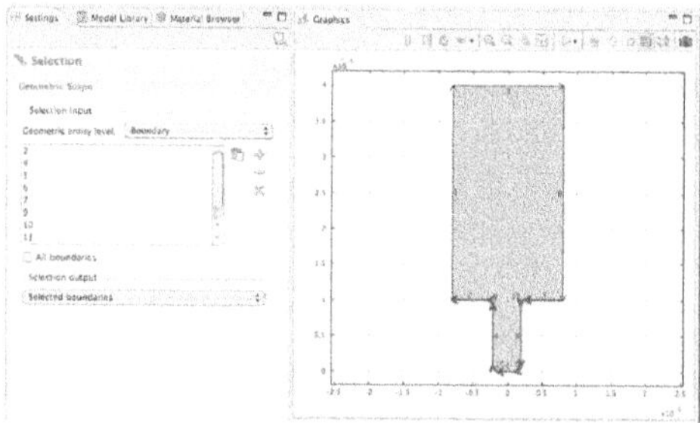

FIGURE 8.11 Cathode Boundary Selection.

Nernst-Planck Equations (chnp)

Click > Model Builder – Model 1 – Twistie for the Nernst-Planck Equations (chnp).

Click > Model Builder – Model 1 – Nernst-Planck Equations (chnp) – Convection, Diffusion, and Migration 1.

Verify that Domain 1 is Selected in Settings – Convection, Diffusion, and Migration – Domains – Selection.

Enter > D_c1 in the Settings – Convection, Diffusion, and Migration – Diffusion D_{c1} edit window.

Enter > D_c2 in the Settings – Convection, Diffusion, and Migration – Diffusion D_{c2} edit window.

Enter > um_c1 in the Settings – Convection, Diffusion, and Migration – Migration in Electric Field $u_{m,c1}$ edit window.

Enter > um_c2 in the Settings – Convection, Diffusion, and Migration – Migration in Electric Field $u_{m,c2}$ edit window.

Enter > z_c1 in the Settings – Convection, Diffusion, and Migration – Migration in Electric Field z_{c1} edit window.

Enter > z_c2 in the Settings – Convection, Diffusion, and Migration – Migration in Electric Field z_{c2} edit window.

See Figure 8.12.

FIGURE 8.12 Nernst-Planck Equation Parameter Filled Edit Windows.

Figure 8.12 shows the Nernst-Planck Equation Parameter filled edit windows.

Initial Values 1

Click > Model Builder – Model 1 – Nernst-Planck Equations (chnp) – Initial Values 1.

NOTE *Since there is only one (1) domain in this model, COMSOL will automatically select domain 1. However, it is important that the modeler always verify that the correct domain for that part of the model has been selected. In this case, the correct domain is domain 1.*

Enter > Cinit in the Settings – Initial Values – Initial Values Concentration c2 edit window.

Enter > (–0.01+0.02*Y/(4e-5[m]))*1[V] in the Settings – Initial Values – Initial Values Electric potential V edit window.

NOTE *The equation (–0.01+0.02*Y/(4e-5[m]))*1[V] inserts a linear potential gradient as the initial condition between Y = 0 and Y = 4.0e-5, with values that range from –0.01 to 0.01.*

See Figure 8.13.

Figure 8.13 shows the Settings – Initial Values edit windows.

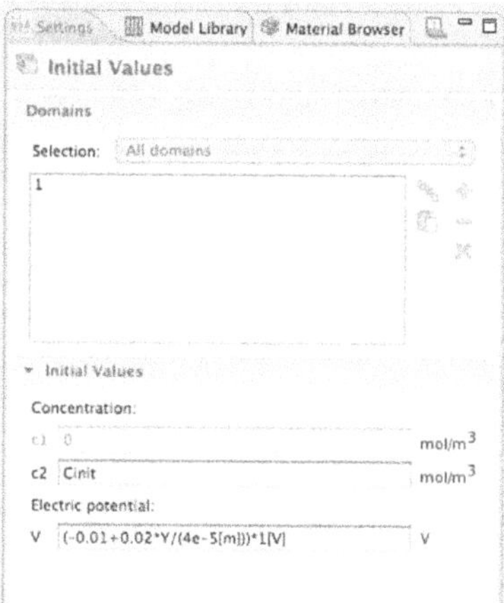

FIGURE 8.13 Settings – Initial Values Edit Windows.

Current Density 1

Right-Click > Model Builder – Model 1 – Nernst-Planck Equations (chnp),

Select > Current Density from the Pop-up menu.

Click > Model Builder – Model 1 – Nernst-Planck Equations (chnp) – Current Density 1.

Click > Settings – Current Density – Boundaries – Selection.

Select > Cathode from the Pull-down menu.

Enter > i_cathode in the Settings – Current Density – Inward Current Density i_0 edit window.

See Figure 8.14.

Figure 8.14 shows the Cathode – Settings – Current Density edit windows.

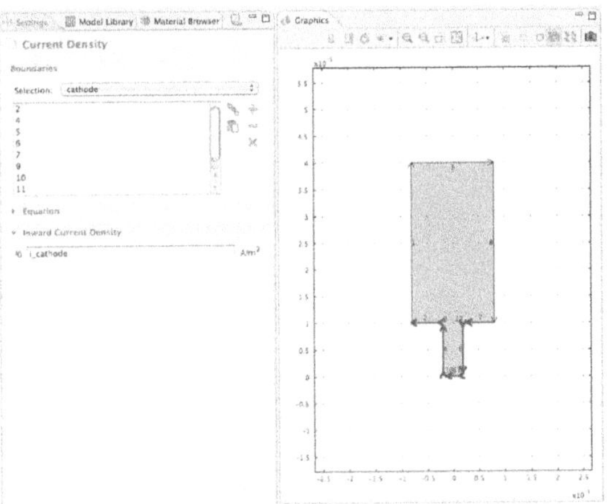

FIGURE 8.14 Cathode - Settings - Current Density Edit Windows.

Current Density 2

Right-Click > Model Builder – Model 1 – Nernst-Planck Equations (chnp).

Select > Current Density from the Pop-up menu.

Click > Model Builder – Model 1 – Nernst-Planck Equations (chnp) – Current Density 2.

Click > Settings – Current Density – Boundaries – Selection.

Select > Anode from the Pull-down menu.

Enter > i_anode in the Settings – Current Density – Inward Current Density i_0 edit window.

See Figure 8.15.

Figure 8.15 shows the Anode – Settings – Current Density edit windows.

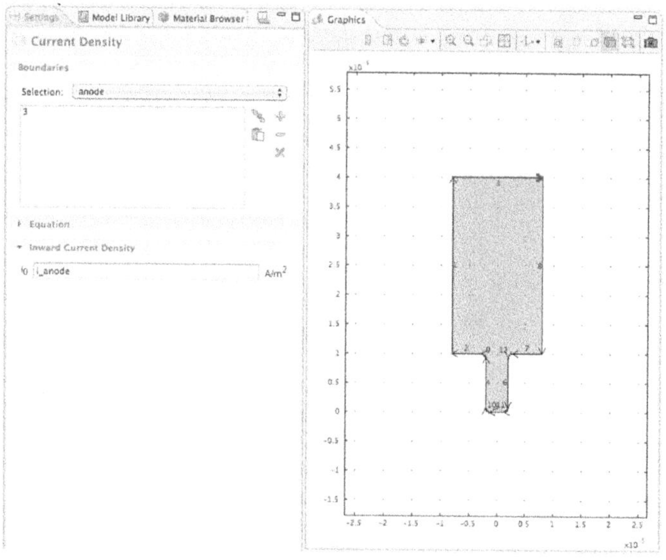

FIGURE 8.15 Anode - Settings - Current Density Edit Windows.

Moving Mesh

Click > Model Builder – Model 1 – Moving Mesh.

Click > Geometry shape order in Settings – Moving Mesh – Frame Settings.

Select > 1 from the Pop-up menu.

See Figure 8.16.

Figure 8.16 shows the Geometry shape order in Settings – Moving Mesh – Frame Settings.

NOTE *By selecting 1, the modeler reduces the probability of distorted shape elements occurring in the mesh.*

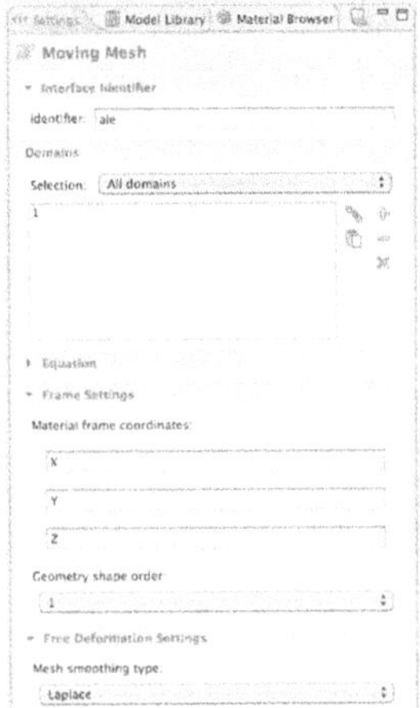

FIGURE 8.16 Geometry Shape Order in Settings - Moving Mesh - Frame Settings.

Moving Mesh – Free Deformation 1

Right-Click > Model Builder – Model 1 – Moving Mesh (ale).

Select > Free Deformation.

Click > Free Deformation 1.

Select > Domain 1 in the Graphics window.

Click > Add to Selection in Settings – Free Deformation – Domains.

See Figure 8.17.

Figure 8.17 shows the Domain Selection in Settings – Free Deformation.

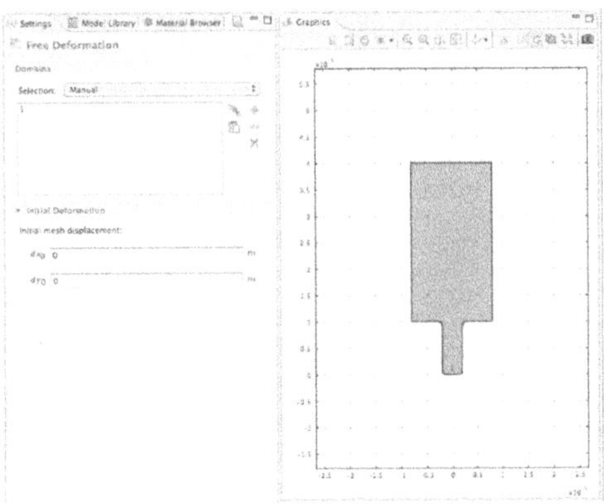

FIGURE 8.17 Domain Selection in Settings - Free Deformation.

Moving Mesh – Prescribed Mesh Velocity 1

Right-Click > Model Builder – Model 1 – Moving Mesh (ale).

Select > Prescribed Mesh Velocity.

Click > Prescribed Mesh Velocity 1.

Click > Selection in Settings – Prescribed Mesh Velocity – Boundaries.

Select > Anode from the Pull-down menu.

Enter > n_growth*chnp.nx in the Settings – Prescribed Mesh Velocity – Prescribed Mesh Velocity v_x edit window.

Enter > n_growth*chnp.ny in the Settings – Prescribed Mesh Velocity – Prescribed Mesh Velocity v_y edit window.

NOTE *The deposition rate of copper on the anode is determined by the product of the growth rate function (n_growth) and the normal vectors {8.15} (chnp.nx and chnp.ny) as is appropriate.*

See Figure 8.18.

Figure 8.18 shows the Prescribed Mesh Velocity 1 Settings in Settings – Prescribed Mesh Velocity.

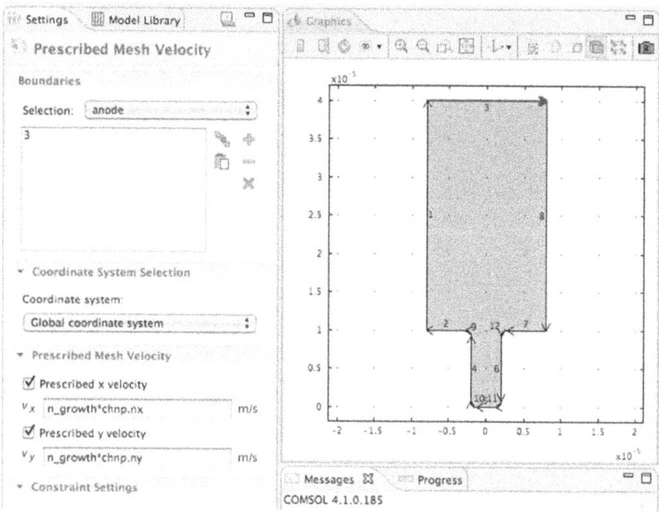

FIGURE 8.18 Prescribed Mesh Velocity 1 Settings in Settings - Prescribed Mesh Velocity.

Moving Mesh – Prescribed Mesh Velocity 2

Right-Click > Model Builder – Model 1 – Moving Mesh (ale).

Select > Prescribed Mesh Velocity.

Click > Prescribed Mesh Velocity 2.

Click > Selection in Settings – Prescribed Mesh Velocity – Boundaries.

Select > Cathode from the Pull-down menu.

Enter > growth*chnp.nx in the Settings – Prescribed Mesh Velocity – Prescribed Mesh Velocity v_x edit window.

Enter > growth*chnp.ny in the Settings – Prescribed Mesh Velocity – Prescribed Mesh Velocity v_y edit window.

NOTE *The deposition rate of copper on the cathode is determined by the product of the growth rate function (growth) and the normal vectors (chnp.nx and chnp.ny) as is appropriate.*

See Figure 8.19.

Figure 8.19 shows the Prescribed Mesh Velocity 2 Settings in Settings – Prescribed Mesh Velocity.

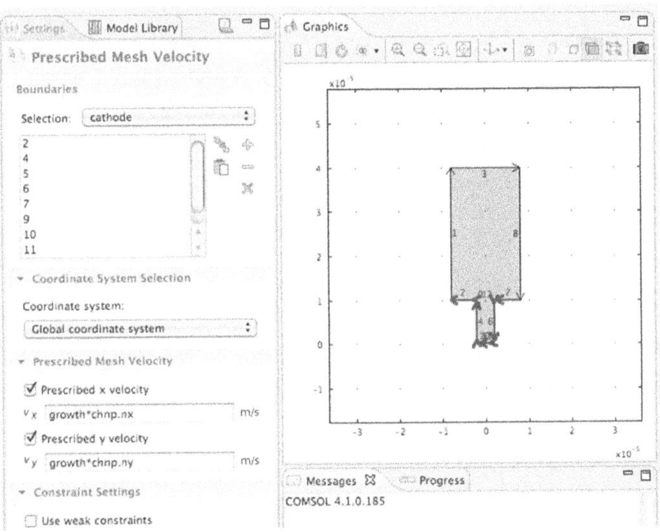

FIGURE 8.19 Prescribed Mesh Velocity 2 Settings in Settings - Prescribed Mesh Velocity.

Moving Mesh – Prescribed Mesh Displacement 2

Right-Click > Model Builder – Model 1 – Moving Mesh (ale).

Select > Prescribed Mesh Displacement.

Click > Prescribed Mesh Displacement 2.

Shift-Click > Boundaries 1 and 8 in the Graphics window.

Click > Add to Selection in Settings – Prescribed Mesh Displacement – Boundaries.

Uncheck > Prescribed y displacement checkbox in Settings – Prescribed Mesh Displacement – Prescribed Mesh Displacement.

See Figure 8.20.

Figure 8.20 shows the Prescribed Mesh Displacement 2 Settings in Settings – Prescribed Mesh Displacement.

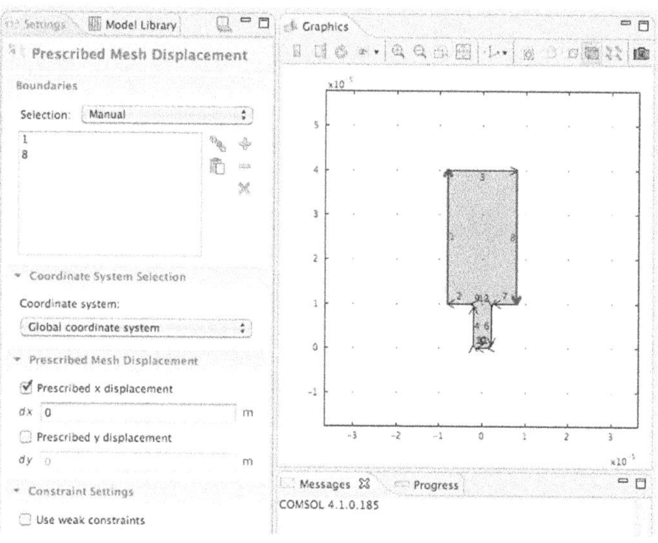

FIGURE 8.20 Prescribed Mesh Displacement 2 Settings in Settings - Prescribed Mesh Displacement.

Mesh 1

Right-Click > Model Builder – Model 1 – Mesh 1.

Select > Edit Physics – Induced Sequence.

Click > Model Builder – Model 1 – Mesh 1 – Size.

Click > Predefined Pull-down list in Settings – Size – Element Size.

Select > Extra fine.

Click > Settings – Size – Element Size Parameters twistie.

Enter > 5.0e-7 in the Settings – Size – Element Size Parameters – Maximum element size edit window.

Enter > 0.9 in the Settings – Size – Element Size Parameters – Resolution of curvature edit window.

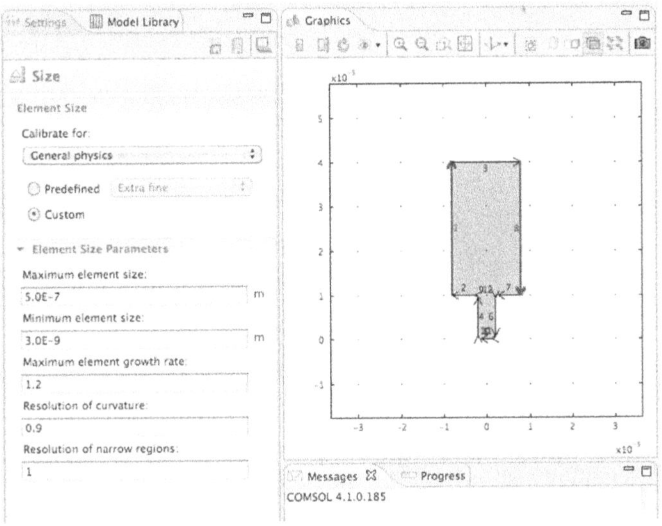

FIGURE 8.21 Settings - Size Parameters.

See Figure 8.21.

Figure 8.21 shows the Settings – Size Parameters.

Click > Build All.

See Figure 8.22.

Figure 8.22 shows the Meshed Model.

NOTE *After the mesh is built, the model should have 5474 elements.*

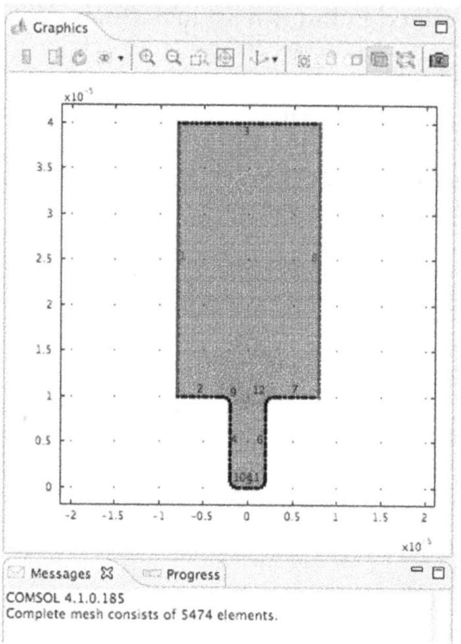

FIGURE 8.22 Meshed Model.

Study 1

Click > Model Builder – Study 1 twistie.

Click > Model Builder – Study 1 – Step 1: Time Dependent.

Click > Range button in Settings – Time Dependent – Study Settings.

Enter > Start = 0, Stop = 5.0, and Step = 0.1.

Click > Replace button.

See Figure 8.23.

Figure 8.23 Settings – Time Dependent – Study Settings edit windows.

In Model Builder, Right-Click Study 1 > Select > Compute.

Computed results, using the default display settings, are shown in Figure 8.24.

Figure 8.24 shows the computed results, using the default display settings.

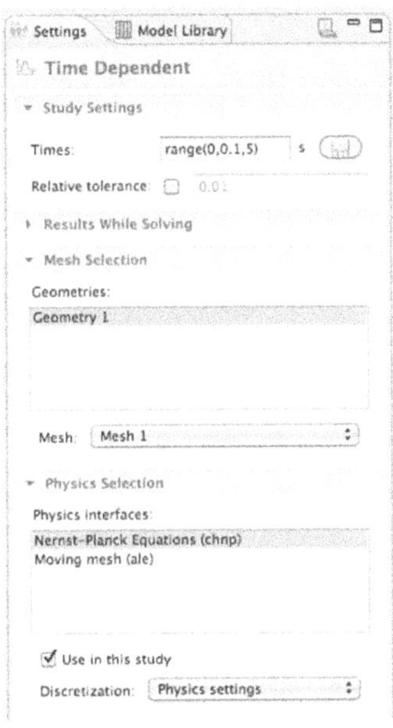

FIGURE 8.23 Settings - Time Dependent - Study Settings Edit Windows.

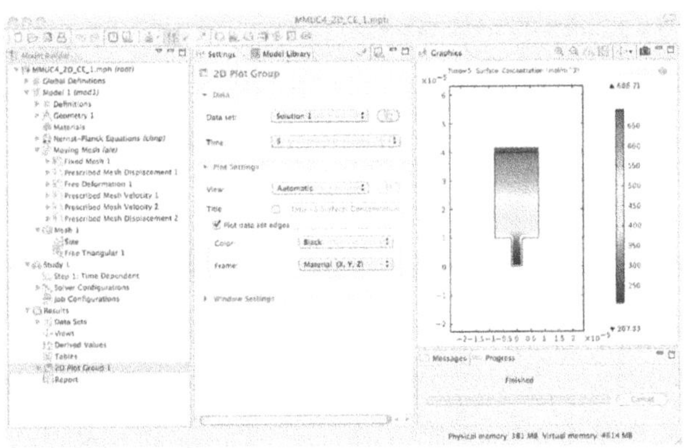

FIGURE 8.24 Computed Results, Using the Default Display Settings.

NOTE *The modeler should note that the computed results, using the default display settings, adequately display the model solution. However, with some additional adjustments in the way that the data are presented, the model solution presentation will be significantly enhanced.*

Results

Click > Model Builder – Results – Data Sets twistie.

Click > Solution 1.

Select > Settings – Solution – Solution – Spatial (x, y, z) from the Frame Pull-down menu.

See Figure 8.25.

Figure 8.25 shows the Settings – Solution – Solution settings.

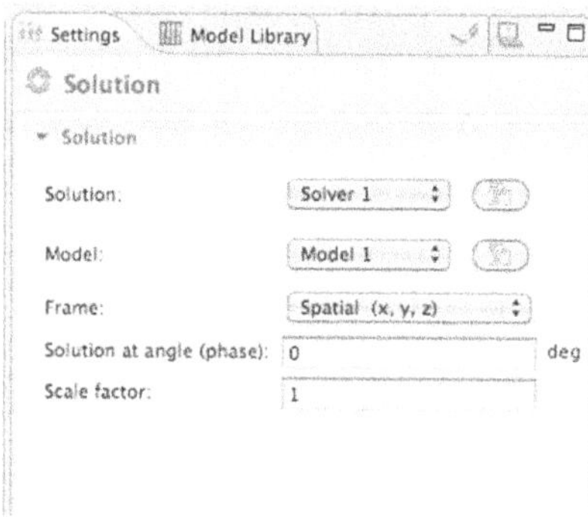

FIGURE 8.25 Settings – Solution – Solution Settings.

2D Plot Group 1

Right-Click > Model Builder – Results – Data Sets.

Select > Solution.

Right-Click > Model Builder – Results – 2D Plot Group 1.

Select > Contour.

Click > Replace Expression in Settings – Contour – Expression.

Select > Nernst-Planck Equations – Electric potential (V) from the Pop-up menu.

Select > Uniform from the Settings – Contour – Color and Style – Coloring Pull-down menu.

Select > Black from the Settings – Contour – Color and Style – Color Pull-down menu.

See Figure 8.26.

Figure 8.26 shows the Settings – Contour settings.

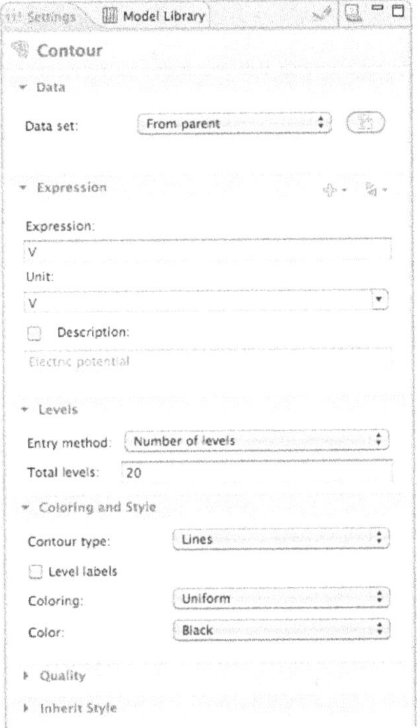

FIGURE 8.26 Settings - Contour Settings.

Right-Click > Model Builder – Results – 2D Plot Group 1.

Select > Streamline.

Click > Replace Expression in Settings – Streamline – Expression.

Select > Nernst-Planck Equations – Species c1 – Total flux from the Pop-up menu.

Select > Magnitude controlled from the Settings – Streamline – Streamline Positioning – Positioning Pull-down menu.

Enter > 15 in the Settings – Streamline – Streamline Positioning – Density edit window.

See Figure 8.27.

Figure 8.27 shows the Settings – Streamline settings.

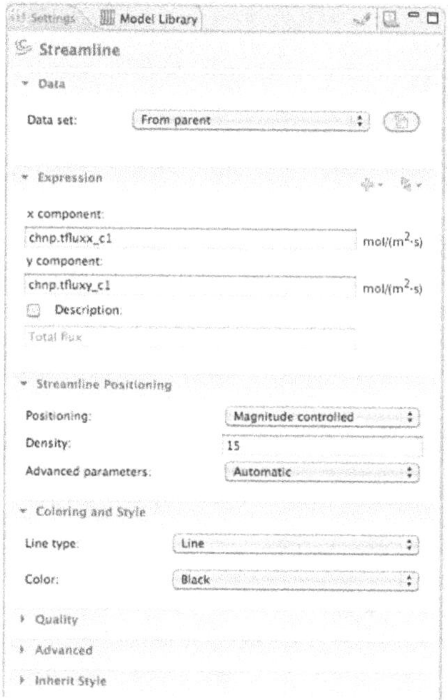

FIGURE 8.27 Settings - Streamline Settings.

Click > Plot.

Click > Zoom Extents.

Right-Click > Model Builder – Results – 2D Plot Group 1.

Select > Line.

Select > Solution 2 from the Settings – Line – Data – Data set Pull-down menu.

Select > Uniform from the Settings – Line – Color and Style – Coloring Pull-down menu.

Select > Black from the Settings – Line – Color and Style – Color Pull-down menu.

See Figure 8.28.

Figure 8.28 shows the Settings – Line settings.

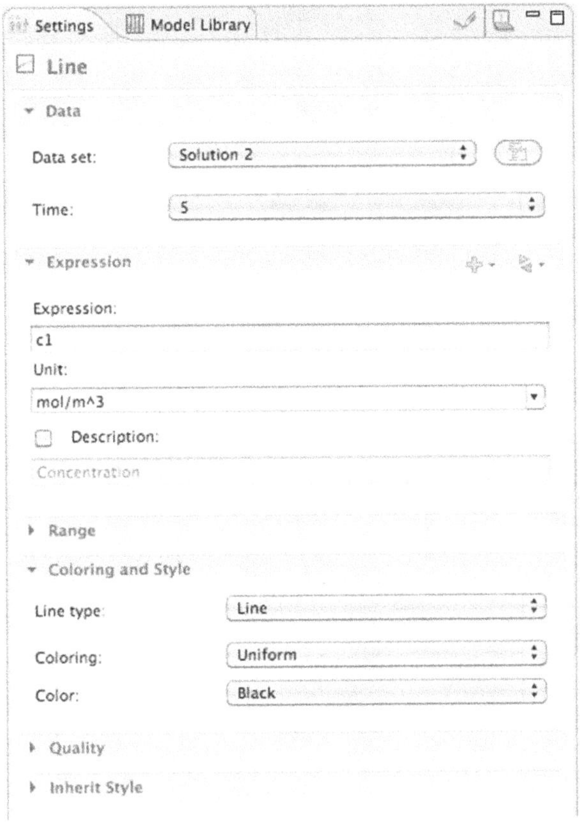

FIGURE 8.28 Settings – Line Settings.

Click > Plot.

Click > Zoom Extents.

See Figure 8.29.

Figure 8.29 shows the Final Copper Deposition Solution Concentration Plot.

More Results

NOTE *Some results from model solution can be displayed to show the thickness of the copper layer (uniformity) as a function of the distance from the bottom of the cathode.*

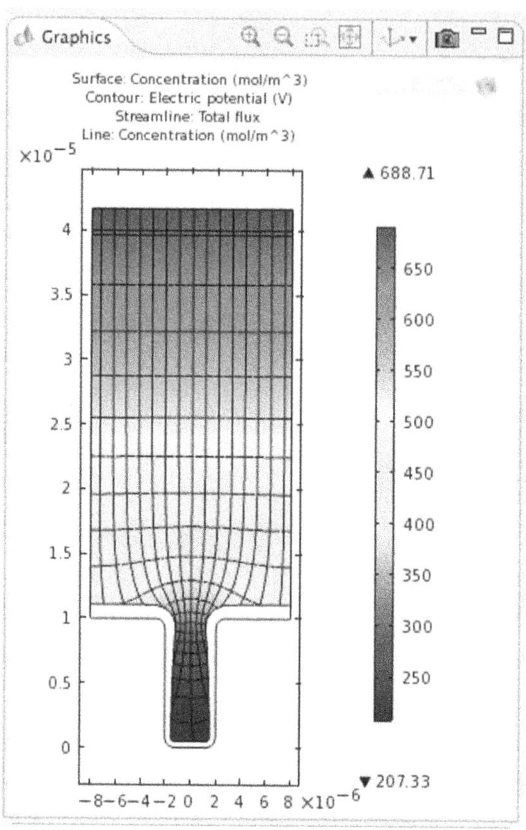

FIGURE 8.29 Final Copper Deposition Solution Concentration Plot.

Right-Click > Model Builder – Results.

Select > 1D Plot Group from the Pop-up menu.

Click > Model Builder – Results – 1D Plot Group 2.

Select > From list in Settings – 1D plot Group – Data – Time selection from the Pop-up menu.

Select > Times 0, 1, 2, 3, 4, 5 from the Settings – 1D plot Group – Data – Times selection list.

Check > Title checkbox, x-axis label checkbox, and the y-axis label checkbox in Settings – 1D plot Group – Plot Settings.

Enter > Thickness of Deposited Copper in the Settings – 1D plot Group – Plot Settings Title edit window.

Enter > Distance from Cavity Bottom in the Settings – 1D plot Group – Plot Settings x- axis label edit window.

Enter > Thickness of Deposited Copper in the Settings – 1D plot Group – Plot Settings y- axis label edit window.

Right-Click > Model Builder – Results – 1D Plot Group 2.

Select > Line Graph from the Pop-up menu.

Select > Boundary 4 only in the Graphics display window.

Click > Add to Selection in Settings – Line Graph – Selection.

Enter > mod1.displ_x in the Settings – Line Graph – Y-Axis Data Expression edit window.

Select > Expression from Settings – Line Graph – X-Axis Data Parameter Pop-up menu.

Enter > Y in the Settings – Line Graph – X-Axis Data Expression edit window.

See Figure 8.30.

FIGURE 8.30 1D Line Graph Settings.

Figure 8.30 shows the 1D Line Graph Settings.

Click > Plot.

See Figure 8.31.

Figure 8.31 shows the Thickness of Deposited Copper Graph.

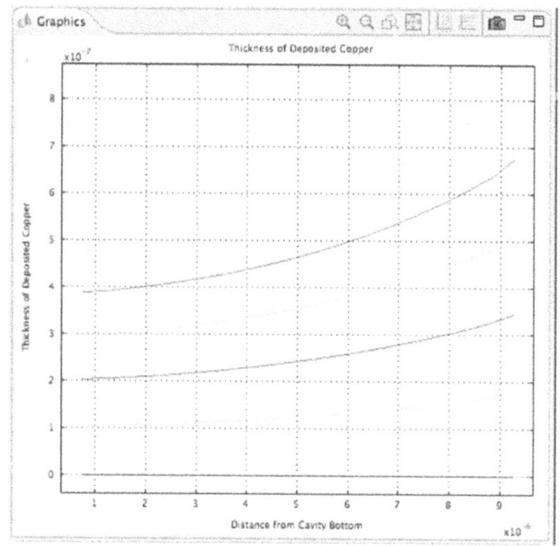

FIGURE 8.31 Thickness of Deposited Copper Graph.

2D Copper Electroplating Model Summary and Conclusions

The 2D Copper Electroplating Model is a powerful modeling tool that can be used to calculate deposition layer thickness in imbedded regions of a block of material. With this model, the modeler can easily vary all of the deposition parameters and optimize the design before the first prototype deposition layer is physically built. The technique has broad application support in industry.

2D Electrocoalescence Oil/Water Separation Model

NOTE

This model is significantly more complex than earlier models. Once built, it will require a relatively long runtime. The model will, however, converge when correctly built. The physics demonstrated in this model is currently widely applied to a diverse collection of practical problems and applied engineering solutions (oil-water separation, meteorological modeling predictions, and many others).

The general definition of the term coalescence specifies the coming together or joining of two or more smaller, similar (compatible) objects (computer code, languages, liquid drops, etc.) to form a larger object {8.16}. In physics, the definition of coalescence has the same implication, however the types of objects considered are restricted to physical objects (liquid drops, particles, etc.) {8.17}.

NOTE *This model introduces calculations employing the Navier-Stokes Laminar Two-Phase Flow Phase Field Interface Equations {8.18} and the Electrostatics Interface {8.19}. In this model, those two areas of physics are coupled through the use of the Maxwell stress tensor {8.20}.*

In the 2D Electrocoalescence Oil/Water Separation Model, the process modeled is the coalescence of water droplets suspended in a flowing oil medium. The solution to the model is obtained by solving the Navier-Stokes Laminar Two-Phase Flow Phase Field Interface Equations and the Electrostatics interface equations simultaneously.

NOTE *In this model, the coalescence of the drops is calculated by tracking the interface between the water droplets and the host oil medium in the presence of an electrostatic field.*

The Navier-Stokes Equations are named after Claude-Louis Navier {8.21} and George Gabriel Stokes. The Navier-Stokes Equations {8.22} apply Newton's Second Law {8.23} to fluid motion with the assumption that the stress is the sum of a viscous term and a pressure term.

Laminar Two-Phase Flow Phase Field Interface Theory

Navier-Stokes Laminar Two-Phase Flow Phase Field Interface equations of fluid motion are as follows:

$$\rho\frac{\partial \mathbf{u}}{\partial t}+\rho(\mathbf{u}\cdot\nabla)\mathbf{u}=\nabla\cdot\left[-p\mathbf{I}+\eta(\nabla\mathbf{u}+(\nabla\mathbf{u})^T)\right]+\mathbf{F}_{ST}+\rho\mathbf{g}+\mathbf{F} \tag{8.9}$$

And:

$$\nabla\cdot\mathbf{u}=0 \tag{8.10}$$

Where: **u** = velocity vector [m/s].

ρ = density [kg/m^3].

p = pressure [Pa].

I = Identity matrix [1] (unitless).

η = dynamic viscosity [Pa·s].

$\mathbf{F}_{ST}$ = surface tension force [N/m].

g = gravity vector [m/s^2].

F = additional volume force [N/m^3].

The phase field method is employed to track the interphase interface (oil/water), as follows:

$$\frac{\partial \phi}{\partial t} + \mathbf{u} \bullet \nabla \phi = \nabla \bullet \frac{3\chi\sigma\varepsilon}{2\sqrt{2}} \nabla \psi \qquad (8.11)$$

And where:

$$\psi = -\nabla \bullet \varepsilon^2 \nabla \phi + (\phi^2 - 1)\phi \qquad (8.12)$$

Where: **u** = velocity vector [m/s].

ϕ = phase field variable (-1 for water and 1 for oil) [1] (unitless).

χ = interface mobility variable [1] (unitless).

ε = numerical parameter for thickness of the interface [m].

$\boldsymbol{\sigma}$ = surface tension force [N/m].

NOTE *The phase field variable varies smoothly from –1 to 1*

See a graphical example of the interphase interface in Figure 8.32.

Figure 8.32 shows an Interphase Interface example.

Electrostatics Interface Theory

When the Electrostatics physics interface is chosen, an equation is set-up for the electric potential [V] as follows:

$$-\nabla \bullet (\varepsilon_0 \varepsilon_r \nabla V) = 0 \qquad (3.13)$$

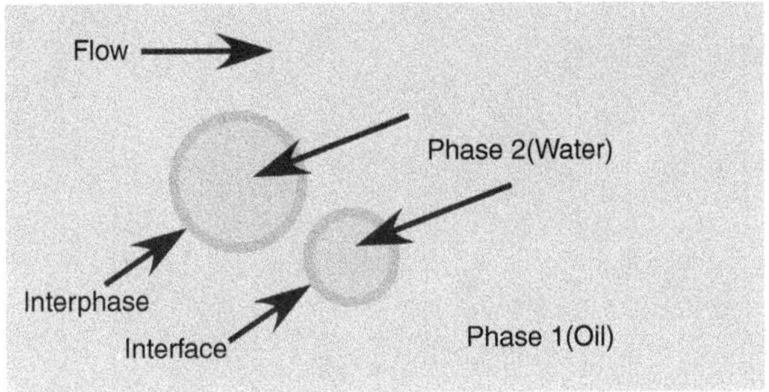

FIGURE 8.32 Interphase Interface Example.

Where: ε_0 = permittivity of vacuum (8.854... E-12) [F/m].
ε_r = relative permittivity.

Coupling Navier-Stokes and Electrostatics

The electrostatic force is given by the divergence of the Maxwell stress tensor:

$$\mathbf{F} = \nabla \cdot \mathbf{T} \tag{8.14}$$

And:

$$\mathbf{T} = \mathbf{E}\mathbf{D}^T - \frac{1}{2}(\mathbf{E} \cdot \mathbf{T})\mathbf{I} \tag{8.15}$$

Where: **E** = Electric field vector [V/m].
D = Electric displacement field [V/m].

And where:

$$\mathbf{E} = -\nabla V \tag{3.16}$$

And:

$$\mathbf{D} = \varepsilon_0 \varepsilon_r \mathbf{E} \tag{3.17}$$

Since this model is 2D, the Maxwell stress tensor **T** is:

$$\mathbf{T} = \begin{bmatrix} T_{xx} & T_{xy} \\ T_{yx} & T_{yy} \end{bmatrix} \tag{3.18}$$

Where:
$$T_{xx} = \varepsilon_0 \varepsilon_r E_x^2 - \frac{1}{2}\varepsilon_0 \varepsilon_r (E_x^2 + E_y^2) \tag{8.19}$$

$$T_{xy} = \varepsilon_0 \varepsilon_r E_x E_y \tag{8.20}$$

$$T_{yx} = \varepsilon_0 \varepsilon_r E_y E_x \tag{8.21}$$

$$T_{yy} = \varepsilon_0 \varepsilon_r E_y^2 - \frac{1}{2}\varepsilon_0 \varepsilon_r (E_x^2 + E_y^2) \tag{8.22}$$

The relative permittivity, at a given point, is determined by the relative volume fraction of the oil and the water, as follows:

$$\varepsilon_r = \varepsilon_{r1} V_{f1} + \varepsilon_{r2} V_{f2} \tag{8.23}$$

Where: ε_{r1} = relative permittivity of oil.
ε_{r2} = relative permittivity of water.
V_{f1} = volume fraction of oil.
V_{f2} = volume fraction of water.

Building the 2D Electrocoalescence Oil/Water Separation Model

Startup 4.x.

Select > 2D.

Click > Next.

Click > Twistie for AC/DC in the Add Physics window.

Click > Electrostatics (es).

Click > Add Selected.

Click > Twistie for Fluid Flow in the Add Physics window.

Click > Twistie for Multiphase Flow in the Add Physics window.

Click > Twistie for Two-Phase Flow, Phase Field in the Add Physics window.

Click > Laminar Two-Phase Flow, Phase Field (tpf).

Click > Add Selected.

Click > Next.

Select > Transient with Initialization in the Select Study Type window.

Click > Finish (Flag).

Click > Save As,

Enter MMUC4_2D_EcOW_1.mph.

See Figure 8.33.

Figure 8.33 shows the Desktop Display for the MMUC4_2D_EcOW_1.mph model.

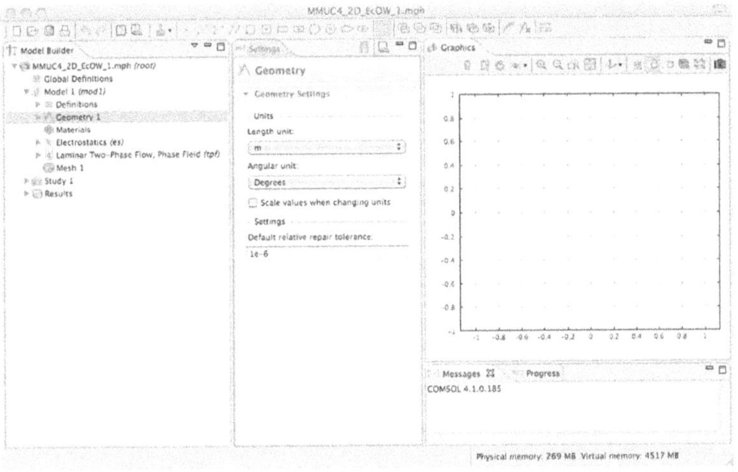

FIGURE 8.33 Desktop Display for the MMUC4_2D_EcOW_1.mph Model.

Geometry 1

In Model Builder – Model 1 (mod1),

Click > Model Builder – Model 1 (mod1) – Geometry 1.

Click > Settings – Geometry – Geometry Settings – Length unit.

Select > mm from the Pull-down menu.

Rectangle 1

In Model Builder – Model 1 (mod1),

Right-Click > Model Builder – Model 1 (mod1) – Geometry 1.

Select > Rectangle from the Pop-up menu.

Enter > 30[mm] in the Settings – Rectangle – Width entry window.

Enter > 10[mm] in the Settings – Rectangle – Height entry window.

Click > Build All.

See Figure 8.34.

Figure 8.34 shows the Settings – Rectangle entry window and built rectangle.

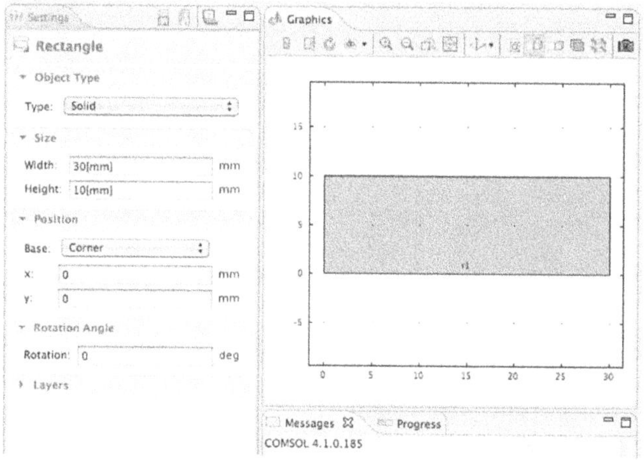

FIGURE 8.34 Settings - Rectangle Entry Window and Built Rectangle.

NOTE *The remaining portions of the geometry will be built without explicitly showing the results incrementally. The total result will be shown, once completed.*

Circle 1

In Model Builder – Model 1 (mod1),

Right-Click > Model Builder – Model 1 (mod1) – Geometry 1.

Select > Circle from the Pop-up menu.

Enter > 1.6[mm] in the Settings – Circle – Size – Radius entry window.

Enter > 4[mm] in the Settings – Circle – Position x entry window.

Enter > 6[mm] in the Settings – Circle – Position y entry window.

Click > Build All.

Circle 2

In Model Builder – Model 1 (mod1),

Right-Click > Model Builder – Model 1 (mod1) – Geometry 1.

Select > Circle from the Pop-up menu.

Enter > 1.2[mm] in the Settings – Circle – Size – Radius entry window.

Enter > 7[mm] in the Settings – Circle – Position x entry window.

Enter > 3.5[mm] in the Settings – Circle – Position y entry window.

Click > Build All.

See Figure 8.35.

Figure 8.35 shows the Electrocoalescence Built Geometry.

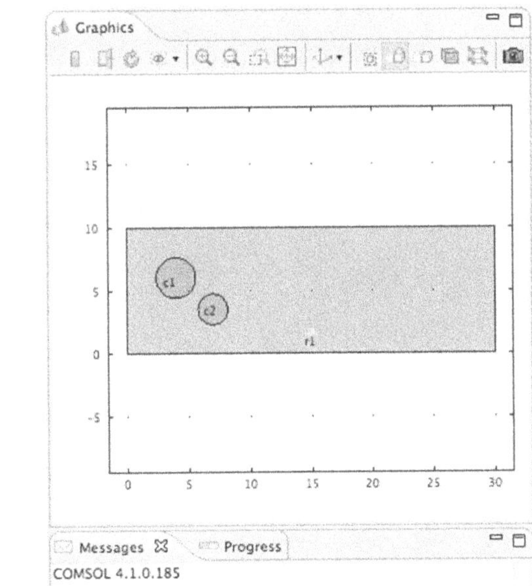

FIGURE 8.35 Electrocoalescence Built Geometry.

Definitions

Selection 1

In Model Builder – Model 1 (mod1),

Right-Click > Model Builder – Model 1 (mod1) – Definitions.

Select > Selection from the Pop-up menu.

Right-Click > Selection 1.

Select > Rename from the Pop-up menu.

Enter > Outlet in the New Name edit window.

Click > OK.

Click > Settings – Selection – Geometric Scope – Geometric entity level.

Select > Boundary from the Pull-down menu.

Click > Boundary 4 in the Graphics window.

Click > Settings – Selection – Geometric Scope – Add to Selection.

See Figure 8.36.

Figure 8.36 shows the Outlet Boundary Selection.

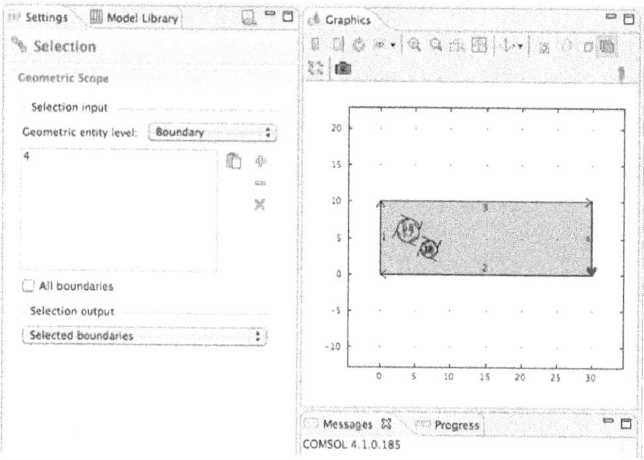

FIGURE 8.36 Outlet Boundary Selection.

Selection 2

In Model Builder – Model 1 (mod1),

Right-Click > Model Builder – Model 1 (mod1) – Definitions.

Select > Selection from the Pop-up menu.

Right-Click > Selection 2.

Select > Rename from the Pop-up menu.

Enter > Inlet in the New Name edit window.

Click > OK.

Click > Settings – Selection – Geometric Scope – Geometric entity level.

Select > Boundary from the Pull-down menu.

Click > Boundary 1 in the Graphics window.

Click > Settings – Selection – Geometric Scope – Add to Selection.

See Figure 8.37.

Figure 8.37 shows the Inlet Boundary Selection.

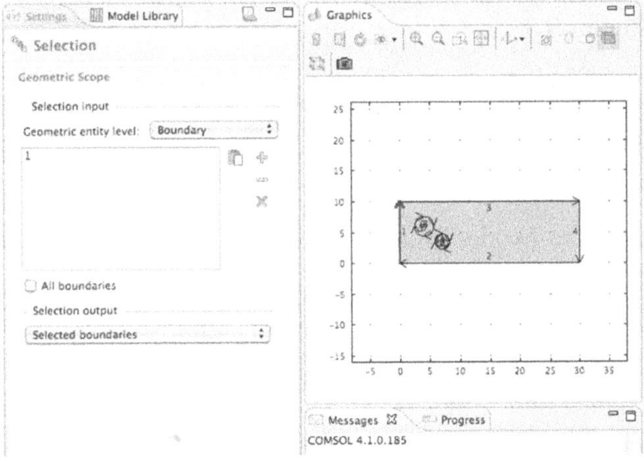

FIGURE 8.37 Inlet Boundary Selection.

Selection 3

In Model Builder – Model 1 (mod1),

Right-Click > Model Builder – Model 1 (mod1) – Definitions.

Select > Selection from the Pop-up menu.

Right-Click > Selection 3.

Select > Rename from the Pop-up menu.

Enter > Oil/Water Interface in the New Name edit window.

Click > OK.

Click > Settings – Selection – Geometric Scope – Geometric entity level.

Select > Boundary from the Pull-down menu.

Click > Settings – Selection – Geometric Scope – All boundaries.

Click > Settings – Selection – Geometric Scope – All boundaries, to uncheck the checkbox.

Shift-Click > Boundaries 1, 2, 3, 4.

Click > Settings – Selection – Geometric Scope – Remove from Selection.

NOTE *That leaves Boundaries 5-12 remaining in the Graphics edit window.*

See Figure 8.38.

Figure 8.38 shows the Oil/Water Interface Boundary Selection.

Global Definitions

Parameters

Right-Click> Model Builder – MMUC4_2D_EcOW_1 – Global Definitions.

Select > Parameters from the Pop-up menu.

Enter the parameters shown in Table 8.3.

NOTE *Numerical parameter values are followed by [units] except when the numerical value is dimensionless. Then the numerical parameter is followed by [1].*

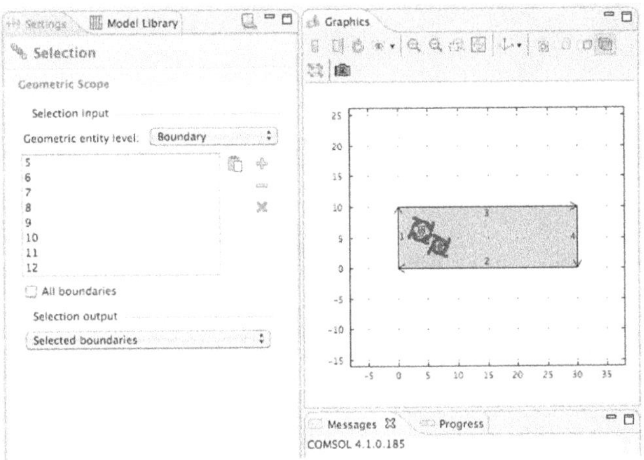

FIGURE 8.38 Oil/Water Interface Boundary Selection.

TABLE 8.3 Parameters

Name	Expression	Description
rho_water	1e3[kg/m^3]	Density, water
mu_water	1.002e-3[Pa*s]	Dynamic viscosity, water
perm_water	8.0e1[1]	Permittivity, water
rho_oil	8.84e2[kg/m^3]	Density, oil
mu_oil	4.74e-1[Pa*s]	Dynamic viscosity, oil
perm_oil	2.2[1]	Permittivity, oil

See Figure 8.39.

Figure 8.39 shows the Parameters entry windows.

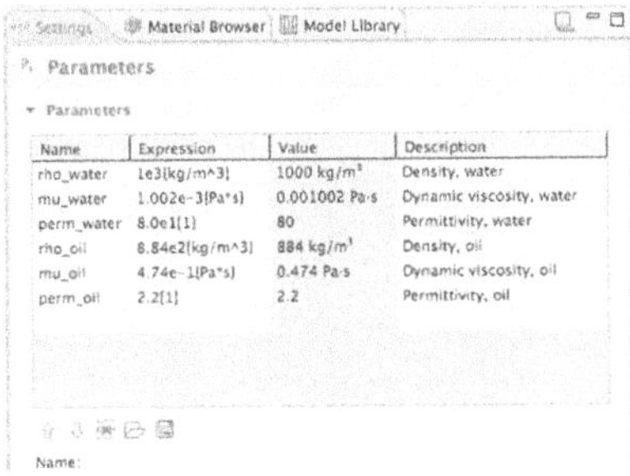

FIGURE 8.39 Parameters Entry Windows.

Definitions

Variables

Right-Click> Model Builder – MMUC4_2D_EcOW_1 – Model 1 – Definitions.

Select > Variables from the Pop-up menu.

Enter the variables shown in Table 8.4.

TABLE 8.4 Variables

Name	Expression	Description
epsilon_r	tpf.Vf1*perm_water+tpf.Vf2*perm_oil	Phase dependent permittivity
Tem11	−epsilon0_const*epsilon_r/2*(es.Ex^2+es.Ey^2) +epsilon0_const*epsilon_r*es.Ex^2	Maxwell stress tensor, 11-component
Tem22	−epsilon0_const*epsilon_r/2*(es.Ex^2+es.Ey^2) +epsilon0_const*epsilon_r*es.Ey^2	Maxwell stress tensor, 22-component
Tem12	epsilon0_const*epsilon_r*es.Ex*es.Ey	Maxwell stress tensor 12-component
Tem21	epsilon0_const*epsilon_r*es.Ex*es.Ey	Maxwell stress tensor, 21-component
Fx	d(Tem11, x)+d(Tem12, y)	Force, x-component
Fy	d(Tem21, x)+d(Tem22, y)	Force, y-component

See Figure 8.40.

Figure 8.40 shows the Variables entry windows.

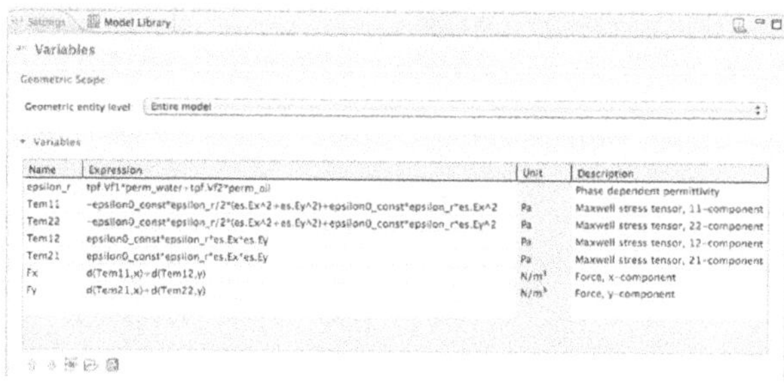

FIGURE 8.40 Variables Entry Windows.

Materials

Water

Right-Click > Model Builder – Model 1 – Materials.

Select > Material from the Pop-up menu.

Click > Domain 1 in the Graphics window.

Click > Remove from Selection in Settings – Material – Geometric Scope.

Right-Click > Material 1.

Select > Rename from the Pop-up menu.

Enter > Water in the New Name edit window.

Click > OK.

Enter > perm_water in the Settings – Material – Material Contents Relative permittivity edit window.

Enter > rho_water in the Settings – Material – Material Contents Density edit window.

Enter > mu_water in the Settings – Material – Material Contents Dynamic viscosity edit window.

Click > Tab.

See Figure 8.41.

Figure 8.41 shows the Water properties entry windows.

Oil

Right-Click> Model Builder – Model 1 – Materials.

Select > Material from the Pop-up menu.

Right-Click > Material 2.

Select > Rename from the Pop-up menu.

Enter > Oil in the New Name edit window.

Click > OK.

Shift-Click > Domain 2 & 3 in the Graphics window.

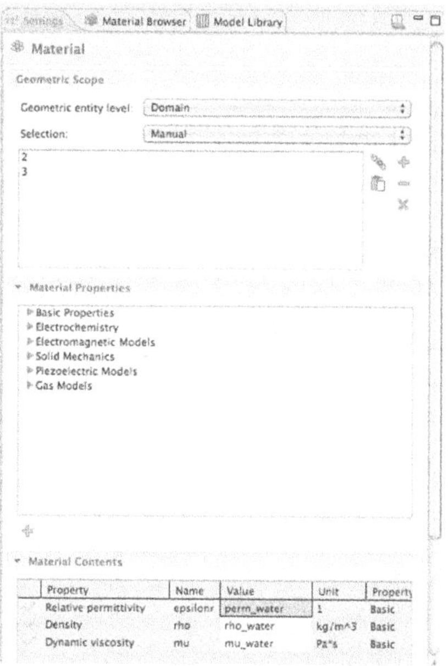

FIGURE 8.41 Water Properties Entry Windows.

Click > Add to Selection in Settings – Material – Geometric Scope.

Enter > perm_oil in the Settings – Material – Material Contents Relative permittivity edit window.

Enter > rho_oil in the Settings – Material – Material Contents Density edit window.

Enter > mu_oil in the Settings – Material – Material Contents Dynamic viscosity edit window.

Click > Tab.

See Figure 8.42.

Figure 8.42 shows the Oil properties entry windows.

Electrostatics (es) Interface

NOTE *The Two-Phase Flow Interface (tpf) employs linear basis functions by default. Thus there is no gain in accuracy level by using quadratic basis functions in the Electrostatics (es) Interface. Thus the basis functions for the Electrostatics (es) Interface needs to be reset to linear basis functions.*

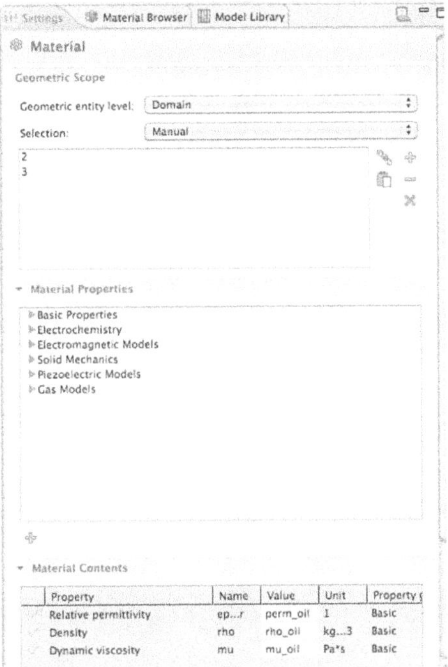

FIGURE 8.42 Oil Properties Entry Windows.

Click> Model Builder – Model 1 – Electrostatics (es).

Click > Settings – Electrostatics – Discretization twistie.

Click > Settings – Electrostatics – Discretization – Electric potential.

Select > Linear from the Pull-down menu.

Charge Conservation 1

Click > Model Builder – Model 1 – Electrostatics (es) twistie.

Click> Model Builder – Model 1 – Electrostatics (es) – Charge Conservation 1.

Click > Settings – Charge Conservation – Electric Field twistie.

Click > Settings – Charge Conservation – Electric Field – Relative permittivity.

Select > User defined from the Pull-down menu.

Enter > epsilon_r in the Relative permittivity edit window.

See Figure 8.43.

Figure 8.43 shows the Charge Conservation Relative Permittivity entry window.

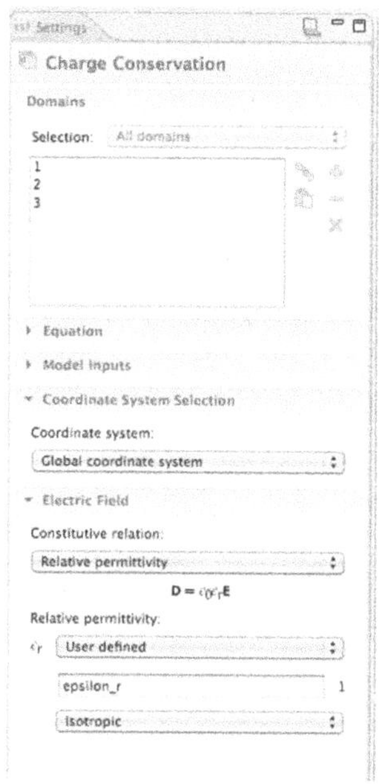

FIGURE 8.43 Charge Conservation Relative Permittivity Entry Window.

Electric Potential 1

Right-Click > Model Builder – Model 1 – Electrostatics (es).

Select > Electric Potential from the Pop-up menu.

Click > Model Builder – Model 1 – Electrostatics – Electric Potential 1.

Click > Boundary 3 in the Graphics window.

Click > Add to Selection in the Settings – Electric potential – Boundaries edit window.

Enter > 5e3[V] in the Settings – Electric Potential – Electric Potential – Voltage edit window.

See Figure 8.44.

Figure 8.44 shows the Electric Potential 1 entry window and Boundary selection.

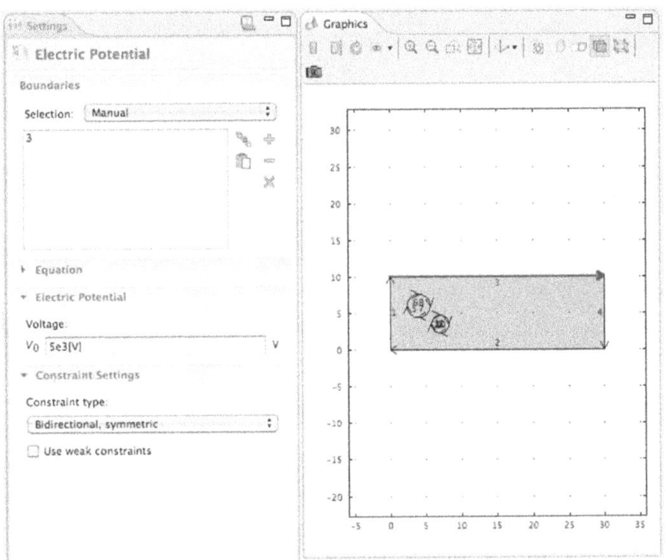

FIGURE 8.44 Electric Potential 1 Entry Window and Boundary Selection.

Electric Potential 2

Right-Click > Model Builder – Model 1 – Electrostatics (es).

Select > Electric Potential from the Pop-up menu.

Click > Model Builder – Model 1 – Electrostatics – Electric Potential 2.

Click > Boundary 2 in the Graphics window.

Click > Add to Selection in the Settings – Electric potential – Boundaries edit window.

Enter > 0[V] in the Settings – Electric Potential – Electric Potential – Voltage edit window.

See Figure 8.45.

Figure 8.45 shows the Electric Potential 2 entry window and Boundary selection.

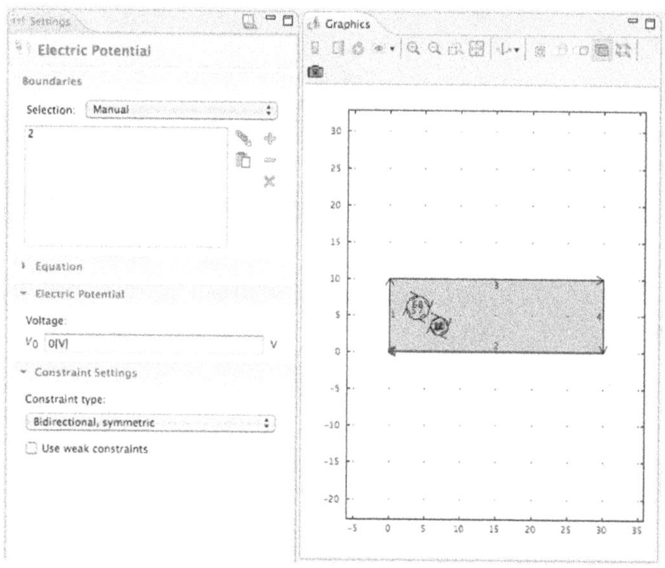

FIGURE 8.45 Electric Potential 2 Entry Window and Boundary Selection.

Laminar Two-Phase Flow, Phase Field (tpf) Interface

Click > Model Builder – Model 1 – Laminar Two-Phase Flow, Phase Field (tpf) twistie.

Click > Model Builder – Model 1 – Laminar Two-Phase Flow, Phase Field (tpf) – Fluid Properties 1.

Fluid 1

Click > Settings – Fluid Properties – Fluid 1 Properties twistie.

Click > Settings – Fluid Properties – Fluid 1 Properties – Fluid 1.

Select > Water (mat1) from the Pull-down menu.

Fluid 2

Click > Settings – Fluid Properties – Fluid 2 Properties twistie.

Click > Settings – Fluid Properties – Fluid 2 Properties – Fluid 2.

Select > Oil (mat2) from the Pull-down menu.

Surface Tension

Click > Settings – Fluid Properties – Surface Tension twistie.

Click > Settings – Fluid Properties – Surface Tension – Surface tension coefficient.

Select > User defined from the Pull-down menu.

Enter > 3.1e-2[N/m] in the Surface tension coefficient edit window.

See Figure 8.46.

Figure 8.46 shows the Fluid Properties and Surface Tension entry windows.

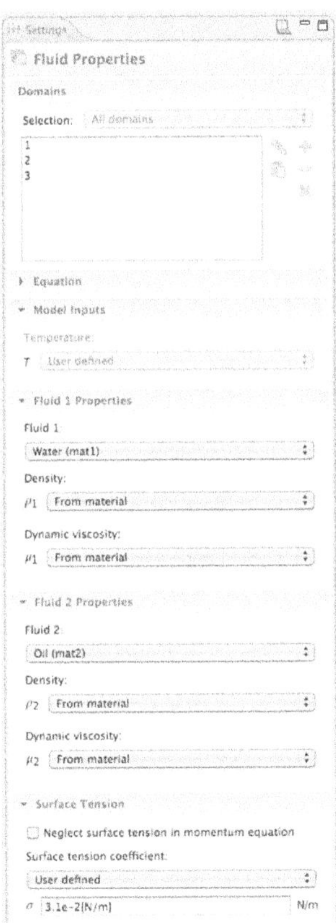

FIGURE 8.46 Fluid Properties and Surface Tension Entry Windows.

Initial Values 1

NOTE *The Initial Values 1 settings do not need to be changed.*

Initial Values 2

Right-Click > Model Builder – Model 1 – Laminar Two-Phase Flow, Phase Field (tpf).

Select > Initial Values from the Pop-up menu.

Click > Initial Values 2.

Click > Fluid 2 button in Settings – Initial Values – Initial Values – Fluid initially in domain.

Click > Domain 1 in the Graphics window.

Click > Add to Selection in Settings – Initial Values – Domains.

See Figure 8.47.

Figure 8.47 shows the Initial Values 2 and Domain Selection.

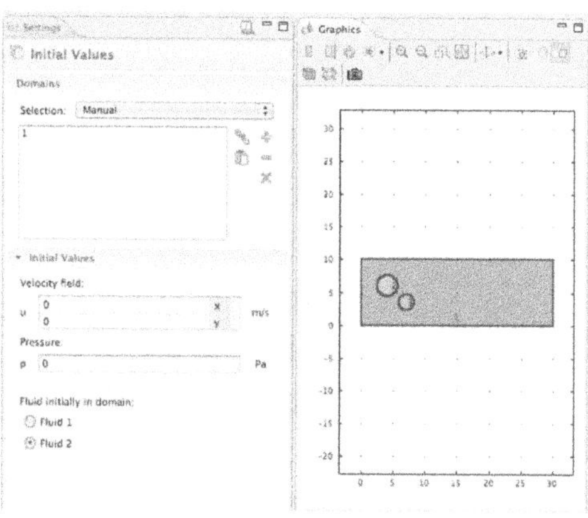

FIGURE 8.47 Initial Values 2 and Domain Selection.

Volume Force 1

Right-Click > Model Builder – Model 1 – Laminar Two-Phase Flow, Phase Field (tpf).

Select > Volume Force from the Pop-up menu.

Click > Model Builder – Model 1 – Laminar Two-Phase Flow, Phase Field (tpf) – Volume Force 1.

Click > Settings – Volume Force – Domains Selection.

Select > All domains from the Pop-up menu.

Enter > Fx in the x edit window.

Enter > Fy in the y edit window.

See Figure 8.48.

Figure 8.48 shows the Volume Force Components.

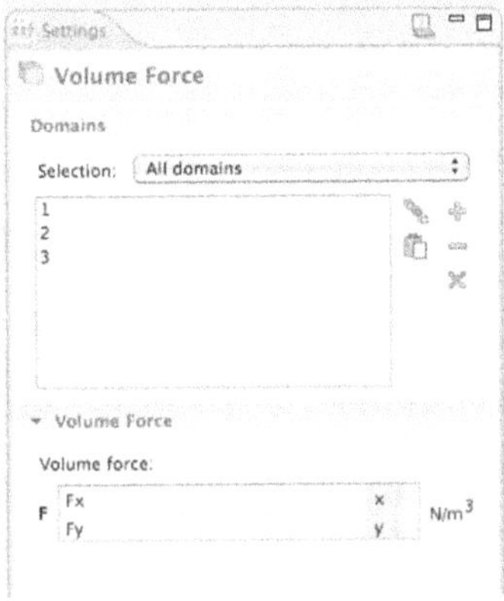

FIGURE 8.48 Volume Force Components.

Inlet 1

Right-Click > Model Builder – Model 1 – Laminar Two-Phase Flow, Phase Field (tpf).

Select > Inlet from the Pop-up menu.

Click > Model Builder – Model 1 – Laminar Two-Phase Flow, Phase Field (tpf) – Inlet 1.

Click > Settings – Inlet – Boundaries – Selection.

Select > Inlet from the Pop-up menu.

Click > Settings – Inlet – Boundary Condition – Volume fraction of fluid 2.

Enter > 1.

Enter > 50[mm/s] in the U_0 Normal inflow velocity edit window.

See Figure 8.49.

Figure 8.49 shows the Inlet Parameter edit windows.

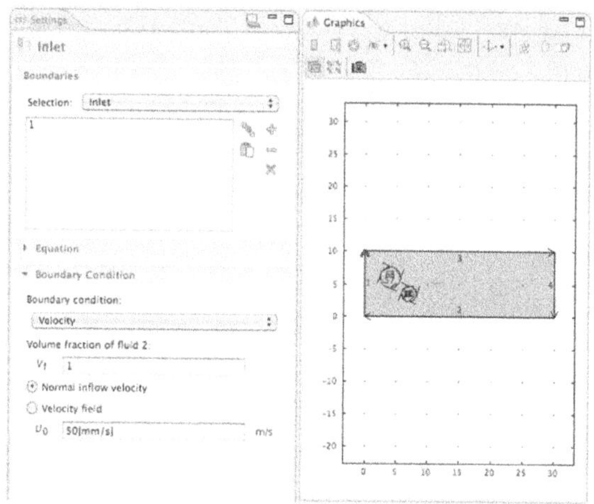

FIGURE 8.49 Inlet Parameter Edit Windows.

Outlet 1

Right-Click > Model Builder – Model 1 – Laminar Two-Phase Flow, Phase Field (tpf).

Select > Outlet from the Pop-up menu.

Click > Model Builder – Model 1 – Laminar Two-Phase Flow, Phase Field (tpf) –

Outlet 1.

Click > Settings – Outlet – Boundaries – Selection.

Select > Outlet from the Pop-up menu.

See Figure 8.50.

Figure 8.50 shows the Outlet Parameter edit windows.

Initial Interface 1

Right-Click > Model Builder – Model 1 – Laminar Two-Phase Flow, Phase Field (tpf).

Select > Initial Interface from the Pop-up menu.

Click > Model Builder – Model 1 – Laminar Two-Phase Flow, Phase Field (tpf) –

Initial Interface 1.

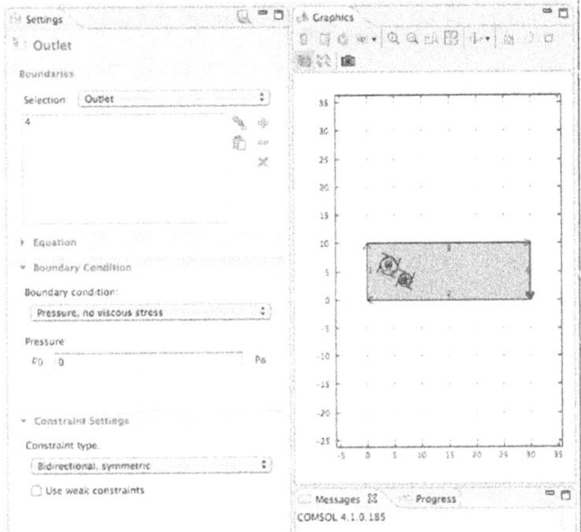

FIGURE 8.50 Outlet Parameter Edit Windows.

Click > Settings – Outlet – Boundaries – Selection.

Select > Oil/Water Interface from the Pop-up menu.

See Figure 8.51.

Figure 8.51 shows the Oil/Water Interface edit window.

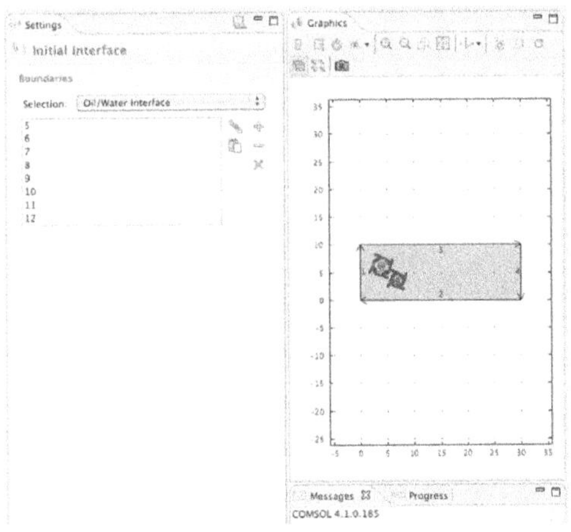

FIGURE 8.51 Oil/Water Interface Edit Window.

Mesh 1

Click > Model Builder – Model 1 – Mesh 1.

Click > Settings – Mesh – Mesh Settings – Element size.

Select > Fine from the Pop-up menu.

Click > Build All button.

After meshing, the modeler should see a message in the message window about the number of elements (10768 elements) in the mesh.

See Figure 8.52.

Figure 8.52 shows the Built Mesh.

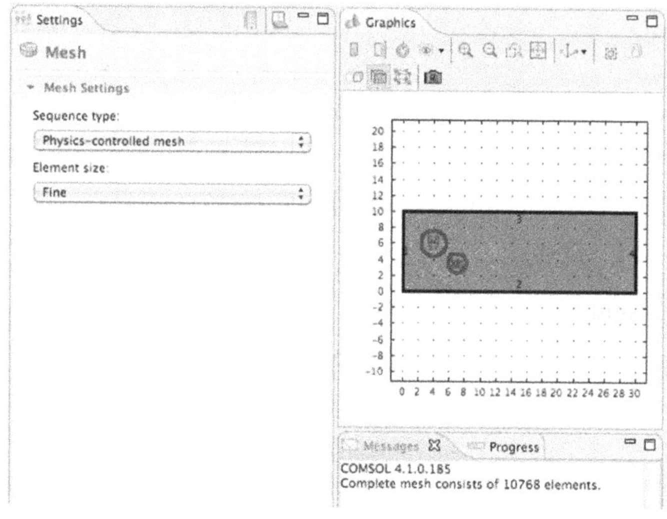

FIGURE 8.52 Built Mesh.

Study 1

Click > Model Builder – Study 1– Twistie.

Click > Study 1 – Step 2: Time Dependent.

Click > Range button in Settings – Time Dependent – Study Settings – Times.

Enter > Start = 0, Stop = 0.3, Step = 0.05 in the Range Pop-up edit window.

Click > Replace button in the Range Pop-up edit window.

See Figure 8.53.

Figure 8.53 shows the Time Dependent Range Settings.

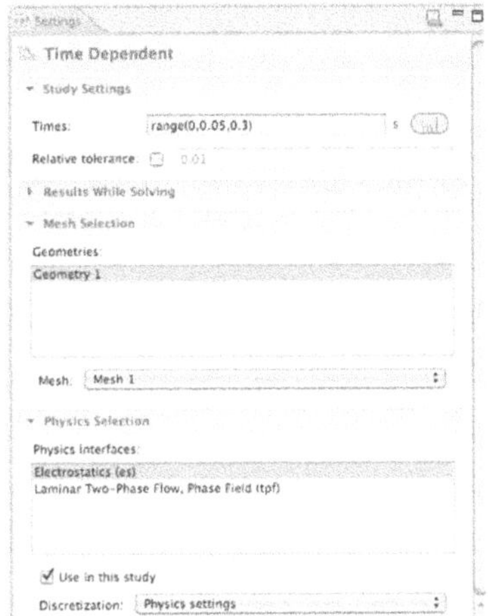

FIGURE 8.53 Time Dependent Range Settings.

Solver 1

NOTE *The Scale factors below that are introduced into the Time-Dependent Solver are chosen to stabilize the solver and aid in a more rapid convergence of the model calculations. For more detailed information see reference {8.25}.*

Right-Click > Model Builder – Study 1.

Select > Show Default Solver from the Pop-up menu.

Click > Model Builder – Study 1 – Solver Configurations twistie.

Click > Model Builder – Study 1 – Solver Configurations – Solver 1 twistie.

Click > Model Builder – Study 1 – Solver Configurations – Solver 1 – Dependent Variables 2.

Click > Settings – Dependent Variables – Scaling – Method.

Select > Manual from the Pop-up menu.
See Figure 8.54.

Figure 8.54 shows the Dependent Variable Scaling Settings.

Click > Model Builder – Study 1 – Solver Configurations – Solver 1 – Dependent Variables 2 twistie.

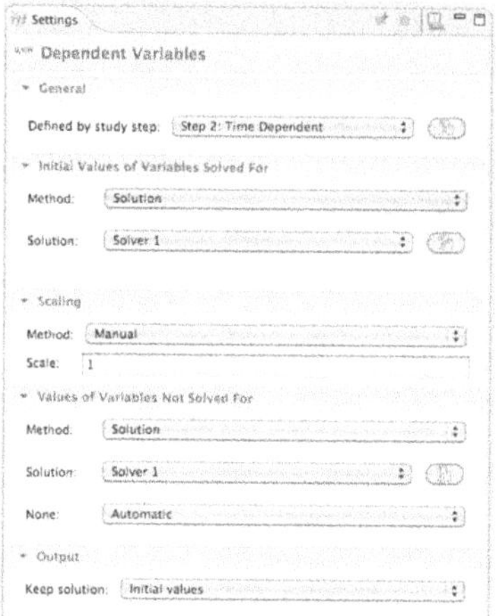

FIGURE 8.54 Dependent Variable Scaling Settings.

modl_V

Click > Model Builder – Study 1 – Solver Configurations – Solver 1 – Dependent Variables 2 – modl_V.

Click > Settings – Field – Scaling.

Select > Manual from the Pull-down menu.

Enter > 1e3 in the Settings – Field – Scaling – Scale edit window.

See Figure 8.55.

Figure 8.55 shows the Field Scaling modl_V settings.

modl_u

Click > Model Builder – Study 1 – Solver Configurations – Solver 1 – Dependent Variables 2 – modl_u.

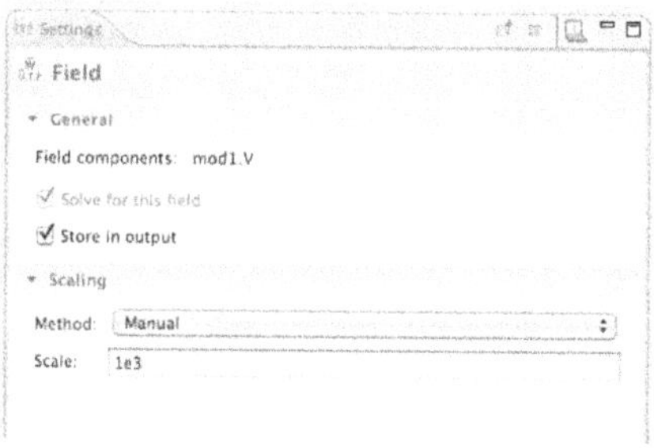

FIGURE 8.55 Field Scaling modl_V Settings.

Click > Settings – Field – Scaling.

Select > Manual from the Pull-down menu.

Enter > 0.1 in the Settings – Field – Scaling – Scale edit window.

See Figure 8.56.

Figure 8.56 shows the Field Scaling modl_u settings.

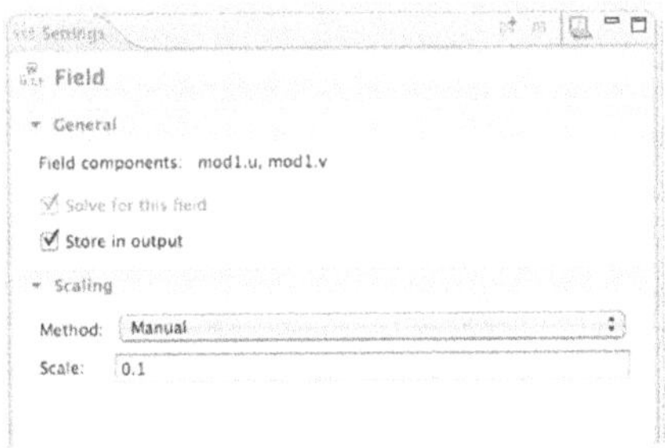

FIGURE 8.56 Field Scaling modl_u Settings.

modl_psi

Click > Model Builder – Study 1 – Solver Configurations – Solver 1 – Dependent Variables 2 – modl_psi.

Click > Settings – Field – Scaling.

Select > Manual from the Pull-down menu.

Enter > 0.1 in the Settings – Field – Scaling – Scale edit window.

See Figure 8.57.

Figure 8.57 shows the Field Scaling modl_psi settings.

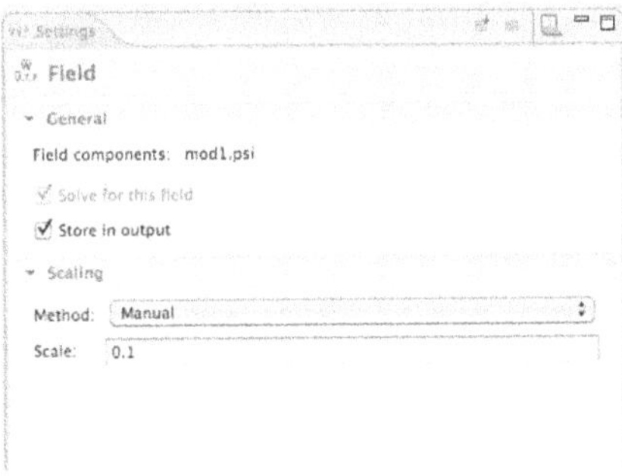

FIGURE 8.57 Field Scaling modl_psi Settings.

modl_p

Click > Model Builder – Study 1 – Solver Configurations – Solver 1 – Dependent Variables 2 – modl_p.

Click > Settings – Field – Scaling.

Select > Manual from the Pull-down menu.

Enter > 1e3 in the Settings – Field – Scaling – Scale edit window.

See Figure 8.58.

Figure 8.58 shows the Field Scaling modl_p settings.

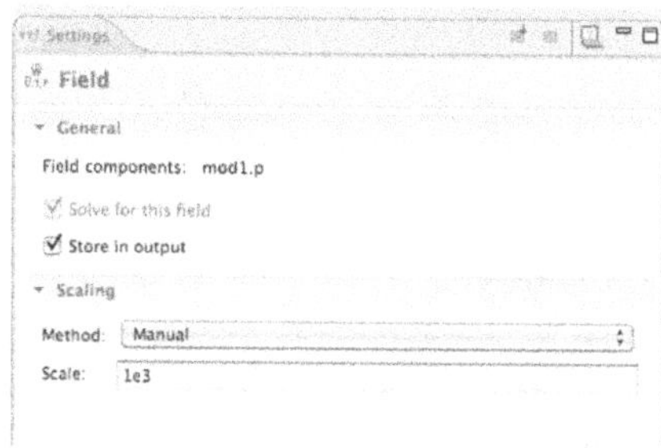

FIGURE 8.58 Field Scaling modl_p Settings.

Time-Dependent Solver 2

Click > Model Builder – Study 1 – Solver Configurations – Solver 1 – Time-Dependent Solver 2.

Click > Settings – Time-Dependent Solver – Advanced twistie.

Click > Settings – Time-Dependent Solver – Advanced – Error estimation.

Select > Exclude algebraic from the Pull-down menu.

In Model Builder, Right-Click Study 1 > Select > Compute.

Results

NOTE *At this point, it is suggested that the modeler save the solved model. Then the modeler should also save the solved model as MMUC4_2D_EcOW_1A.mph. Thus, if errors are made when postprocessing changes incorporated into the model for display purposes, not all the previous work is lost.*

Click > Zoom Extents.

Computed results, using the default display settings, plus Zoom Extents, are shown in Figure 8.59.

Figure 8.59 shows the 2D Electrocoalescence Oil/Water Separation Model results, using the default display settings, plus Zoom Extents.

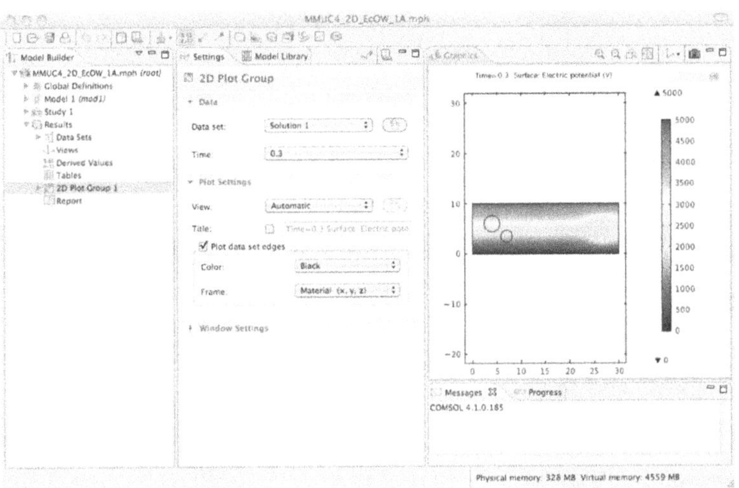

FIGURE 8.59 2D Electrocoalescence Oil/Water Separation Model Results, Using the Default Display Settings, Plus Zoom Extents.

NOTE *In order to display the electrocoalescence process as it occurs, it is necessary to modify the display settings.*

Click > Model Builder – Results – 2D Plot Group 1 twistie.

Click > Model Builder – Results – 2D Plot Group 1 – Surface 1.

Click > Settings – Surface – Expression – Replace Expression.

Select > Laminar Two-Phase Flow, Phase Field > Volume fraction of fluid 1 (tpf.Vf1) from the Pop-up menus.

Click > Uncheck the Color legend checkbox in Settings – Surface – Coloring and Style.

See Figure 8.60.

Figure 8.60 shows the Oil/Water Volume Fraction Plot.

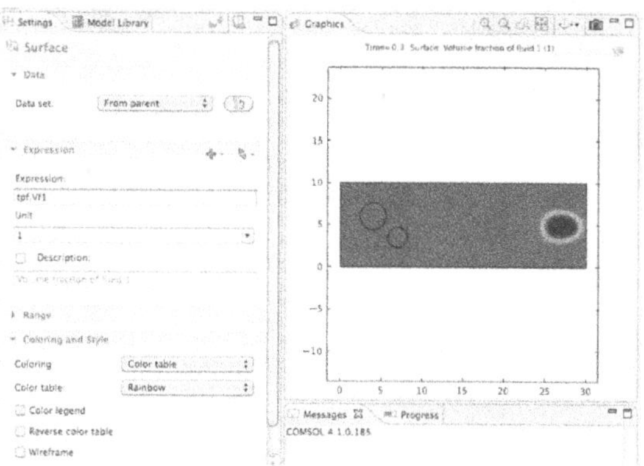

FIGURE 8.60 Oil/Water Volume Fraction Plot.

Right-Click > Model Builder – Results – 2D Plot Group 1.

Select > Contour from the Pop-up menu.

Click > Uncheck the Color legend checkbox in Settings – Contour – Coloring and Style.

Click > Zoom Extents.

See Figure 8.61.

Figure 8.61 shows the Oil/Water Volume Fraction Plot with Contours.

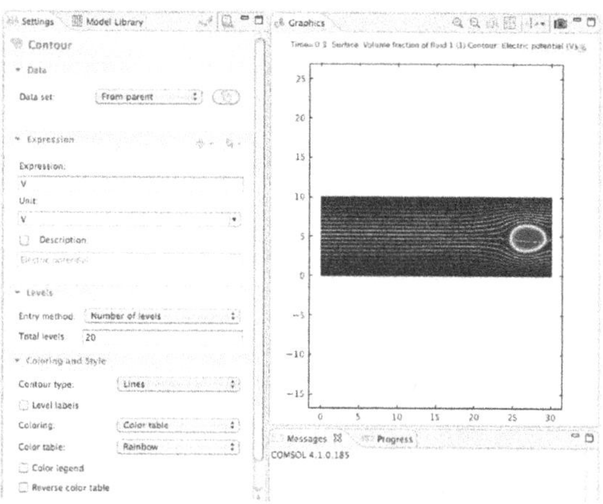

FIGURE 8.61 Oil/Water Volume Fraction Plot with Contours.

Results Plots

Plot at Time 0

Click > Model Builder – Results – 2D Plot Group 1.

Click > Settings – 2D Plot Group – Data – Time.

Select > 0 from the Pull-down menu.

Click > Plot.

See Figure 8.62.

Figure 8.62 shows the Electrocoalescence of Oil/Water Time 0.

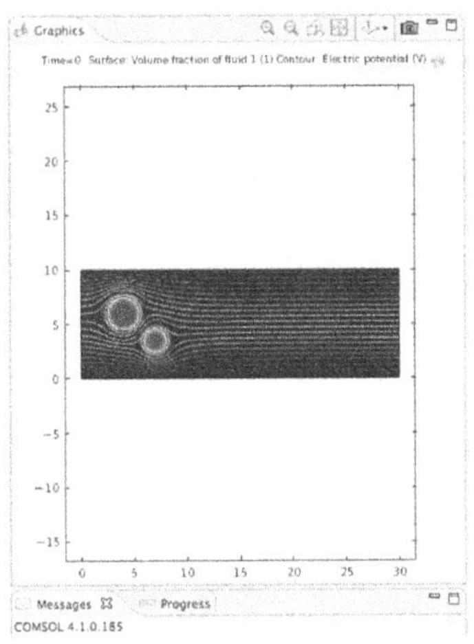

FIGURE 8.62 Electrocoalescence of Oil/Water Time 0.

Plot at Time 0.1

Click > Model Builder – Results – 2D Plot Group 1.

Click > Settings – 2D Plot Group – Data – Time.

Select > 0.1 From the Pull-down menu.

Click > Plot.

See Figure 8.63.

Figure 8.63 shows the Electrocoalescence of Oil/Water Time 0.1.

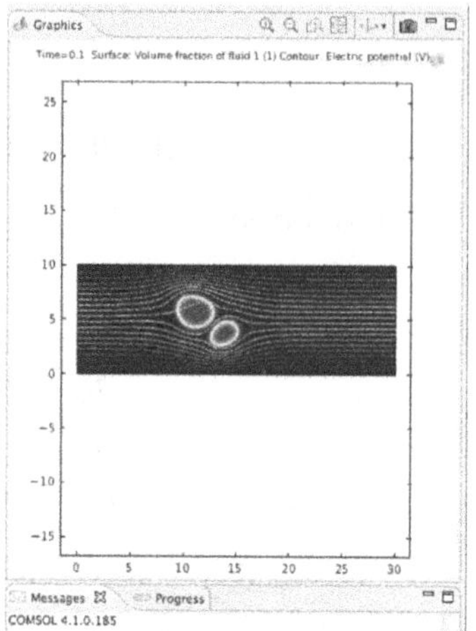

FIGURE 8.63 Electrocoalescence of Oil/Water Time 0.1.

Plot at Time 0.2

Click > Model Builder – Results – 2D Plot Group 1.

Click > Settings – 2D Plot Group – Data – Time.

Select > 0.2 From the Pull-down menu.

Click > Plot.

See Figure 8.64.

Figure 8.64 shows the Electrocoalescence of Oil/Water Time 0.2.

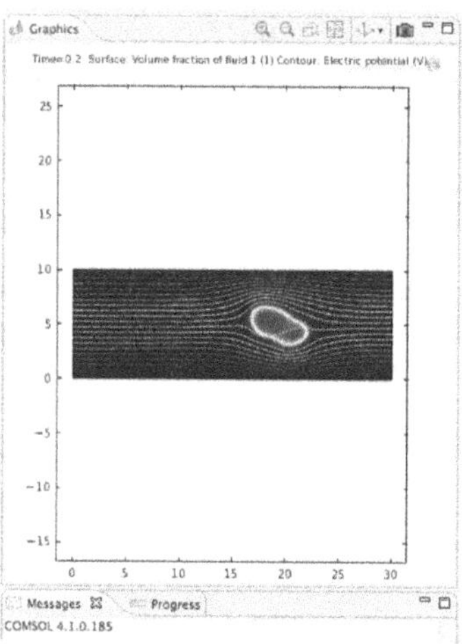

FIGURE 8.64 Electrocoalescence of Oil/Water Time 0.2.

Plot at Time 0.3

Click > Model Builder – Results – 2D Plot Group 1.

Click > Settings – 2D Plot Group – Data – Time.

Select > 0.3 From the Pull-down menu.

Click > Plot.

See Figure 8.65.

Figure 8.65 shows the Electrocoalescence of Oil/Water Time 0.3.

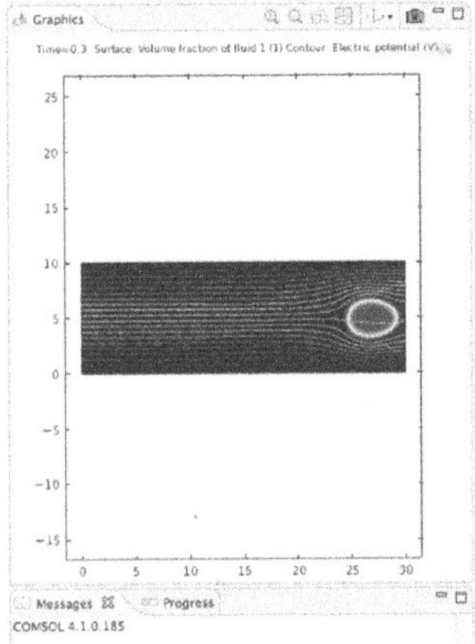

FIGURE 8.65 Electrocoalescence of Oil/Water Time 0.3.

2D Electrocoalescence Oil/Water Separation Model Summary and Conclusions

The 2D Electrocoalescence Oil/Water Separation Model is a powerful tool that can be used to develop the design of numerous different models for fluid separation problems. With this model, the modeler can easily vary all of the machine's parameters and optimize the design to obtain the best fluid separation process before the first prototype is physically built. This technique works best when the physical contrast in the properties of the materials is as large as possible within the model's constraints. This technique can be used either statically or dynamically and is a broadly applied solution in various industries.

FIRST PRINCIPLES AS APPLIED TO 2D COMPLEX MIXED MODE MODEL DEFINITION

First Principles Analysis derives from the fundamental laws of nature. In the case of models using this Classical Physics Analysis approach, the laws of conservation in physics require that what goes in (as mass, energy, charge, etc.) must come out (as mass, energy, charge, etc.) or must accumulate within the boundaries of the model.

The careful modeler must be knowledgeable of the implicit assumptions and default specifications that are normally incorporated into the COMSOL Multiphysics software model when a model is built using the default settings.

Consider, for example, the two 2D models developed in this chapter. In these models, it is implicitly assumed there are no thermally related changes (mechanical, electrical, etc.). It is also assumed the materials are homogeneous and isotropic, except as specifically indicated and there are no thin insulating contact barriers at the thermal junctions. None of these assumptions are typically true in the general case. However, by making such assumptions, it is possible to easily build a 2D Complex Mixed Mode First Approximation Model.

NOTE

A First Approximation Model is one that captures all the essential features of the problem that needs to be solved, without dwelling excessively on all of the small details. A good First Approximation Model will yield an answer that enables the modeler to determine if he needs to invest the time and the resources required to build a more highly detailed model.

Also, the modeler needs to remember to name model parameters carefully as pointed out in Chapter 1.

REFERENCES

8.1 COMSOL Multiphysics Users Guide, Version 4.1, pp. 288–292

8.2 http://en.wikipedia.org/wiki/Electroplating

8.3 http://en.wikipedia.org/wiki/Coalescer

8.4 COMSOL Chemical Reaction Engineering Module Users Guide, p. 94

8.5 COMSOL CFD Module Users Guide, p. 189

8.6 http://en.wikipedia.org/wiki/Electrochemistry

8.7 http://en.wikipedia.org/wiki/John_Frederic_Daniell

8.8 http://en.wikipedia.org/wiki/Michael_Faraday

8.9 N. Kanani, Ed., *Electroplating and Electroless Plating of Copper & its Alloys,* ASM International, Materials Park, Ohio, ISBN: 0-904477-26-6

8.10 N. Kanani, Ed., *Electroplating and Electroless Plating of Copper & its Alloys,* ASM International, Materials Park, Ohio, ISBN: 0-904477-26-6, pp 34–42

8.11 http://en.wikipedia.org/wiki/Nernst–Planck_equation

8.12 http://en.wikipedia.org/wiki/Fick%27s_laws_of_diffusion

8.13 http://en.wikipedia.org/wiki/Butler-Volmer_equation

8.14 Z. Chen and S. Liu, "Simulation of Copper Electroplating Fill Process of Through Silicon Via", 11th International conference on electronic Packaging Technology & High Density Packaging, 2010, pp. 433–437

8.15 COMSOL Multiphysics Users Guide, Version 4.1, pp. 101–102

8.16 http://en.wikipedia.org/wiki/Coalescence_(disambiguation)

8.17 http://en.wikipedia.org/wiki/Coalescence_(physics)

8.18 COMSOL CFD Module Users Guide, p. 192

8.19 COMSOL ACDC Module Users Guide, p. 78

8.20 COMSOL ACDC Module Users Guide, p. 33

8.21 http://en.wikipedia.org/wiki/Claude-Louis_Navier

8.22 http://en.wikipedia.org/wiki/Sir_George_Stokes,_1st_Baronet

8.23 http://en.wikipedia.org/wiki/Newton%27s_laws_of_motion

8.24 http://en.wikipedia.org/wiki/Sir_George_Stokes,_1st_Baronet

8.25 COMSOL Multiphysics Reference Guide, p. 365

Suggested Modeling Exercises

1. Build, mesh, and solve the 2D Copper Electroplating Model as presented earlier in this chapter.
2. Build, mesh, and solve the 2D Electrocoalescence Oil/Water Separation Model as presented earlier in this chapter.
3. Change the values of the materials parameters and then build, mesh, and solve the 2D Copper Electroplating Model as an example problem.
4. Change the values of the materials parameters and then build, mesh, and solve the 2D Electrocoalescence Oil/Water Separation Model as an example problem.
5. Change the value of the electrical parameters and then build, mesh, and solve the 2D Copper Electroplating Model as an example problem.
6. Change the value of the geometries of the drops and the tube and then solve the 2D Electrocoalescence Oil/Water Separation Model as an example problem.

CHAPTER 9

3D MODELING USING COMSOL MULTIPHYSICS 4.X

In This Chapter

- Guidelines for 3D Modeling in 4.x
  - 3D Modeling Considerations
- 3D Models
  - 3D Spiral Coil Microinductor Model
  - 3D Linear Microresistor Beam Model
- First Principles as Applied to 3D Model Definition
- References
- Suggested Modeling Exercises

GUIDELINES FOR 3D MODELING IN 4.X

NOTE

In this chapter, 3D models will be presented. Such 3D models are typically more conceptually and physically complex than the models that were presented in earlier chapters of this text. 3D models have proven to be very valuable to the science and engineering communities as first-cut evaluations of potential systemic physical behavior under the influence of mixed external stimuli. 3D model responses and other such ancillary information can be gathered and screened early in a project for a first-cut evaluation of the physical behavior of planned prototype. The calculated model (simulation) information can be used in the prototype fabrication stage as guidance in the selection of prototype geometry and materials.

Since the models in this and subsequent chapters are more complex and more difficult to solve than the models presented thus far, it is important that the modeler have available the tools necessary to most easily utilize the powerful capabilities of the 4.x software. In order to do that, if you have not done this previously, the modeler should go to the main 4.x toolbar, Click > Options – Preferences – Model builder. When the Preferences – Model builder edit window is shown, Select > Show equation view checkbox and Show more options checkbox. Click > Apply {9.1}.

3D Modeling Considerations

3D models are typically difficult, simply based on their mathematical size, if nothing else. The modeler needs to ensure that all the appropriate assumptions have been made and that the correct materials properties have been incorporated in all dimensions. 3D models also allow the use of vector and tensor materials properties {9.2, 9.3}. Thus, depending upon the nature of the materials properties, the behavior of the material in the model may not be linear or homogeneous or isotropic or well behaved, in any sense of the word. However, if the materials are properly characterized and the modeler is sufficiently patient, a first approximation model can be built and converged to yield a first approximation answer (solution) to a difficult, non-analytic problem.

NOTE *An Analytic Function has a Closed-form Solution which can be written as a bounded number of well-known functions {9.4}. A non-analytic function cannot be so written.*

In compliance with the laws of physics, a 3D model initially implicitly assumes that energy flow, materials properties, environment, and all other conditions and variables that are of interest are homogeneous, isotropic, and/or constant, unless otherwise specified, throughout the entire domain of interest both within the model and through the boundary conditions and in the environs of the model.

It is the responsibility of the modeler to ensure that any implicit assumptions that need to be adjusted are. Especially in 3D models, the modeler needs to bear the above stated conditions in mind and carefully ensure that all of the modeling conditions and associated parameters (default settings) in each model created are properly considered, defined, verified, and/or set to the appropriate values.

It is always mandatory that the modeler be able to accurately anticipate the expected results of the model and accurately specify the manner in which those results will be presented. Never assume that any of the default values that are present when the model is created necessarily satisfy the needs or conditions of a particular model.

NOTE

Always verify that any parameters employed in the model are of the correct value needed for that model. Calculated solutions that significantly deviate from the anticipated solution or from a comparison of values to those measured in an experimentally derived realistic model are probably indicative of one or more modeling errors either in the original model design, in the earlier model analysis, in the understanding of the underlying physics, or are simply due to human error.

3D Coordinate System

In a 3D model, if the parameters can only vary as a function of the position, the (x), (y), and (z) coordinates, then such a 3D model represents the parametric condition of the model in a time independent mode (stationary). In a time-dependent study or frequency domain study model, parameters can vary both with position in (x), (y), and/or (z) and with time (t).

See Figure 9.1.

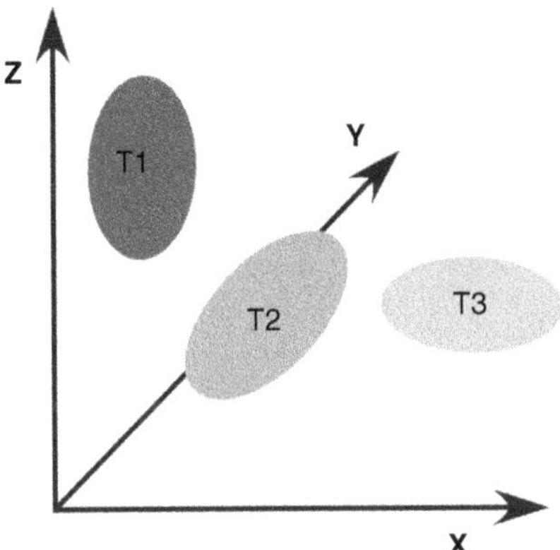

FIGURE 9.1 Parametric Variations in a 3D Coordinate System, Plus Time.

Figure 9.1 shows parametric variations in a 3D coordinate system, plus time.

In this chapter, the modeler is introduced to three new modeling concepts: the Terminal boundary condition, lumped parameters {9.5}, and coupled thermal, electrical, and structural multiphysics analysis. The Terminal boundary condition and the lumped parameter concepts are employed in the solution of the 3D Spiral Coil Microinductor Model. The fully coupled multiphysics solution is employed in the 3D Linear Microresistor Beam Model.

NOTE

The lumped parameter (lumped element) modeling approach approximates a spatially distributed collection of diverse physical elements by a collection of topologically (series and/or parallel) connected discrete elements. This technique is commonly employed for first approximation models in electrical, electronic, mechanical, heat transfer, acoustic, and other physical systems.

Inductance Theory

Michael Faraday {9.6} discovered electromagnetic induction {9.7} in 1831. Joseph Henry {9.8} independently discovered electromagnetic induction during approximately the same era; however he chose not to publish his discovery until later. Oliver Heaviside coined the name inductance in 1886 {9.9}.

Inductance (a lumped parameter) is the physical property associated with electrical components configured in specific circuit configurations. There are basically two commonly employed configurations for inductance: self inductance and mutual inductance. Self inductance (L) is defined as the inductance of an individual coil in a circuit and is mathematically defined as follows:

$$v = L\frac{di}{dt} \tag{9.1}$$

Where: v = induced voltage in volts [V].

L = inductance in henries [H].

i = current flowing in the circuit in amperes [A].

The configuration for a self inductive coil is as shown in Figure 9.2.

Figure 9.2 shows a Self Inductance circuit.

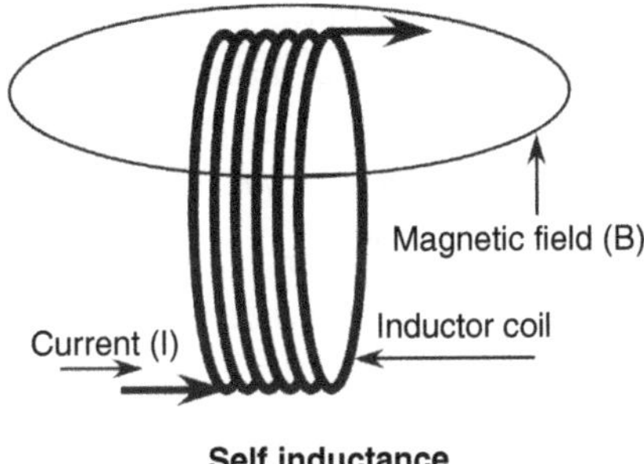

FIGURE 9.2 Self Inductance Circuit.

Mutual inductance is somewhat more complex. Mutual inductance involves the coupling between at least two discrete coils through an intermediate medium. The mutual inductance for two coils is typically defined as follows:

$$M_{\alpha\beta} = k_{\alpha\beta}(L_{\alpha} \bullet L_{\beta})^{1/2} \qquad (9.2)$$

Where: $M_{\alpha\beta}$ = the mutual inductance in henries [H].

$k_{\alpha\beta}$ = coupling coefficient through the intermediate media $0 \le k_{\alpha\beta} \le 1$ and is unitless [1].

L_{α} = inductance of the first coil in henries [H].

L_{β} = inductance of the second coil in henries [H].

The configuration for two mutual inductive coils is as shown in Figure 9.3.

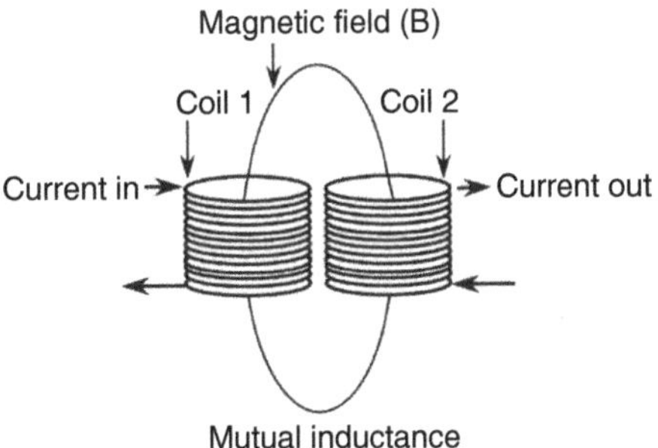

FIGURE 9.3 Mutual Inductance Circuit.

Figure 9.3 shows a Mutual Inductance circuit.

NOTE

Inductance can be calculated in the general case by using Maxwell's equations {9.10}. Expressions (formulas) for the inductance of most common inductor configurations (geometries) have been calculated and are available in the form of tables and handbooks {9.11}.

The easiest modeling method to employ in the solution of a lumped parameter is the use of an impedance matrix {9.12, 9.13}, which is merely an expansion of Ohm's Law {9.14}.

When current flows in an electrical circuit, energy is stored in the magnetic field {9.15}. The energy stored in the magnetic field of an inductor {9.16, 9.17} is proportional to the inductance and to the square of the current as follows:

$$E_m = \frac{1}{2} L I^2 \tag{9.3}$$

Where: E_m = stored magnetic energy in joules [J].

L = inductance in henries [H].

I = current in amperes [A].

NOTE

In a circuit that contains an inductive component L, the diagonal term L_{11} of the impedance matrix is the magnitude of the self-inductance L.

3D MODELS

3D Spiral Coil Microinductor Model

Building the 3D Spiral Coil Microinductor Model

Startup 4.x.

Select > 3D.

Click > Next.

Click > Twistie for AC/DC Interface.

Click > Magnetic and Electric Fields (*mef*).

Click > Add Selected.

Click > Next.

Select > Preset Studies – Stationary.

Click > Finish (Flag).

Click > Save As.

Enter MMUC4_3D_SCM_1.mph.

Click > Save.

See Figure 9.4.

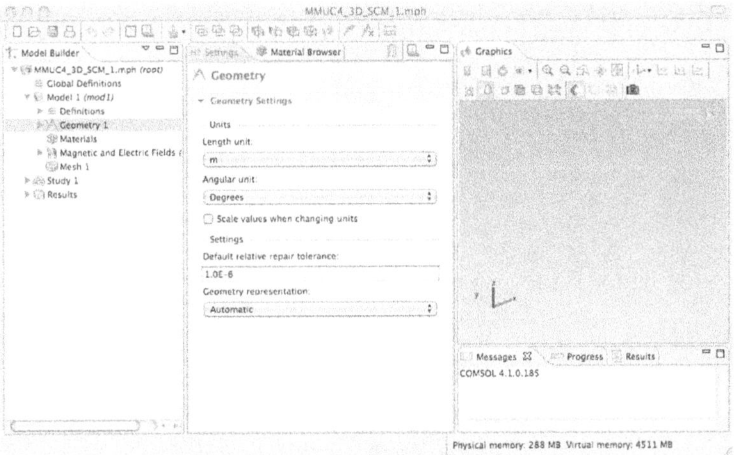

FIGURE 9.4 Desktop Display for the MMUC4_3D_SCM_1.mph Model.

Figure 9.4 shows the Desktop Display for the MMUC4_3D_SCM_1.mph model.

Geometry

Spiral Coil

Block 1

In Model Builder – Model 1 (mod1),

Right-Click > Model Builder – Model 1 (mod1) – Geometry 1.

Select > Block from the Pop-up menu.

Enter > 60E-6[m] in the Settings – Block – Size and Shape – Width entry window.

Enter > 20E-6[m] in the Settings – Block – Size and Shape – Depth entry window.

Enter > 20E-6[m] in the Settings – Block – Size and Shape – Height entry window.

Select > Corner from the Settings – Block – Position – Base Pull-down menu.

Enter > 0E-6[m] in the Settings – Block – Position – x entry window.

Enter > 0E-6[m] in the Settings – Block – Position – y entry window.

Enter > 0E-6[m] in the Settings – Block – Position – z entry window.

Click > Build All.

See Figure 9.5.

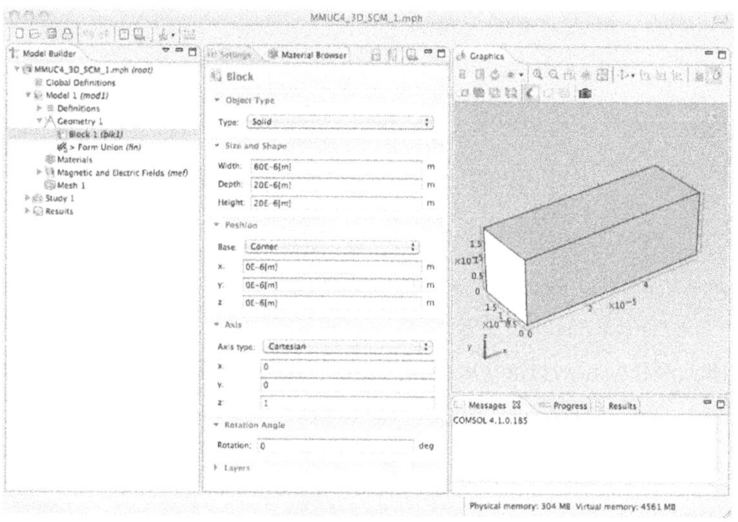

FIGURE 9.5 Settings - Block 1 Entry Window and the Built Block 1.

Figure 9.5 shows the Settings – Block 1 entry window and the built Block 1.

NOTE *The remaining portions of the geometry will be built without explicitly showing the results incrementally. The total result will be shown, once completed.*

Build the remainder of the Spiral Coil Blocks 2 – 15, by following the procedure employed to build Block 1.

See Table 9.1 for the Block build values.

TABLE 9.1 Spiral Coil Elements

Block Dimensions				Corner Position		
Block	**Width**	**Depth**	**Height**	x	y	z
2	20E-6	80E-6	20E-6	40E-6	20E-6	0E-6
3	200E-6	20E-6	20E-6	40E-6	100E-6	0E-6
4	20E-6	140E-6	20E-6	220E-6	-40E-6	0E-6
5	160E-6	20E-6	20E-6	80E-6	-60E-6	0E-6
6	20E-6	100E-6	20E-6	80E-6	-40E-6	0E-6
7	120E-6	20E-6	20E-6	80E-6	60E-6	0E-6
8	20E-6	60E-6	20E-6	180E-6	0E-6	0E-6
9	80E-6	20E-6	20E-6	120E-6	-20E-6	0E-6
10	20E-6	20E-6	20E-6	120E-6	0E-6	0E-6
11	40E-6	20E-6	20E-6	120E-6	20E-6	0E-6
12	20E-6	20E-6	20E-6	140E-6	20E-6	20E-6
13	140E-6	20E-6	20E-6	140E-6	20E-6	40E-6
14	20E-6	20E-6	20E-6	260E-6	20E-6	20E-6
15	40E-6	20E-6	20E-6	260E-6	20E-6	0E-6

Click > Zoom Extents after each block is built.

Right-Click > Model Builder – Model 1 (mod1) – Geometry 1.

Select > Boolean Operations – Union from the Pop-up menu.

Uncheck > Keep interior boundaries Checkbox.

Click > Select Box in the Graphics Toolbar.

Select > All Blocks.

Click > Add to Selection in Settings – Union – Union.

Click > Build All.

See Figure 9.6.

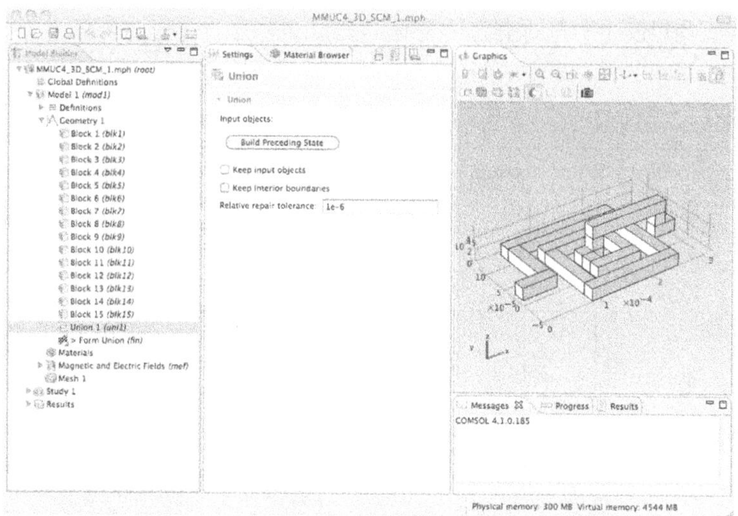

FIGURE 9.6 Graphics Window - Built Spiral Coil.

Figure 9.6 shows the Graphics Window – Built Spiral Coil.

The Spiral Coil needs to have an environment as part of the model. The next step adds a Block to define the Spiral Coil environment.

Surrounding Environment

Right-Click > Model Builder – Model 1 (mod1) – Geometry 1.

Select > Block from the Pop-up menu.

Enter > 300E-6[m] in the Settings – Block – Size and Shape – Width entry window.

Enter > 300E-6[m] in the Settings – Block – Size and Shape – Depth entry window.

Enter > 200E-6[m] in the Settings – Block – Size and Shape – Height entry window.

Select > Corner from the Settings – Block – Position Pull-down menu.

Enter > 0E-6[m] in the Settings – Block – Position – x entry window.

Enter > –120E-6[m] in the Settings – Block – Position – y entry window.

Enter > –100E-6[m] in the Settings – Block – Position – z entry window.

Click > Build All.

Click > Zoom Extents.

Click > Wireframe Rendering in the Graphics Window Toolbar.

See Figure 9.7.

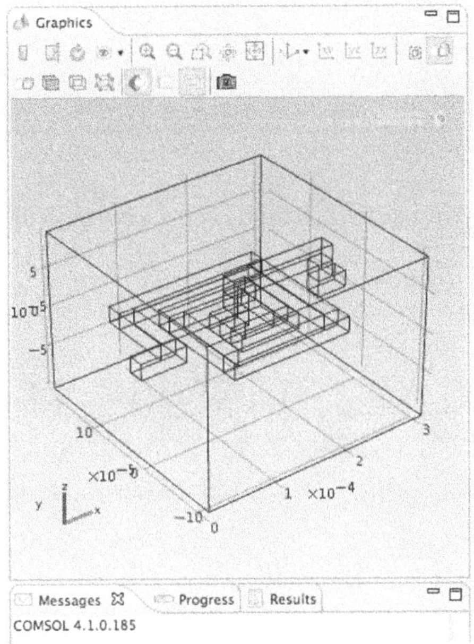

FIGURE 9.7 Graphics Window - Built Spiral Coil Plus Environmental Box.

Figure 9.7 shows the Graphics Window – Built Spiral Coil plus Environmental Box.

NOTE *The Spiral Coil now has the necessary environment to complete the Geometry of the model. The next step defines the materials of the Spiral Coil and the Environmental Box.*

Materials

Material 1

NOTE *In this model, the materials properties will be entered by the modeler.*

Right-Click > Model Builder – Model 1 (mod1) – Materials.

Select > Material from the Pop-up menu.

Right-Click > Material 1.

Select > Rename from the Pop-up menu.

Enter > Conductor in the New Name edit window.

Click > OK.

Click > 1 in Settings – Material – Geometric Scope – Selection edit window.

Click > Remove from Selection (Minus Sign) in Settings – Material – Geometric Scope.

Enter > 1E6 in Settings – Material – Material Contents – sigma Value edit window.

Enter > 1E0 in Settings – Material – Material Contents – epsilonr Value edit window.

Enter > 1E0 in Settings – Material – Material Contents – mur Value edit window.

See Figure 9.8.

Figure 9.8 shows the Settings – Material Contents (Properties) Conductor Windows.

Material 2

Right-Click > Model Builder – Model 1 (mod1) – Materials.

Select > Material from the Pop-up menu.

Right-Click > Material 2.

Select > Rename from the Pop-up menu.

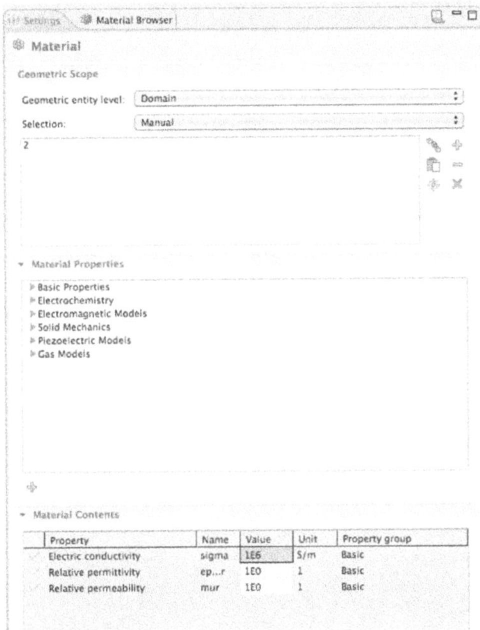

FIGURE 9.8 Settings - Material Contents (Properties) Conductor Windows.

Enter > Air in the New Name edit window.

Click > OK.

Click > Domain 1 (Environmental Box) in the Graphics window.

Click > Add to Selection (Plus Sign) in Settings – Material – Geometric Scope.

Enter > 1E-6 in Settings – Material – Material Contents – sigma Value edit window.

Enter > 1E0 in Settings – Material – Material Contents – epsilonr Value edit window.

Enter > 1E0 in Settings – Material – Material Contents – mur Value edit window.

NOTE *The value chosen for sigma Air (1E-6) needs to be sufficiently small, but not zero (0), in order for the model to converge to solution without incurring a divide by zero problem.*

See Figure 9.9.

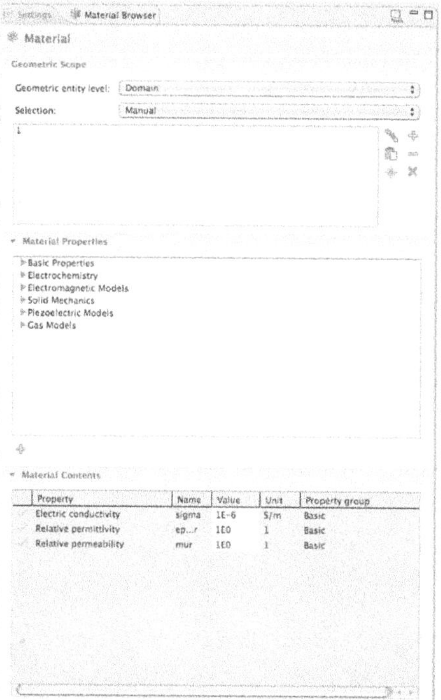

FIGURE 9.9 Settings - Material Contents (Properties) Air Windows.

Figure 9.9 shows the Settings – Material Contents (Properties) Air Windows.

Magnetic and Electric Fields (mef)

Terminal 1

Click > Model Builder – Model 1 – Twistie for the Magnetic and Electric Fields (mef).

Right-Click > Model Builder – Model 1 – Magnetic and Electric Fields (mef) – Magnetic Insulation 1.

Select > Terminal from the Pop-up menu.

Click > Terminal 1.

Select > 1-4 in the Settings – Terminal Boundaries edit window.

Click > Remove from Selection.

Select > 6-82 in the Settings – Terminal Boundaries edit window.

Click > Remove from Selection.

Enter > 1E0 in Settings – Terminal – Terminal I_0 edit window.

See Figure 9.10.

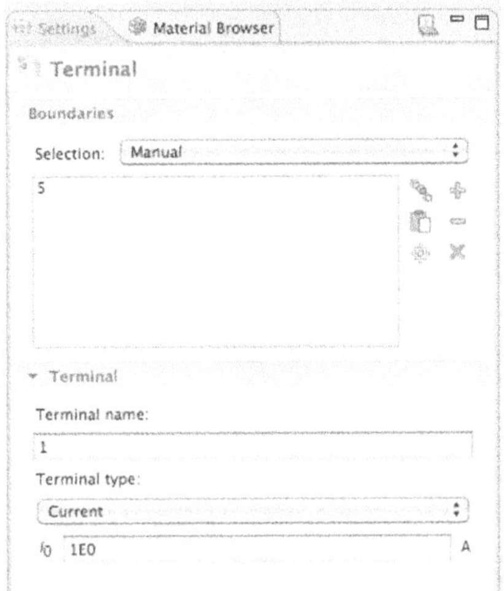

FIGURE 9.10 Settings - Terminal 1 Edit Windows.

Figure 9.10 shows the Settings – Terminal 1 edit windows.

Ground 1

Right-Click > Model Builder – Model 1 – Magnetic and Electric Fields (mef) – Magnetic Insulation 1.

Select > Ground from the upper section of the Pop-up menu.

Click > Ground 1.

Select > 1-80 Settings – Ground Boundaries edit window.

Click > Remove from Selection.

See Figure 9.11.

Figure 9.11 shows the Settings – Ground 1 edit window.

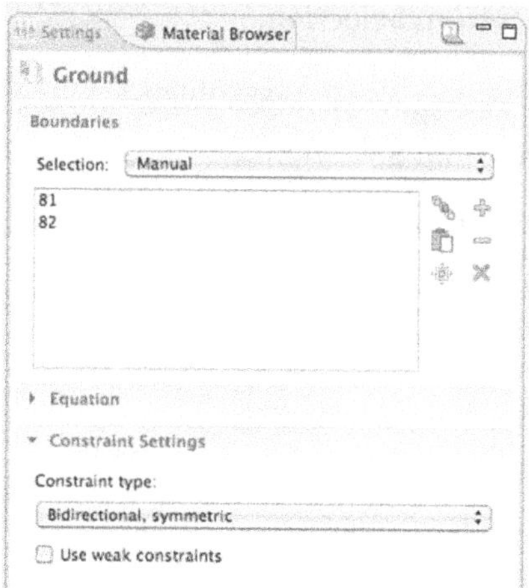

FIGURE 9.11 Settings - Ground 1 Edit Window.

Mesh 1

Right-Click > Model Builder – Model 1 – Mesh 1.

Select > Free Tetrahedral from the Pop-up menu.

Click > Model Builder – Model 1 – Mesh 1 – Size.

Click > Predefined Pull-down list in Settings – Size – Element Size.

Select > Coarse.

See Figure 9.12.

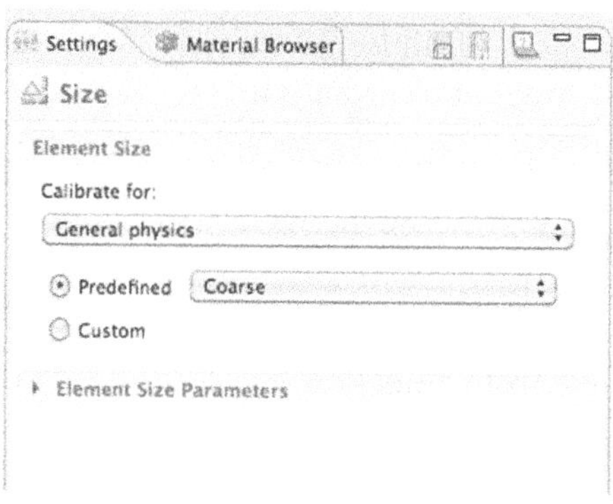

FIGURE 9.12 Settings - Size Pull-Down List.

Figure 9.12 shows the Settings – Size Pull-down list.

Click > Build All.

See Figure 9.13.

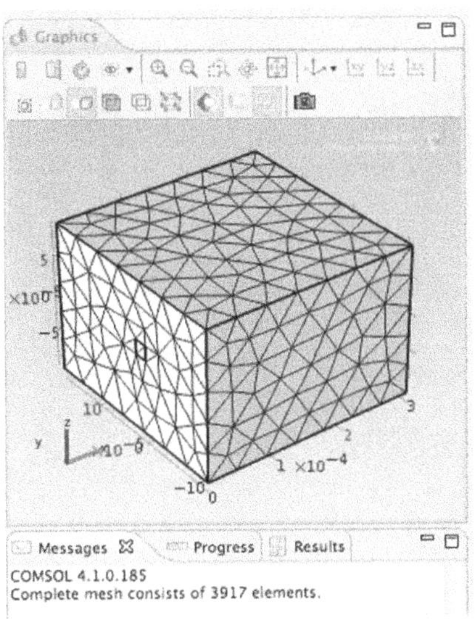

FIGURE 9.13 Graphics Window, Mesh.

Figure 9.13 shows the Graphics Window, Mesh.

NOTE *After the mesh is built, the model should have 3917 elements.*

Study 1

In Model Builder, Right-Click Study 1 > Select > Compute.

Computed results, using the default display settings, are shown in Figure 9.14.

Figure 9.14 shows the Graphics Window, Mesh.

NOTE *The modeler should note that the computed results, using the default display settings, initially display the model solution. However, with some additional display parameter adjustments in the way that the data are presented, the model solution presentation will be significantly enhanced.*

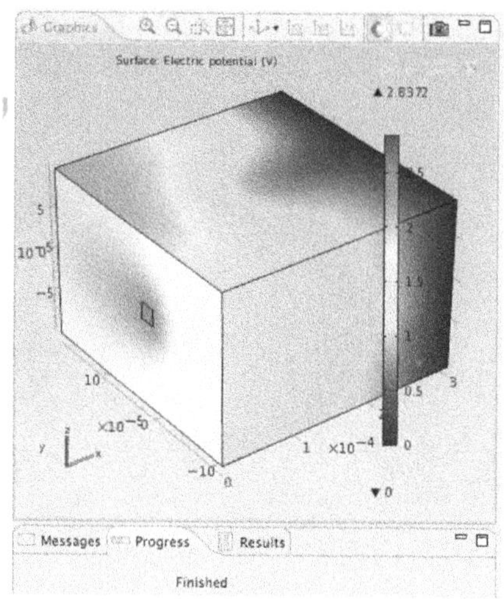

FIGURE 9.14 Graphics Window, Mesh.

Results

Color and Style

Click > Model Builder – Results – 3D Plot Group 1 – Surface 1.

Select > Thermal from the Settings – Surface – Coloring and Style – Color table Pull-down menu.

See Figure 9.15.

Figure 9.15 shows the Color and Style Plot Settings.

Data Sets

Click > Model Builder – Results – Data Sets twistie.

Right-Click > Model Builder – Results – Data Sets – Solution 1.

Select > Add Selection from the Pop-up menu.

NOTE *To select all the boundaries of the Spiral Coil, first, select all of the boundaries of the model and then Remove from Selection all of the boundaries of the Air Box.*

Click > Settings – Selection – Geometric Scope – Geometric entity level.

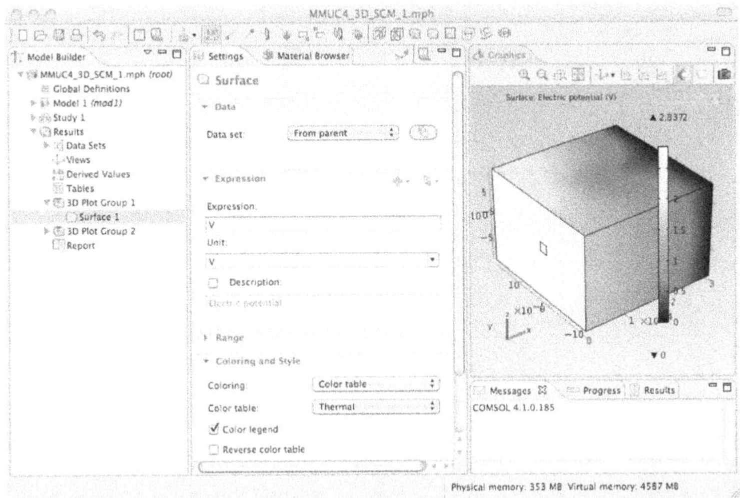

FIGURE 9.15 Color and Style Plot Settings.

Select > Boundary from the Pop-up menu.

Click > Settings – Selection – Geometric Scope – Selection.

Select > All boundaries from the Pop-up menu.

Select > Boundaries 1, 2, 3, 4, 10, 81.

Click > Remove from Selection in Settings – Selection – Geometric Scope – Selection.

NOTE *When selected boundaries are removed from the selection window, the process becomes manual.*

See Figure 9.16.

Figure 9.16 shows the Selected Spiral Coil Boundaries.

3D Plot Group 1

Right-Click > Model Builder – Results – 3D Plot Group 1.

Select > Streamline from the Pop-up menu.

Click > Replace Expression in Settings – Streamline – Expression.

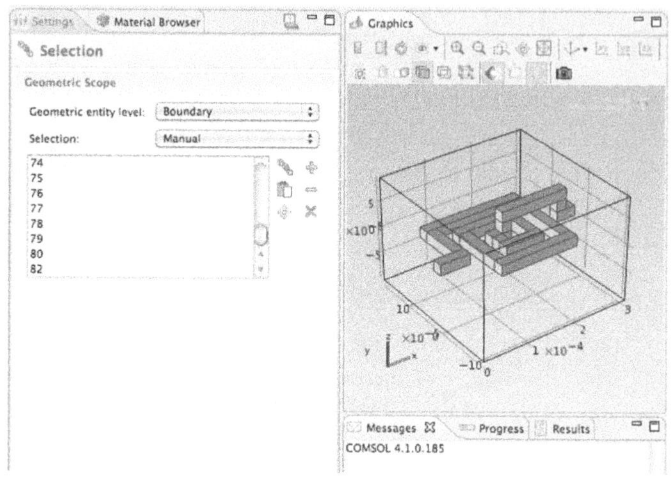

FIGURE 9.16 Selected Spiral Coil Boundaries.

Select > Magnetic and Electric Fields – Magnetic Flux Density (mef.Bx, mef.By, mef.Bz) from the Pop-up menu.

Enter > 3 in the Settings – Streamline – Streamline Positioning – Points edit window.

Click > Settings – Streamline – Coloring and Style – Line type.

Select > Tube from the Pull-down menu.

Enter > 1E-6 in the Settings – Streamline – Coloring and Style – Tube radius expression edit window.

Click > Settings – Streamline – Quality twistie.

Click > Settings – Streamline – Quality – Resolution.

Select > Finer from the Pop-up menu.

See Figure 9.17.

Figure 9.17 shows the Streamline Settings.

FIGURE 9.17 Streamline Settings.

Click > Plot.

See Figure 9.18.

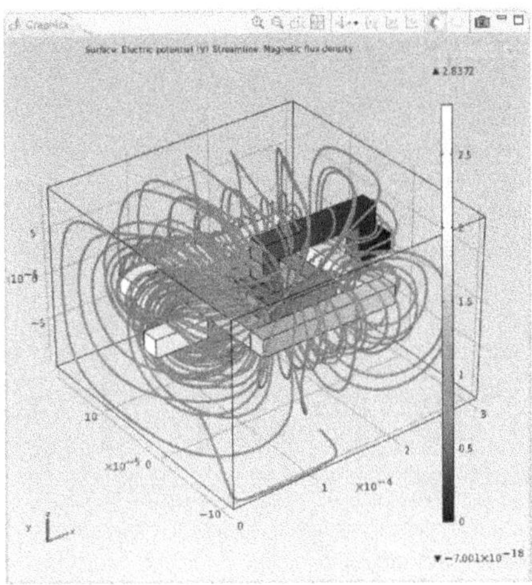

FIGURE 9.18 Spiral Coil Magnetic Field.

Figure 9.18 shows the Spiral Coil Magnetic Field.

NOTE *Because the Spiral Coil is contained within the Air Box and the Air Box is magnetically insulated, the field lines follow the contour of the box. The limited size of the box introduces a small error in the calculated value of the inductance. The error can be roughly estimated from the relative magnitude of the magnetic field at the boundary of the box.*

Right-Click > Model Builder – Results – 3D Plot Group 1 – Streamline 1.

Select > Color Expression from the Pop-up menu.

Click > Replace Expression in Settings – Color Expression – Expression.

Select > Magnetic and Electric Fields – Magnetic flux density norm (mef.normB).

See Figure 9.19.

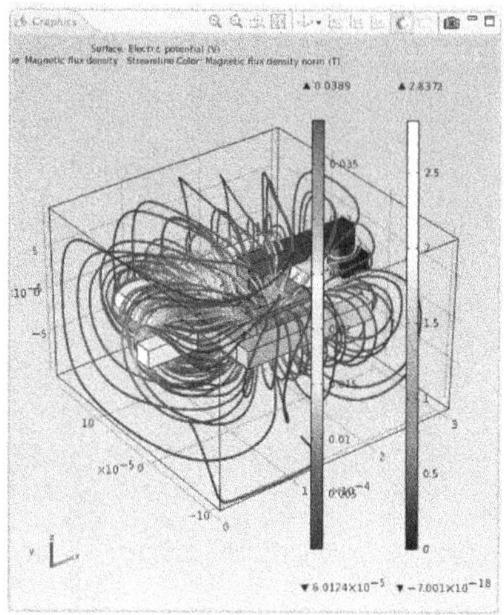

FIGURE 9.19 Spiral Coil Magnetic Field Intensity.

Figure 9.19 shows the Spiral Coil Magnetic Field Intensity.

NOTE *The modeler should note that the magnetic field intensity in the center of the Spiral Coil is approximately 650 times stronger than the magnetic field intensity at the boundary of the Air Box (ratio of the strongest magnetic field in the box to that of the weakest magnetic field in the box).*

Calculating Inductance

Right-Click > Model Builder – Results – 3D Plot Group 1 – Streamline 1.

Select > Plot from the Pop-up menu.

NOTE *The Plot command was just executed to insure that the software was in the correct place to execute the next step.*

Right-Click > Model Builder – Results – Derived Values.

Select > Global Evaluation.

Click > Settings – Global Evaluation – Expression – Replace Expression.

Select > Magnetic and Electric Fields (Electric Currents) – Inductance (mef.L11).

Click > Evaluate (Equals Sign (=)) in the Settings Toolbar.

NOTE *The calculated value for the Spiral Coil Inductance is shown in the Results window below the Graphics window. The value of L for this model is 0.52556E-9H (nanoHenry [nH]).*

See Figure 9.20.

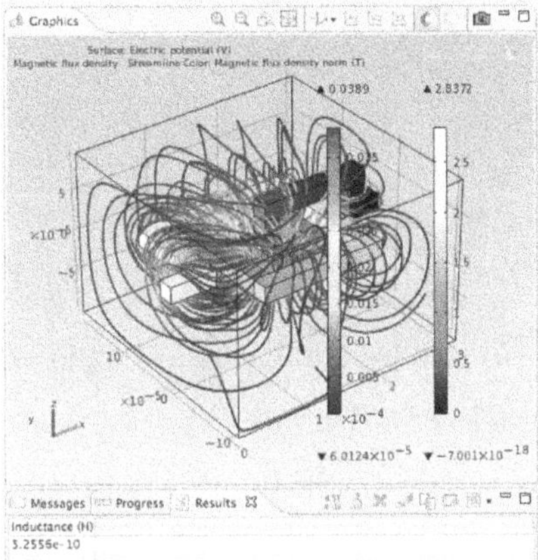

FIGURE 9.20 Spiral Coil Inductance.

Figure 9.20 shows the Spiral Coil Inductance.

3D Spiral Coil Microinductor Model Summary and Conclusions

The 3D Spiral Coil Microinductor Model is a powerful modeling tool that can be used to calculate the inductance of spiral coils of various different sizes. With this model, the modeler can easily vary all of the deposition parameters and optimize the design before the first prototype is physically built. These types of spiral coils are widely used in industry.

3D Linear Microresistor Beam Model

In this model, three different areas of physical analysis are combined and solved simultaneously. Those three areas of physics are electrical, thermal, and structural analysis. The analysis of this problem begins explicitly with the use of Ohm's Law {9.18}.

Which is:

$$V = IR \tag{9.4}$$

Where: V = Applied voltage [V].

I = Current in amperes [A].

R = Resistance in ohms [ohms].

When current flows through a resistance, power is dissipated and then Joule's First Law {9.19} applies. James Prescott Joule {9.20} discovered his power (first) law in the 1840s.

That Law is:

$$Q = I^2 \bullet R \bullet t \tag{9.5}$$

Where: Q = Heat Generated in Joules [J].

I = Current in Amperes [A] (assumed to be constant).

R = Resistance in Ohms [ohms] (assumed to be constant).

t = Time in Seconds [s].

Using Ohm's Law and substituting for the resistance R, yields:

$$Q = I^2 \bullet \frac{V}{I} \bullet t = I \bullet V \bullet t \tag{9.6}$$

Where: Q = Heat Generated in joules [J].

I = Current in amperes [A].

V = Voltage in volts [V].

t = Time in seconds [s].

Dividing the heat generated by the time yields:

$$\frac{Q}{t} = I \bullet V = P \tag{9.7}$$

Where: Q = Heat Generated in joules [J].

I = Current in amperes [A].

V = Voltage in volts [V].

t = Time in seconds [s].
P = Power in watts [J/s].

Where P is the instantaneous rate of power dissipation.

It has long been observed that most materials change volume in either a positive (expand) or a negative (shrink) fashion when heated or cooled. Quantification of this anecdotal evidence began in approximately 1650 {9.21} when Otto von Guericke {9.22} built the first vacuum pump {9.23}. Subsequent work by Nicolas Leonard Sadi Carnot {9.24}, William Thomson (Lord Kelvin) {9.25}, and many others have brought thermodynamics to a very high level of development. Of interest in this particular model is the change in the resistance and shape of the copper beam, due to thermal expansion, from the resistive heating caused by current flow through the beam.

NOTE *As current flows through the beam, heat is deposited. As heat is deposited, the copper expands. As the copper expands, the beam bends upward. The upward bending is caused by the beam becoming longer with the beam ends fixed in place.*

Building the 3D Linear Microresistor Beam Model

Startup 4.x.

Select > 3D.

Click > Next.

Click > Twistie for Structural Mechanics Interface.

Click > Joule Heating and Thermal Expansion (tem).

Click > Add Selected.

Click > Next.

Select > Preset Studies – Stationary.

Click > Finish (Flag).

Click > Save As.

Enter MMUC4_3D_MRTE_1.mph.

Click > Save.

See Figure 9.21.

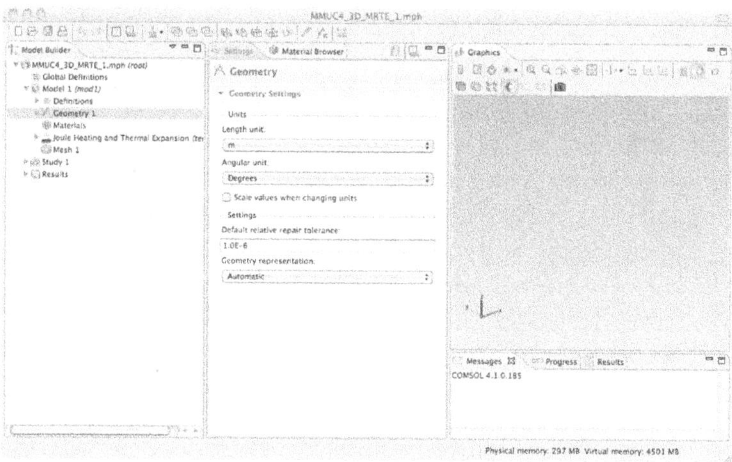

FIGURE 9.21 Desktop Display for the MMUC4_3D_MRTE_1.mph Model.

Figure 9.21 shows the Desktop Display for the MMUC4_3D_MRTE_1.mph model.

Global Definitions – Parameters

Right-Click > Model Builder – Global Definitions.

Select > Parameters from the Pop-up menu.

Enter > Parameters from Table 9.2 into the Settings – Parameters – Parameters edit windows.

TABLE 9.2 Microresistor Beam Parameters

Name	Expression	Description
V0	0.2[V]	Applied voltage
T0	323[K]	Base temperature
Text	298[K]	Room temperature
ht	5[W/(m^2*K)]	Heat transfer coefficient

See Figure 9.22.

Figure 9.22 shows the Filled Settings – Parameters – Parameters edit window.

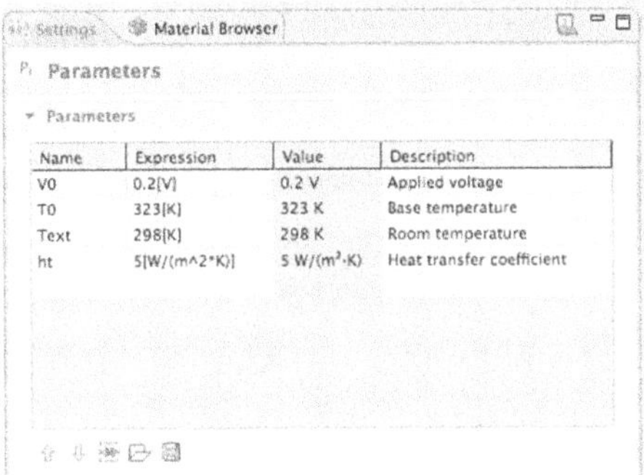

FIGURE 9.22 Filled Settings - Parameters - Parameters Edit Window.

Geometry

Geometry 1

Click > Model Builder – Model 1 (mod1) – Geometry 1.

Click > Length unit in Settings – Geometry – Geometry Settings – Units.

Select > μm from the list on the Pull-down menu.
See Figure 9.23.

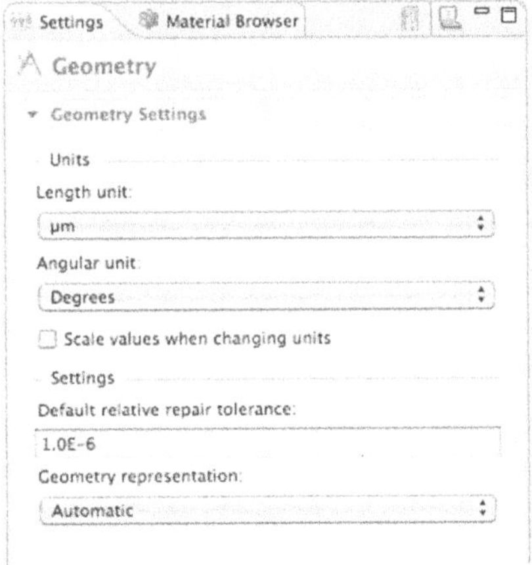

FIGURE 9.23 Settings - Geometry - Geometry Settings - Length Unit Setting.

Figure 9.23 shows the Settings – Geometry – Geometry Settings – Length unit setting.

In the following portion of the development of this model, a powerful tool is introduced. That tool is the Work Plane. The Work Plane allows the modeler to create the outline of a 3D object in 2D and then to add the third dimension by extrusion {9.26}.

Work Plane 1

Right-Click > Model Builder – Model 1 (mod1) – Geometry 1.

Select > Work Plane from the Pop-up menu.

Bézier Polygon 1

Right-Click > Model Builder – Model 1 (mod1) – Geometry 1 – Work Plane 1 – Geometry.

Select > Bézier Polygon from the Pop-up menu.

Click > Add Linear button in Settings – Bézier Polygon – Polygon Segments.

See Figure 9.24.

Figure 9.24 shows the Settings – Bézier Polygon edit window.

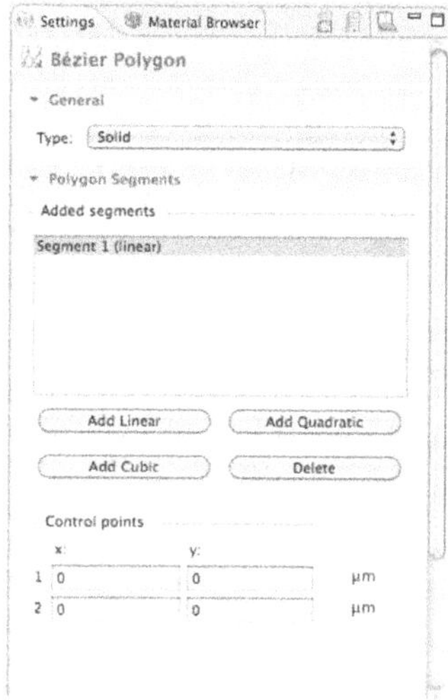

FIGURE 9.24 Settings - Bézier Polygon Edit Window.

Enter the Control point values shown in Table 9.3 in the Bézier Polygon edit window and Click the Add Linear button once, as indicated.

TABLE 9.3 Bézier Polygon Control Points

Segment	Row #	x	y	Click
1	1	0	0	(no Click)
1	2	5	1.5	Add Linear
2	2	5	1.5	Add Linear
3	2	18	1.5	Add Linear
4	2	23	0	Add Linear
5	2	23	4	Add Linear
6	2	18	2.5	Add Linear
7	2	5	2.5	Add Linear
8	2	0	4	Add Linear

NOTE *Closing the curve, as indicated below, adds the 8th segment that completes (closes) the curve.*

Click > Close Curve button in Settings – Bézier Polygon – Polygon Segments.

Click > Build All.

Click > Zoom Extents.
See Figure 9.25.

Figure 9.25 shows the Bézier Polygon Settings and Built Curve.

Extrude 1

Right-Click > Model Builder – Model 1 (mod1) – Geometry 1 – Work Plane 1.

Select > Extrude from the Pop-up menu.

Enter > 3 in Settings – Extrude – Distances from Work Plane – Distances edit window.

Click > Build All.

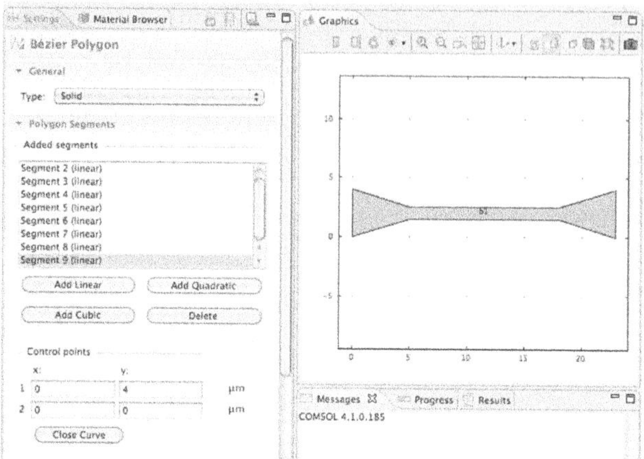

FIGURE 9.25 Bézier Polygon Settings and Built Curve.

See Figure 9.26.

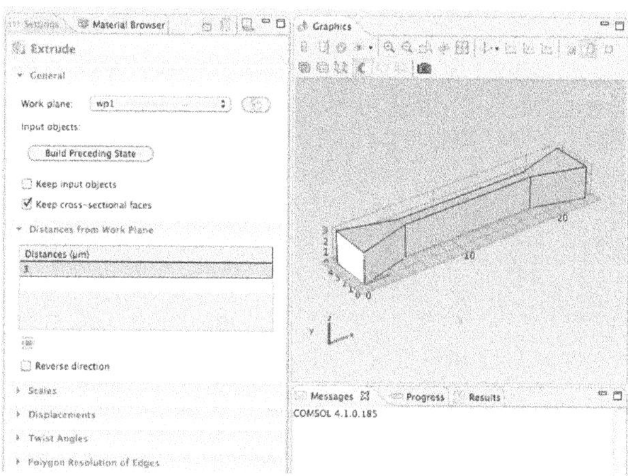

FIGURE 9.26 Bézier Polygon Settings and Built Curve Extruded.

Figure 9.26 shows the Bézier Polygon Settings and Built Curve Extruded.

Work Plane 2

Right-Click > Model Builder – Model 1 (mod1) – Geometry 1.

Select > Work Plane from the Pop-up menu.

Click > Plane type in Settings – Work Plane – Work Plane.

Select > Face parallel from the Pop-up menu.

Click > Boundary 6 on ext 1 in the Graphics window (see Figure 9.27).

Figure 9.27 shows the Boundary 6 of ext 1 in Graphics window.

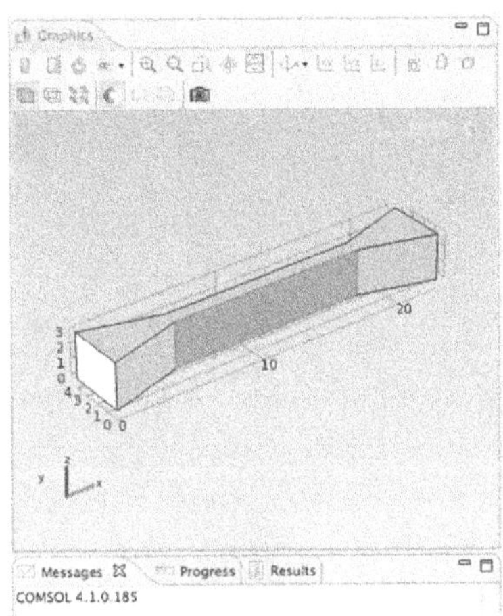

FIGURE 9.27 Boundary 6 of Ext 1 in Graphics Window.

Click > Add to Selection (Plus sign) in Settings – Work Plane – Work Plane.

See Figure 9.28.

Figure 9.28 shows Work Plane 2 before offset.

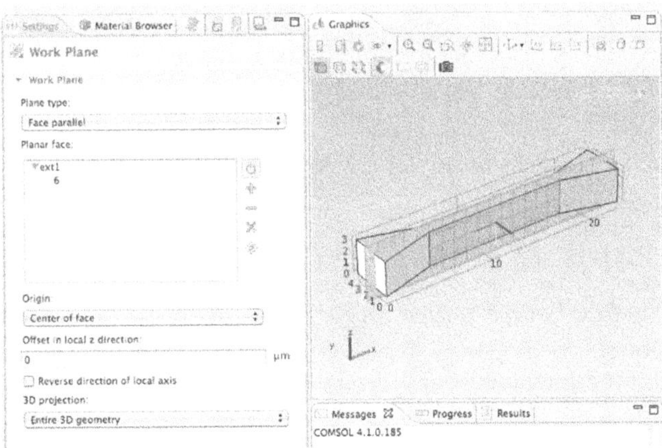

FIGURE 9.28 Work Plane 2 Before Offset.

NOTE *In the following model building steps, the location of the work plane and the local axis will be changed. These changes are made so that when the final solid figure is extruded, it will be created in the correct location.*

Enter > -1.5 in Settings – Work Plane – Work Plane – Offset in local z direction.

Click > Reverse direction of local axis check box in Settings – Work Plane – Work Plane.

See Figure 9.29.

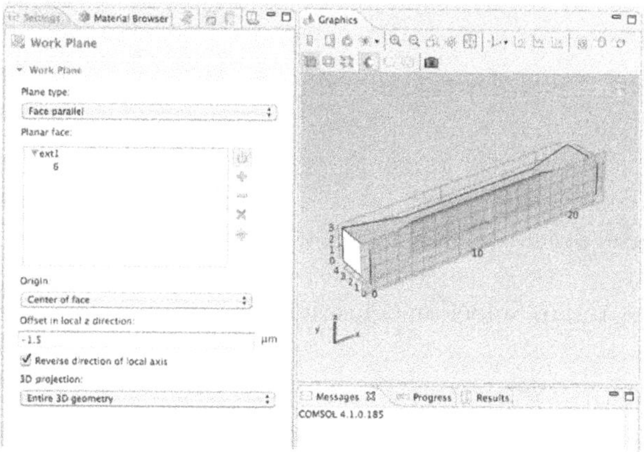

FIGURE 9.29 Work Plane 2 After Offset.

Figure 9.29 shows Work Plane 2 after offset.

Bézier Polygon 1

Right-Click > Model Builder – Model 1 (mod1) – Geometry 1 – Work Plane 2 –Geometry.

Select > Bézier Polygon from the Pop-up menu.

Click > Add Linear button in Settings – Bézier Polygon – Polygon Segments.

See Figure 9.30.

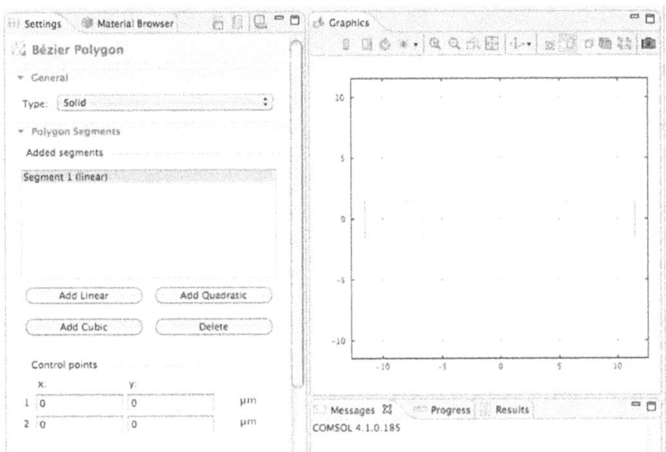

FIGURE 9.30 Settings - Bézier Polygon Edit Window.

Figure 9.30 shows the Settings – Bézier Polygon edit window.

Enter the Control point values shown in Table 9.4 in the Bézier Polygon edit window and Click the Add Linear button once, as indicated.

TABLE 9.4 Bézier Polygon Control Points

Segment #	Row #	x	y	Click
1	1	-11.5	-1.5	(no Click)
1	2	-6.3	-1.5	Add Linear
2	2	-6.3	0.5	Add Linear
3	2	6.3	0.5	Add Linear
4	2	6.3	-1.5	Add Linear
5	2	11.5	-1.5	Add Linear
6	2	6.5	1.5	Add Linear
7	2	-6.5	1.5	Add Linear

NOTE *Closing the curve, as indicated below, adds the 8thsegment that completes (closes) the curve.*

Click > Close Curve button in Settings – Bézier Polygon – Polygon Segments.

Click > Build All.

Click > Zoom Extents.

See Figure 9.31.

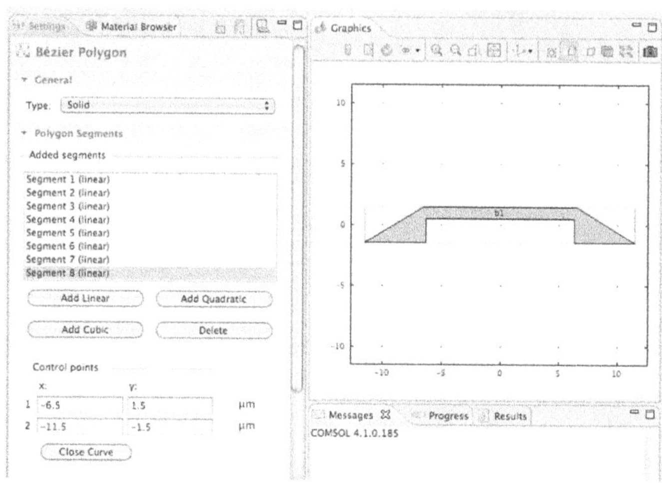

FIGURE 9.31 Settings - Bézier Polygon Edit Window and Closed Bézier Polygon.

Figure 9.31 shows the Settings – Bézier Polygon edit window and Closed Bézier Polygon.

Fillet 1

Right-Click > Model Builder – Model 1 (mod1) – Geometry 1 – Work Plane 2 – Geometry.

Select > Fillet from the Pop-up menu.

Select > Points 4 and 6 in the Graphics window.

Click > Add to Selection in Settings – Fillet – Points.

Enter > 0.3 in Settings – Fillet – Radius.

Click > Build All.

See Figure 9.32.

Figure 9.32 shows the Settings – Bézier Polygon with Fillets.

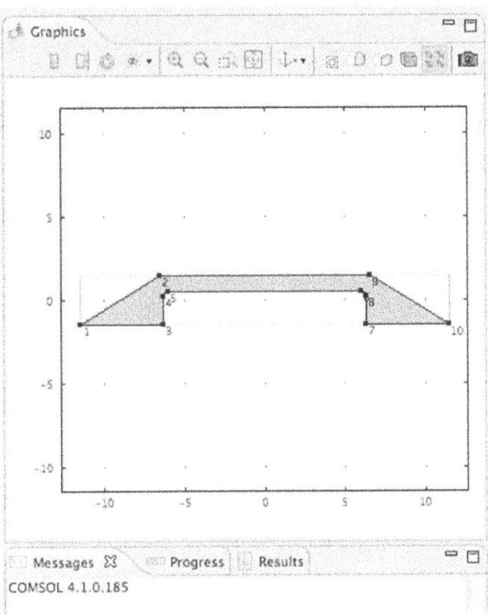

FIGURE 9.32 Settings - Bézier Polygon with Fillets.

Extrude 2

Right-Click > Model Builder – Model 1 (mod1) – Geometry 1 – Work Plane 2.

Select > Extrude from the Pop-up menu.

Enter > 4 in Settings – Extrude – Distances from Work Plane – Distances edit window.

Click > Build All.

See Figure 9.33.

Figure 9.33 shows the Work Plane 2 Settings and Extrusion.

Intersection 1

Right-Click > Model Builder – Model 1 (mod1) – Geometry 1.

Select > Boolean Operations – Intersection from the Pop-up menu.

Select > Objects ext1 and ext2 in the Graphics window.

Click > Add to Selection in Settings – Intersection – Intersection – Input objects (Plus sign).

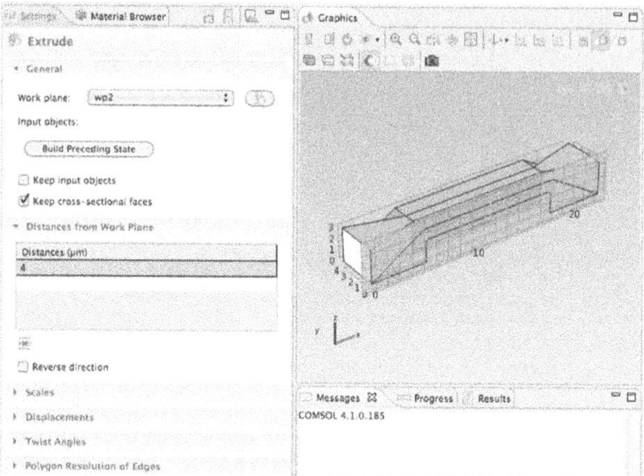

FIGURE 9.33 Work Plane 2 Settings and Extrusion.

See Figure 9.34.

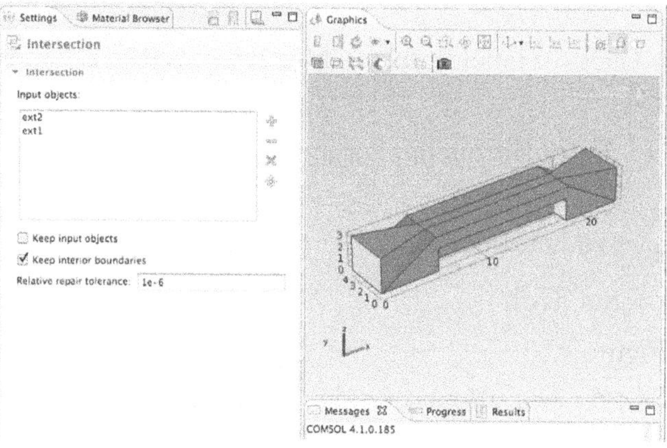

FIGURE 9.34 Work Plane 2 Settings and Extrusion.

Figure 9.34 shows the Work Plane 2 Settings and Extrusion.

Click > Build All.

Form Union

Right-Click > Model Builder – Model 1 (mod1) – Geometry 1 – Work Plane 2 – Form Union.

Select > Build Selected.

See Figure 9.35.

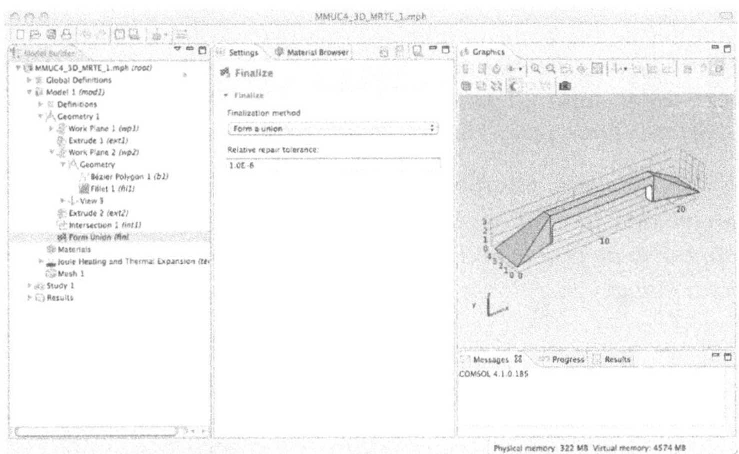

FIGURE 9.35 Desktop Display of the Completed Microresistor Beam.

Figure 9.35 shows the Desktop Display of the Completed Microresistor Beam.

Definitions

Now that the Microresistor Beam has been built, some definitions of collections of boundaries are going to be added to make the setting of boundary conditions easier later in the model.

Selection 1

Right-Click > Model Builder – Model 1 (mod1) – Definitions.

Select > Selection.

Right-Click > Selection 1.

Select > Rename.

Enter > connector1 in the Rename Selection – New name edit window.

Click > OK.

Click > Geometric entity level in Settings – Selection – Geometric Scope.

Select > Boundary from the Pull-down menu.

NOTE

The easiest method of picking a single boundary or a small set of boundaries is by Clicking the All boundaries check box first and then Clicking the All boundaries check box a second time. All the modeler then has to do is remove (Remove from Selection (Minus sign)) the boundaries that are not needed from the edit window.

If the modeler wants to view the selected boundary in the Graphics window to verify his choice and the selected boundary does not appear to be visible in that view, rotate the graphic in the Graphics window.

Select > Boundary 1 only.

Selection 2

Right-Click > Model Builder – Model 1 (mod1) – Definitions.

Select > Selection.

Right-Click > Selection 2.

Select > Rename.

Enter > connector2 in the Rename Selection – New name edit window.

Click > OK.

Click > Geometric entity level in Settings – Selection – Geometric Scope.

Select > Boundary from the Pull-down menu.

Select > Boundary 13 only.

Selection 3

Right-Click > Model Builder – Model 1 (mod1) – Definitions.

Select > Selection.

Right-Click > Selection 3.

Select > Rename.

Enter > connectors in the Rename Selection – New name edit window.

Click > OK.

Click > Geometric entity level in Settings – Selection – Geometric Scope.

Select > Boundary from the Pull-down menu.

Select > Boundaries 1 and 13 only.

Materials

Material 1

Right-Click > Model Builder – Model 1 (mod1) – Materials.

Select > Open Material Browser from the Pop-up menu.

Click > Settings – Material Browser – MEMS – Metals – Twistie.

Click > Settings – Material Browser – MEMS – Metals – Cu.

Right-Click > Settings – Material Browser – MEMS – Metals – Cu.

Select > Add Material to Model.

Material - Cu

Click > Model Builder – Model 1 (mod1) – Materials - Cu.

Enter > 1 in the Settings – Material – Material Contents Relative Permittivity (epsilonr) edit window.

See Figure 9.36.

Figure 9.36 shows the Cu Settings – Material.

Click > Settings – Material – Material Properties – Electromagnetic Models – Twistie.

Click > Settings – Material – Material Properties – Electromagnetic Models – Linearized Resistivity – Twistie.

Right-Click > Settings – Material – Material Properties – Electromagnetic Models – Linearized Resistivity – Reference Resistivity (rho0).

Select > Add to Material (plus sign).

Enter > Values as indicated in Table 9.5.

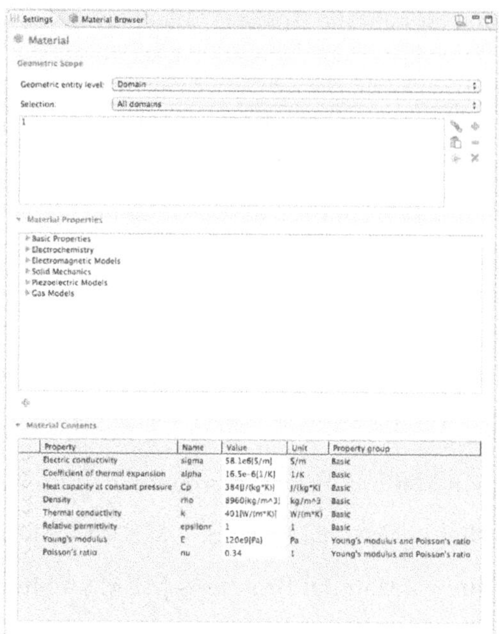

FIGURE 9.36 Cu Settings - Material.

TABLE 9.5 Cu Reference Resistivity Parameters

Property	Name	Value
Reference Resistivity	rho0	1.72E-8[ohm*m]
Resistivity temperature coefficient	alpha	3.9E-3[1/K]
Reference temperature	Tref	293[K]

See Figure 9.37.

Figure 9.37 shows the Linearized Cu Settings – Material.

Joule Heating and Thermal Expansion

Click > Model Builder – Model 1 (mod1) – Joule Heating and Thermal Expansion (tem) – Twistie.

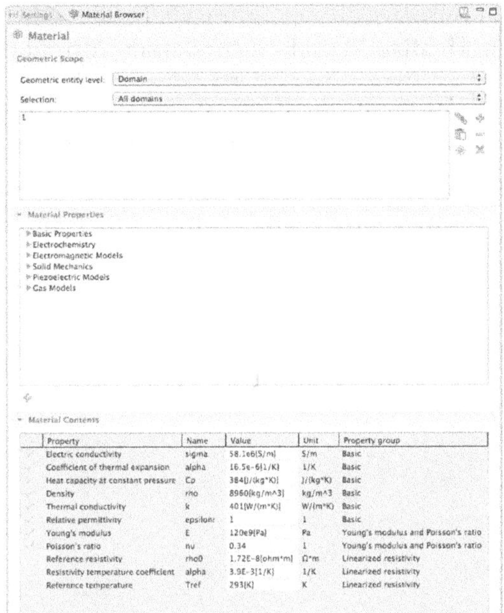

FIGURE 9.37 Linearized Cu Settings – Material.

Click > Model Builder – Model 1 (mod1) – Joule Heating and Thermal Expansion (tem) – Joule Heating Model 1.

Click > Electrical conductivity in Settings – Joule Heating Model – Conduction Current.

Select > Linearized resistivity from the Pop-up menu.

NOTE *The model is initially solved for a temperature independent resistivity to allow comparison of the temperature dependent and temperature independent resistive behavior.*

Click > Resistivity temperature coefficient in Settings – Joule Heating Model – Conduction Current.

Select > User defined from the Pop-up menu.

Enter > 0 in Resistivity temperature coefficient edit window in Settings – Joule Heating Model – Conduction Current (default value).

See Figure 9.38.

Figure 9.38 shows the Joule Heating Model 1 Settings.

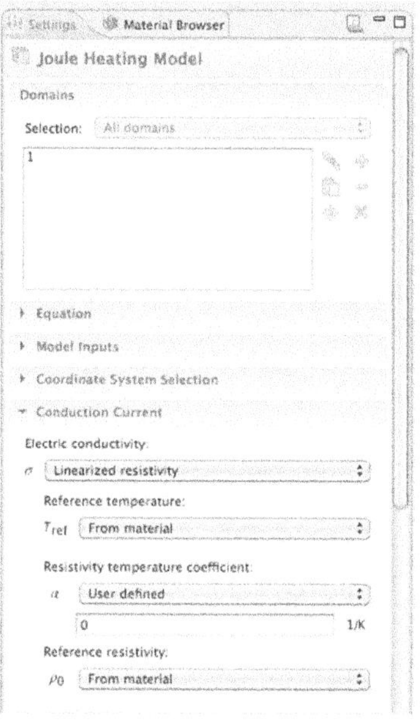

FIGURE 9.38 Joule Heating Model 1 Settings.

Thermal Linear Elastic 1

Click > Model Builder – Model 1 (mod1) – Joule Heating and Thermal Expansion (tem) – Thermal Linear Elastic 1.

Enter > Text (a Global Parameter) in Settings – Thermal Linear Elastic – Thermal Expansion – Strain reference temperature Tref edit window.

See Figure 9.39.

Figure 9.39 shows the Thermal Linear Elastic 1 Settings.

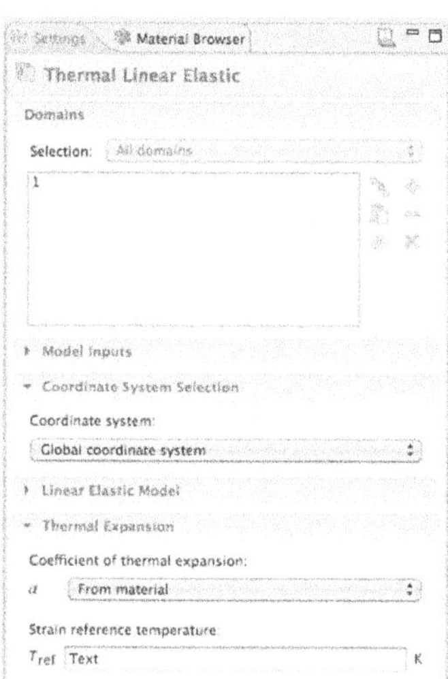

FIGURE 9.39 Thermal Linear Elastic 1 Settings.

Initial Values 1

Click > Model Builder – Model 1 (mod1) – Joule Heating and Thermal Expansion (tem) – Initial Values 1.

Enter > T0 in Settings – Initial Values – Initial Values temperature T edit window.

See Figure 9.40.

Figure 9.40 shows the Initial Values 1 Settings.

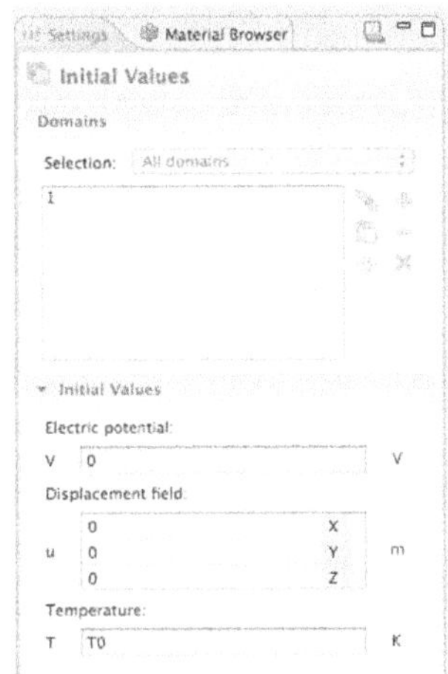

FIGURE 9.40 Initial Values 1 Settings.

Ground 1

The Pop-up menu for the Physics Interfaces is divided into hierarchical sections {9.27}. First are Domains. Second are Boundaries, etc.

Right-Click > Model Builder – Model 1 (mod1) – Joule Heating and Thermal Expansion (tem).

Select > Electric Currents – Ground from the boundary condition section of the Pop-up menu.

Click > Ground 1.

Click > Selection in Settings – Ground – Boundaries – Selection.

Select > connector2 from the Pop-up menu.

FIGURE 9.41 Ground Boundary Setting.

See Figure 9.41.

Figure 9.41 shows the Ground Boundary Setting.

Electric Potential 1

Right-Click > Model Builder – Model 1 (mod1) – Joule Heating and Thermal Expansion (tem).

Select > Electric Currents – Electric Potential from the boundary condition section of the Pop-up menu.

Click > Electric Potential 1.

Click > Selection in Settings – Electric Potential – Boundaries – Selection.

Select > connector1 from the Pop-up menu.

Enter > V0 in the Settings – Electric Potential – Electric Potential – Voltage V_0 edit window.

See Figure 9.42.

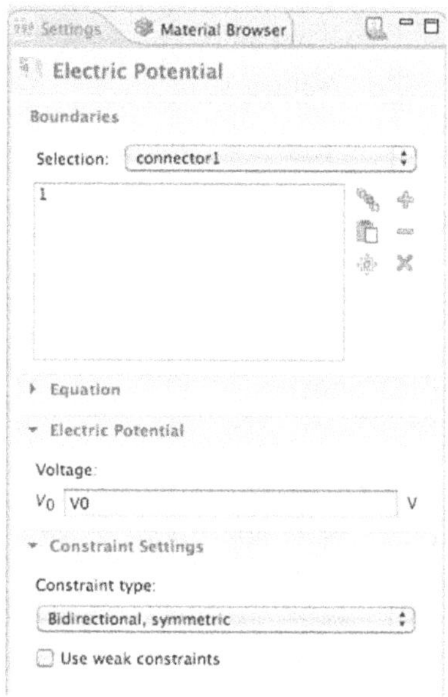

FIGURE 9.42 Electric Potential Boundary Setting.

Figure 9.42 shows the Electric Potential Boundary Setting.

Heat Flux 1

Right-Click > Model Builder – Model 1 (mod1) – Joule Heating and Thermal Expansion (tem).

Select > Heat Transfer – Heat Flux from boundary condition section of the Pop-up menu.

Click > Heat Flux 1.

Click > Selection in Settings – Heat Flux – Boundaries – Selection.

Select > All boundaries from the Pop-up menu.

Click > Settings – Heat Flux – Heat Flux – Inward heat flux.

Enter > ht in the Settings – Heat Flux – Heat Flux – Heat transfer coefficient h edit window.

Enter > Text in the Settings – Heat Flux – Heat Flux – External temperature Text edit window.

See Figure 9.43.

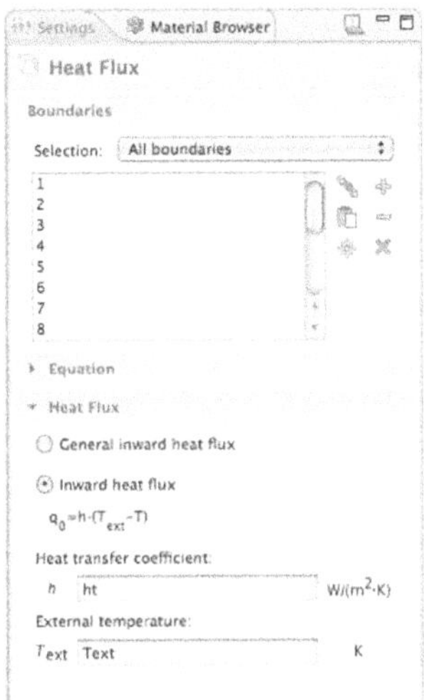

FIGURE 9.43 Heat Flux Boundary Settings.

Figure 9.43 shows the Heat Flux Boundary Settings.

Temperature 1

Right-Click > Model Builder – Model 1 (mod1) – Joule Heating and Thermal Expansion (tem).

Select > Heat Transfer – Temperature from boundary condition section of the Pop-up menu.

Click > Temperature 1.

Click > Selection in Settings – Temperature – Boundaries – Selection.

Select > connectors from the Pop-up menu.

Enter > T0 in the Settings – Temperature – Temperature – Temperature T_0 edit window.

See Figure 9.44.

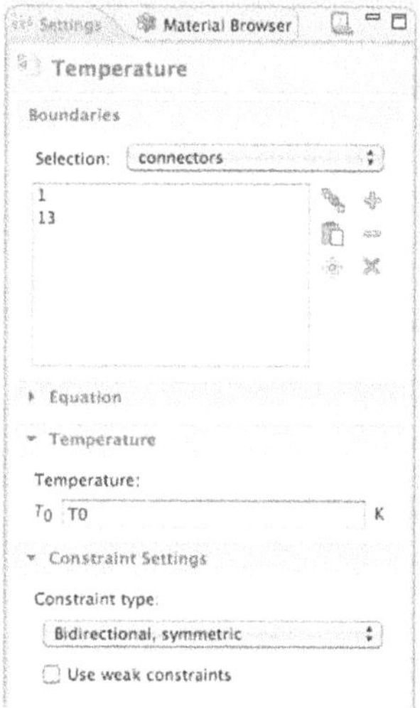

FIGURE 9.44 Temperature 1 Boundary Settings.

Figure 9.44 shows the Temperature 1 Boundary Settings.

Fixed Constraint 1

Right-Click > Model Builder – Model 1 (mod1) – Joule Heating and Thermal Expansion (tem).

Select > Solid Mechanics – Fixed Constraint from boundary condition section of the Pop-up menu.

Click > Fixed Constraint 1.

Click > Selection in Settings – Fixed Constraint – Boundaries – Selection.

Select > connectors from the Pop-up menu.

See Figure 9.45.

Figure 9.45 shows the Fixed Constraint 1 Boundary Settings.

Mesh 1

Right-Click > Model Builder – Model 1 – Mesh 1.

Select > Free Tetrahedral from the Pop-up menu.

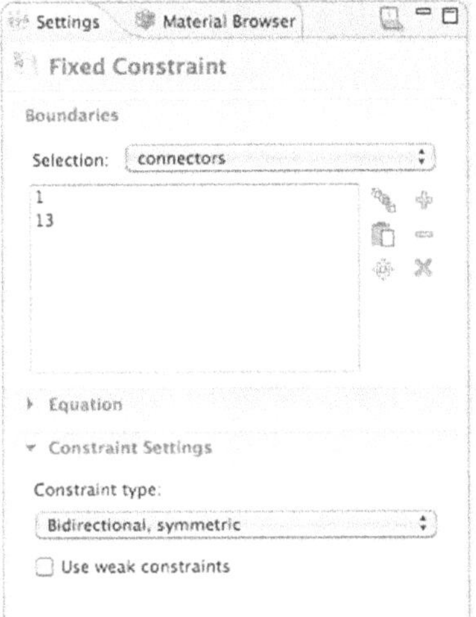

FIGURE 9.45 Fixed Constraint 1 Boundary Settings.

Click > Model Builder – Model 1 – Mesh 1 – Size.

Click > Predefined Pull-down list in Settings – Size – Element Size.

Select > Finer.

See Figure 9.46.

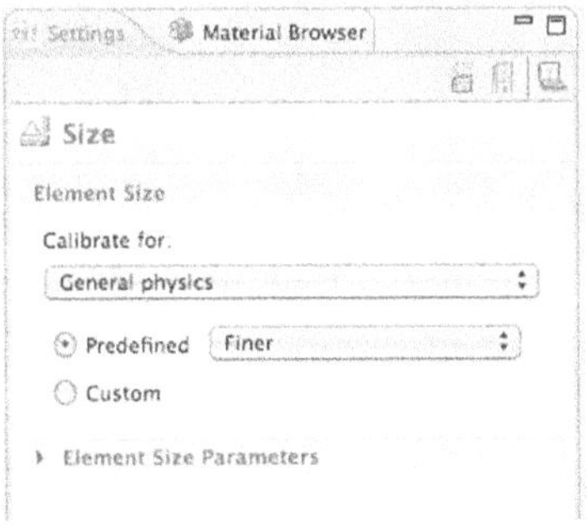

FIGURE 9.46 Mesh Size Settings.

Figure 9.46 shows the Mesh Size Settings.

Click > Build All.

See Figure 9.47.

Figure 9.47 shows the Graphics Window, Mesh.

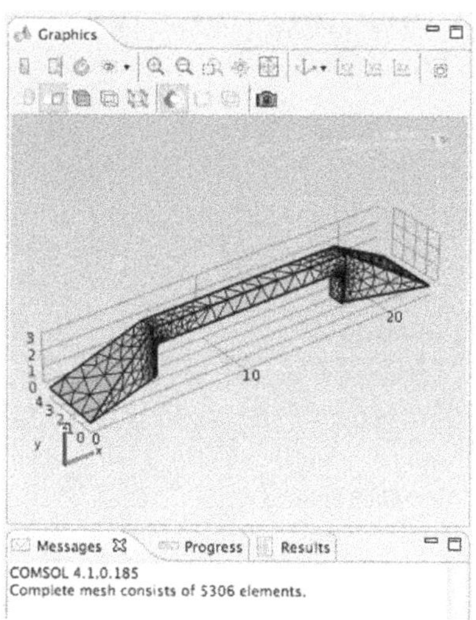

FIGURE 9.47 Graphics Window, Mesh.

NOTE *After the mesh is built, the model should have 5306 elements.*

Study 1

In Model Builder,

Right-Click > Study 1.

Select > Compute from the Pop-up menu.

Computed results, using the default display settings, are shown in Figure 9.48.

Figure 9.48 shows the Converged Microresistor Beam Model using the Default Plot Parameters.

NOTE *The modeler should note that the computed results, using the default display settings, initially display the model solution. However, with some additional display parameter adjustments in the way that the data are presented, the model solution presentation will be significantly enhanced.*

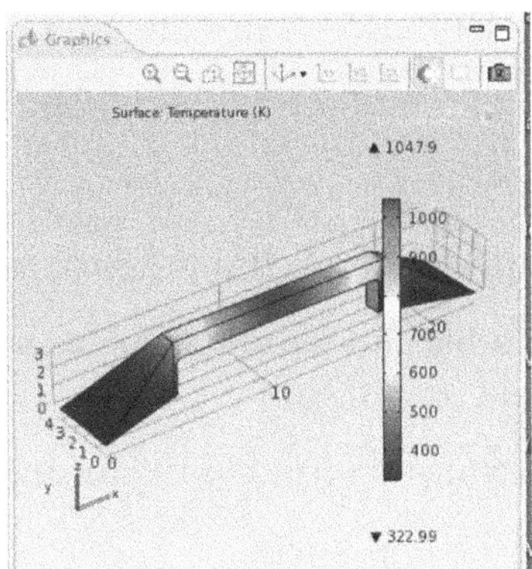

FIGURE 9.48 Converged Microresistor Beam Model Using the Default Plot Parameters.

Results

3D Plot Group 1

NOTE *The modeler should note that the computed results for the maximum temperature, using the copper linearized resistivity, is approximately 1048 K.*

Click > Model Builder – Results – Twistie.

Click > Model Builder – Results – 3D Plot Group 1 – Twistie.

NOTE *The modeler executes the following steps to ensure that the model is properly initialized, before proceeding to make modifications in the postprocessing display parameters.*

Right-Click > Model Builder – Results – 3D Plot Group 1 – Surface 1.

Select > Plot from the Pop-up menu.

3D Plot Group 2

Click > Model Builder – Results – 3D Plot Group 2 – Twistie.

Right-Click > Model Builder – Results – 3D Plot Group 2 – Slice 1.

Select > Delete from the Pop-up menu.

Click > Yes button on the Pop-up window.

Right-Click > Model Builder – Results – 3D Plot Group 2.

Select > Surface.

Click > Surface 1.

Click Replace Expression in Settings – Surface – Expression.

Select > Joule Heating and Thermal Expansion (Solid Mechanics) – Total Displacement (tem.disp).

Click > Unit in Settings – Surface – Expression.

Select > nm from the Pop-up list.
See Figure 9.49.

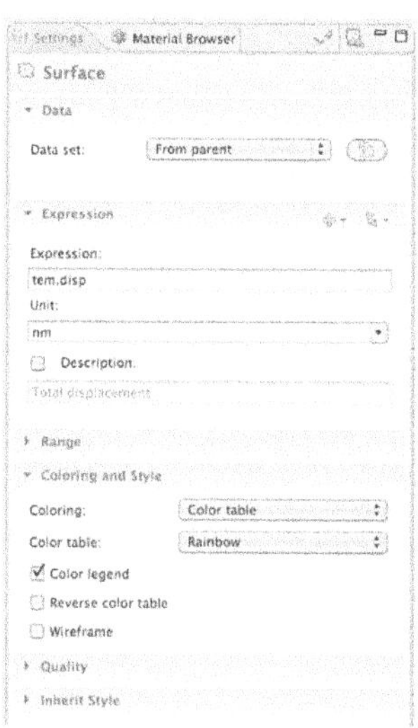

FIGURE 9.49 Thermal Expansion Parameters.

Figure 9.49 shows the Thermal Expansion Parameters.

Click > Plot in the Settings Toolbar.

NOTE

The modeler should note that the maximum thermal expansion under the linearized resistivity condition is approximately 88 nm.

Next, it is interesting to plot the displacement and the surface deformation.

Right-Click > Model Builder – Results – 3D Plot Group 2 – Surface 1.

Select > Deformation from the Pop-up menu.

See Figure 9.50.

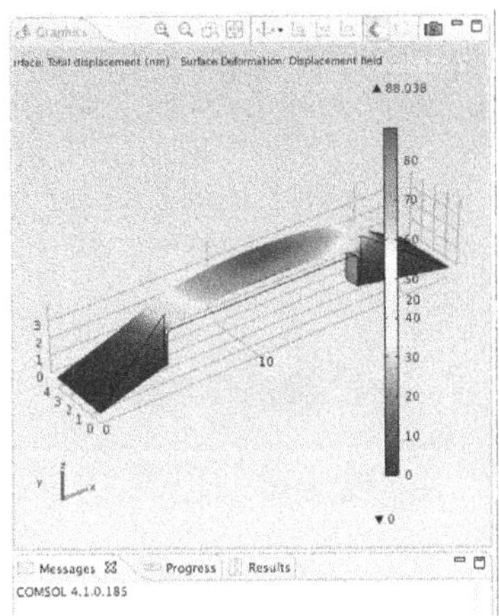

FIGURE 9.50 Thermal Deformation Graphics Plot.

Figure 9.50 shows the Thermal Deformation Graphics Plot.

Now that the modeler has computed the model with the linearized resistivity, it will be interesting to do the same model with a thermally dependent resistivity.

Joule Heating Model 1

Click > Model Builder – Model 1 (mod1) – Joule Heating and Thermal Expansion (tem) – Joule Heating Model 1.

Click > Resistivity temperature coefficient (a) in Settings – Joule Heating Model – Conduction Current.

Select > From material from the Pop-up menu.

See Figure 9.51.

Figure 9.51 shows the Resistivity Temperature Coefficient Settings.

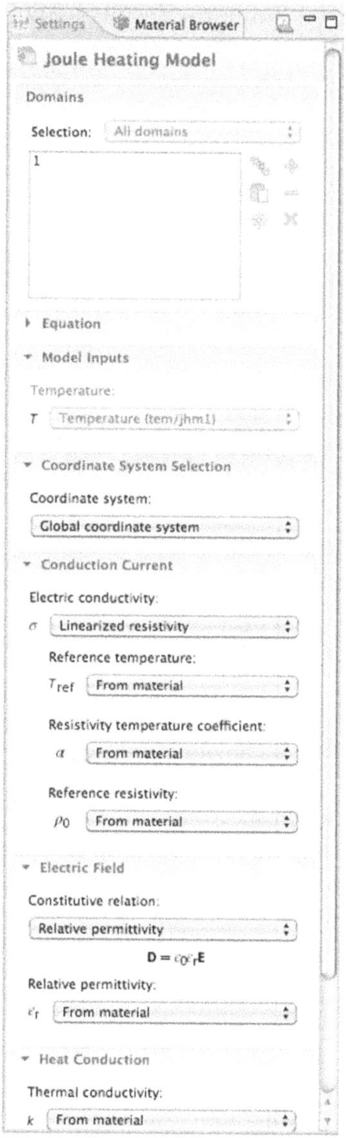

FIGURE 9.51 Resistivity Temperature Coefficient Settings.

Study 1

In Model Builder,

Right-Click > Study 1.

Select > Compute from the Pop-up menu.

Results

3D Plot Group 1

Computed results, using the thermally dependent resistivity settings, are shown in Figure 9.52.

Figure 9.52 shows the Converged Thermally Dependent Microresistor Beam Model.

NOTE *The modeler should note that the maximum temperature has dropped from approximately 1048 K to approximately 710 K.*

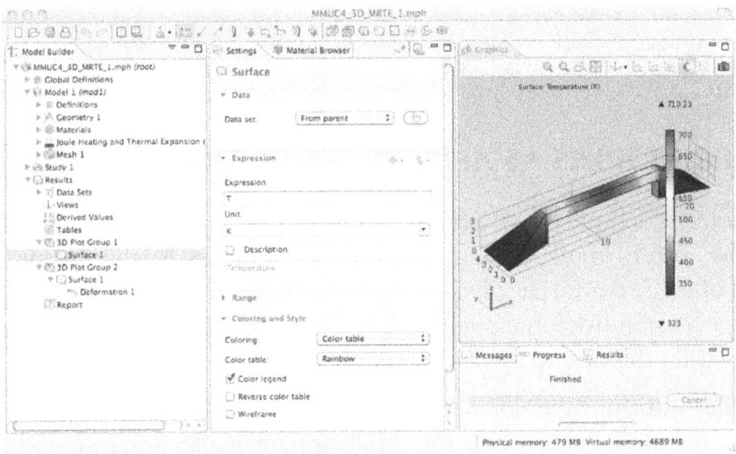

FIGURE 9.52 Converged Thermally Dependent Microresistor Beam Model.

3D Plot Group 2

Computed deformation results, using the thermally dependent resistivity settings, are shown in Figure 9.53.

Figure 9.53 shows the Converged Thermally Deformation in the Microresistor Beam Model.

NOTE *The modeler should note that the maximum deformation has dropped from approximately 88 nm to approximately 48 nm.*

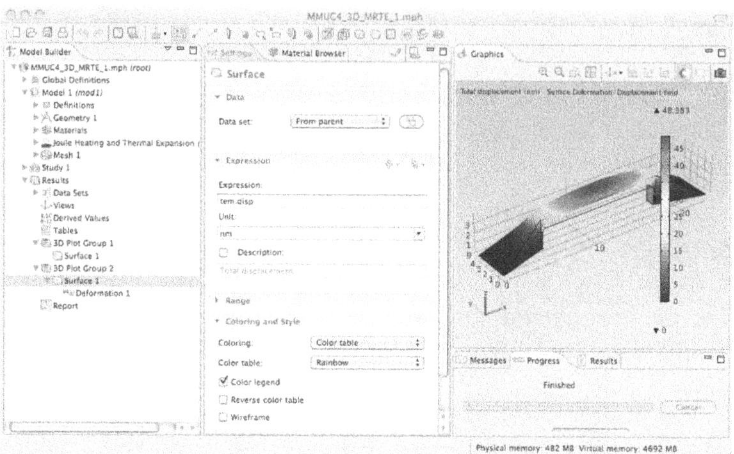

FIGURE 9.53 Converged Thermally Deformation in the Microresistor Beam Model.

FIRST PRINCIPLES AS APPLIED TO 3D MODEL DEFINITION

First Principles Analysis derives from the fundamental laws of nature. In the case of models using this Classical Physics Analysis approach, the laws of conservation in physics require that what goes in (as mass, energy, charge, etc.) must come out (as mass, energy, charge, etc.) or must accumulate within the boundaries of the model.

The careful modeler must be knowledgeable of the implicit assumptions and default specifications that are normally incorporated into the COMSOL Multiphysics software model when a model is built using the default settings.

Consider, for example, the two 3D models developed in this chapter. In these models, it is implicitly assumed that there are no thermally related changes (mechanical, electrical, etc.), except as specified. It is also assumed the materials are homogeneous and isotropic, except as specifically indicated and there are no thin insulating contact barriers at the thermal junctions. None of these assumptions are typically true in the general case. However, by making such assumptions, it is possible to easily build 3D First Approximation Models.

NOTE

A First Approximation Model is one that captures all the essential features of the problem that needs to be solved, without dwelling excessively on all of the small details. A good First Approximation

Model will yield an answer that enables the modeler to determine if he needs to invest the time and the resources required to build a more highly detailed model.

Also, the modeler needs to remember to name model parameters carefully as pointed out in Chapter 1.

REFERENCES

9.1 COMSOL Multiphysics Users Guide, Version 4.1, pp. 288–292

9.2 http://en.wikipedia.org/wiki/Vector_(mathematics_and_physics)

9.3 http://en.wikipedia.org/wiki/Tensor

9.4 http://en.wikipedia.org/wiki/Analytic_solution

9.5 http://en.wikipedia.org/wiki/Lumped_element_model

9.6 http://en.wikipedia.org/wiki/Michael_Faraday

9.7 http://en.wikipedia.org/wiki/Electromagnetic_induction

9.8 http://en.wikipedia.org/wiki/Joseph_Henry

9.9 http://en.wikipedia.org/wiki/Oliver_Heaviside

9.10 http://en.wikipedia.org/wiki/Maxwell%27s_equations

9.11 http://en.wikipedia.org/wiki/Inductance/Details_for_some_circuit_types

9.12 http://en.wikipedia.org/wiki/Impedance_parameters

9.13 MEMS Module Users Guide, Version 4.1, p. 46

9.14 http://en.wikipedia.org/wiki/Ohm%27s_law

9.15 http://en.wikipedia.org/wiki/Magnetic_field

9.16 http://en.wikipedia.org/wiki/Magnetic_energy

9.17 http://en.wikipedia.org/wiki/Inductor

9.18 http://en.wikipedia.org/wiki/Ohm%27s_law

9.19 http://en.wikipedia.org/wiki/Joule%27s_Laws

9.20 http://en.wikipedia.org/wiki/James_Prescott_Joule

9.21 http://en.wikipedia.org/wiki/Thermodynamics

9.22 http://en.wikipedia.org/wiki/Otto_von_Guericke

9.23 http://en.wikipedia.org/wiki/Vacuum_pump

9.24 http://en.wikipedia.org/wiki/Nicolas_Léonard_Sadi_Carnot

9.25 http://en.wikipedia.org/wiki/William_Thomson,_1st_Baron_Kelvin

9.26 COMSOL Multiphysics Users Guide, Version 4.1, p. 214

9.27 COMSOL Multiphysics Users Guide, Version 4.1, p. 293

Suggested Modeling Exercises

1. Build, mesh, and solve the 3D Spiral Coil Microinductor Model as presented earlier in this chapter.
2. Build, mesh, and solve the 3D Linear Microresistor Beam Model as presented earlier in this chapter.
3. Change the values of the materials parameters and then build, mesh, and solve the 3D Spiral Coil Microinductor Model as an example problem.
4. Change the values of the materials parameters and then build, mesh, and solve the 3D Linear Microresistor Beam Model as an example problem.
5. Change the value of the electrical parameters and then build, mesh, and solve the 3D Spiral Coil Microinductor Model as an example problem.
6. Change the value of the geometries of the microresistor and then solve the 3D Linear Microresistor Beam Model as an example problem.

CHAPTER 10

PERFECTLY MATCHED LAYER MODELS USING COMSOL MULTIPHYSICS 4.X

In This Chapter

GUIDELINES FOR PERFECTLY MATCHED LAYER (PML) MODELING IN 4.X

NOTE

In this chapter, two 2D PML models will be presented. PML models have proven to be very valuable in the study and application of wave propagation for the science and engineering communities, both in the past and currently. Such models serve as first-cut evaluations of potential systemic physical behavior under the influence of complex external stimuli. PML model responses and other such ancillary information can be gathered and screened early in a project for a first-cut evaluation of the physical behavior of a planned prototype. PML models are typically more conceptually and physically complex than the models that were presented in earlier chapters of this text. The calculated model (simulation) information can be used in the prototype fabrication stage as guidance in the selection of prototype geometry and materials.

Since the models in this and subsequent chapters are more conceptually complex and are potentially more difficult to solve than the models presented thus far, it is important that the modeler have available the tools necessary to most easily utilize the powerful capabilities of the 4.x software. In order to do that, if you have not done this previously, the modeler should go to the main 4.x toolbar, Click > Options – Preferences – Model builder. When the Preferences – Model builder edit window is shown, Select > Show equation view checkbox and Show more options checkbox. Click > Apply {10.1}.

Perfectly Matched Layer (PML) Modeling Guidelines and Coordinate Considerations

PML Theory

One of the fundamental difficulties underlying electromagnetic wave equation calculations (Maxwell's Equations {10.2}) is dealing with a propagating wave after the wave interacts with a boundary (reflection). If the boundary of a model domain is terminated in the typical fashion {10.3}, unwanted reflections will typically be incorporated into the solution, potentially creating undesired and possibly erroneous model solution values. Fortunately, for the modeler of today, there is a methodology that works sufficiently well that it essentially eliminates reflection problems at the domain boundary. That methodology is the Perfectly Matched Layer.

The Perfectly Matched Layer (PML) {10.4} is an approximation methodology originally developed in 1994 by Jean-Pierre Berenger for use with FDTD {10.5} (Finite-Difference Time-Domain) electromagnetic modeling calculations. The PML technique has now been adapted and applied to other calculational techniques that have similar domain mediated needs (e.g. FEM and others) {10.6}. The PML methodology can be applied to a large variety of diverse wave equation problems {10.7}. Herein, however, it is only applied to electromagnetic problems within the context of the COMSOL RF Module {10.8}.

NOTE *For more detailed applications and a history of the PML methodology in other types of wave problems, the modeler is referred to the literature.*

The PML methodology functions by adding anisotropic attenuating domains (layers) outside the modeled domain. The anisotropic attenuating domains (PMLs) create for the modeled domain a set of essentially reflectionless boundaries.

Examples of modeling domains with PMLs can be found in the COMSOL Multiphysics literature {10.9}. The coordinate systems employed with the domain structures are those that are associated with their respective geometries.

In order to achieve the desired behavior of the wave equation PDE, the entire model domain, including the Perfectly Matched Layers, is transformed to a complex coordinate system. For a Cartesian System (x, y, z), the transformation occurs as follows:

$$\frac{\partial}{\partial x}\to\frac{1}{1+\frac{i\sigma(x)}{\omega}}\frac{\partial}{\partial x};\quad \frac{\partial}{\partial y}\to\frac{1}{1+\frac{i\sigma(y)}{\omega}}\frac{\partial}{\partial y};\quad \frac{\partial}{\partial z}\to\frac{1}{1+\frac{i\sigma(z)}{\omega}}\frac{\partial}{\partial z} \tag{10.1}$$

NOTE *Where: $\sigma(x, y, z)$ = is a step function that is zero inside the solution domain and a positive real number or an appropriate function of the designated coordinate variable (x, y, z), outside the solution domain and inside the PML.*

The transformation of the PDE in the above fashion results in a solution with a multiplicative term that is, in general, as follows:

$$F(x,y,z)=f(x,y,z)*e^{-\frac{k\sigma(x,y,z)}{\omega}} \tag{10.2}$$

Where: $F(x,y,z) = f(x,y,z)*e^{-0}$ *(The Solution inside the domain).*

$F(x,y,z) = f(x,y,z)*e^{-\kappa\sigma(x,y,z)/\omega}$ *(The Decaying Solution within the PML domain).*

At the outer PML boundary, the preferred boundary condition is the Scattering boundary condition. However, if the attenuation of the propagating wave, at the outer boundary of the PML, is sufficient, then the particular boundary condition invoked is largely irrelevant. This pertains, since the amplitude of the reflected wave will be sufficiently small as not to contribute to the final solution.

Figures 10.1 and 10.2 show examples of wavefront behavior within a domain and within a PML. For an example of the wavefront inside the modeling domain see Figure 10.1. For an example of the wavefront inside the PML domain see Figure 10.2.

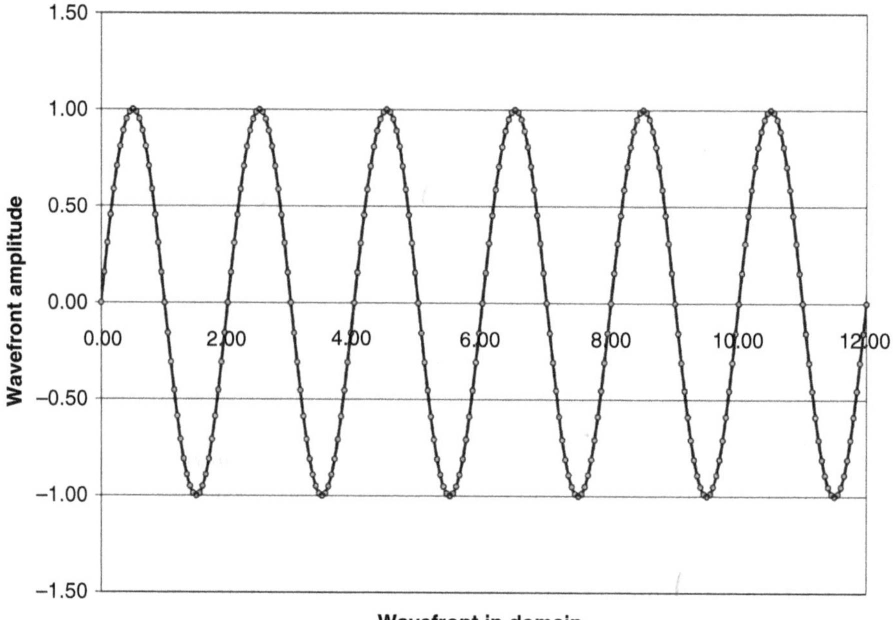

FIGURE 10.1 Wave Equation Solution Example Inside the Modeling Domain.

Figure 10.1 shows the wave equation solution example inside the modeling domain.

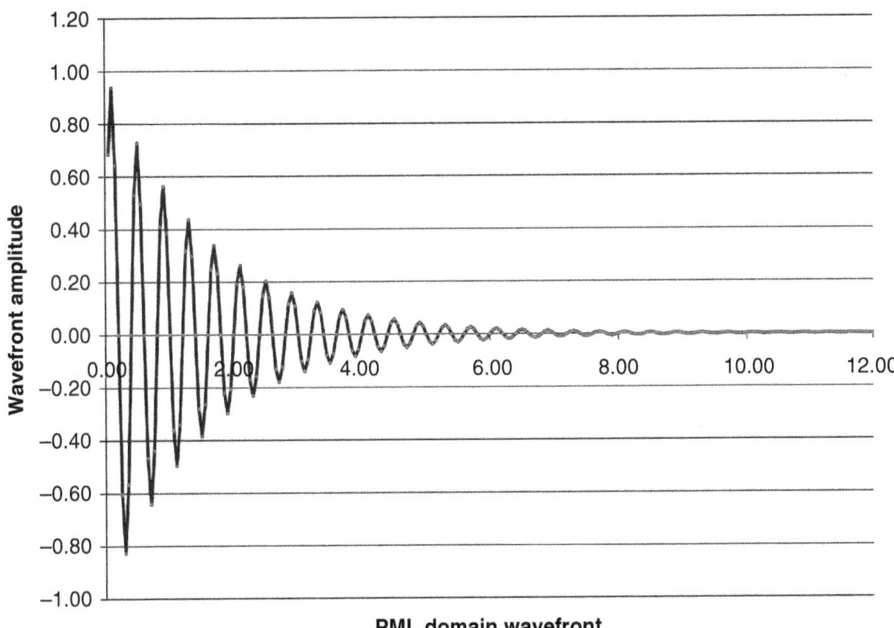

FIGURE 10.2 Wave Equation Solution Example Inside the PML Domain.

Figure 10.2 shows the wave equation solution example inside the PML domain.

PERFECTLY MATCHED LAYER MODELS

NOTE *The two models presented in this chapter are derived from the COMSOL PML tutorial model "Radar Cross Section."*

2D Concave Metallic Mirror PML Model

NOTE *The Concave Metallic Mirror is a concept widely utilized in Optical Physics. In this application, the principles of Optics are applied to lower frequency electromagnetic waves to focus an impinging wavefront into the region of a sensor. The act of focusing the wavefront effectively increases the magnitude of the impinging signal (i.e. adds Gain {10.10}) in the region of the sensor, thus making the focused signal more easily detectable.*

The focusing concept can also be utilized to concentrate large-area, diffuse energy from a renewable energy source (solar), into a smaller area, higher energy density source for more convenient application. An example of that concept will be demonstrated in the 2D Energy Concentrator PML Model.

Startup 4.x.

Select > 2D.

Click > Next.

Click > Twistie for the Radio Frequency interface.

Click > Electromagnetic Waves (emw).

Click > Add Selected.

Click > Next.

Select > Preset Studies – Frequency Domain.

Click > Finish (Flag).

Click > Save As.

Enter MMUC4_2D_PML_CMM_1.mph.

Click > Save.

See Figure 10.3.

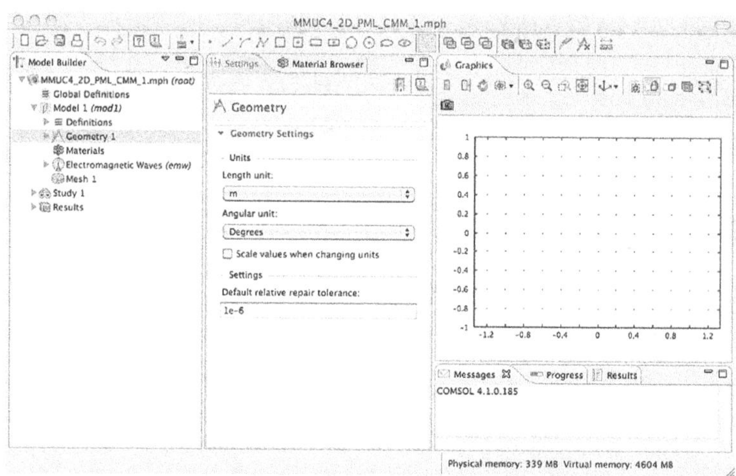

FIGURE 10.3 Desktop Display for the MMUC4_2D_PML_CMM_1.mph Model.

Figure 10.3 shows the Display for the MMUC4_2D_PML_CMM_1.mph model.

Global Definitions

Parameters

In Model Builder – Model 1 (mod1),

Right-Click > Model Builder – Global Definitions.

Select > Parameters from the Pop-up menu.

Enter all the Parameters, as shown, in Table 10.1.

NOTE

The speed of light in free space (vacuum) is actually 299,792,458 m/s (exactly) {10.11}. However, for use in this model, the approximate value of 3.0e8 m/s will be used, as a good first approximation.

The wave number is proportional to the reciprocal of wavelength {10.12}.

TABLE 10.1 CMM Parameters

Name	Expression	Description
f_a	100[MHz]	Frequency
c_L	3.0e8[m/s]	Speed of light in free space
k_0	2*pi*f_a/c_L	Free space wave number
E_z	1[V/m]	Electric field amplitude
me_s	c_L/f_a/6	Maximum element size

See Figure 10.4.

Figure 10.4 shows the Settings – Parameters – Parameters edit window.

Variables

In Model Builder – Model 1 (mod1),

Right-Click > Model Builder – Global Definitions.

Select > Variables from the Pop-up menu.

Enter all the Variables, as shown, in Table 10.2.

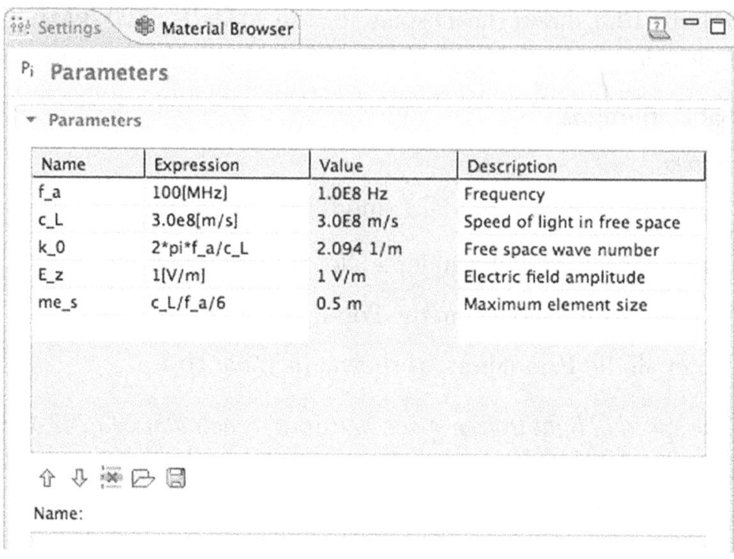

FIGURE 10.4 Settings - Parameters - Parameters Edit Window.

TABLE 10.2 CMM Variables

Name	Expression	Description
phideg	0	Initial angle
phi	phideg*pi/180	Angle of incidence, rad
E_b	exp(j*k_0*(x*cos(phi)+y*sin (phi)))*E_z	Electric field
dm_x	x*cos(phi)	Destination map x expression
dm_y	x*sin(phi)	Destination map y expression
Efar	1	Reciprocal backscatter

See Figure 10.5.

Figure 10.5 shows the Settings – Variables – Variables edit window.

Geometry

PML Domain

In Model Builder – Model 1 (mod1),

Right-Click > Model Builder – Model 1 (mod1) – Geometry 1.

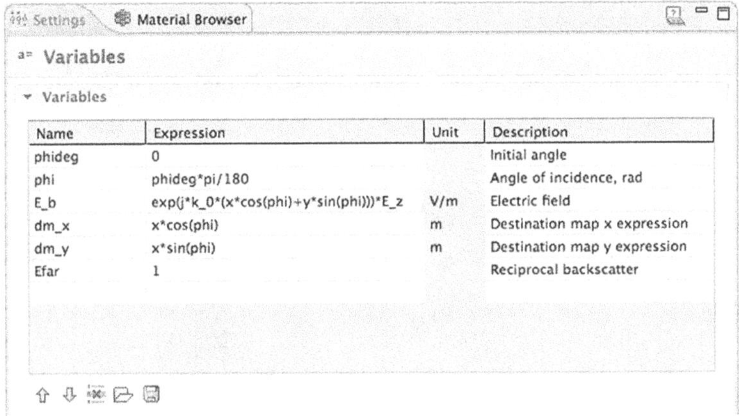

FIGURE 10.5 Settings - Variables - Variables Edit Window.

Select > Circle from the Pop-up menu.

Enter the coordinates shown in Table 10.3.

Click as instructed.

Repeat the sequence until completed.

TABLE 10.3 PML Domain

Circle	Radius	Click
1	15[m]	Build Selected
2	12[m]	Build All

See Figure 10.6.

Figure 10.6 shows the PML domain in the Graphics edit window.

CMM Domain

In Model Builder – Model 1 (mod1),

Right-Click > Model Builder – Model 1 (mod1) – Geometry 1.

Select > Circle from the Pop-up menu.

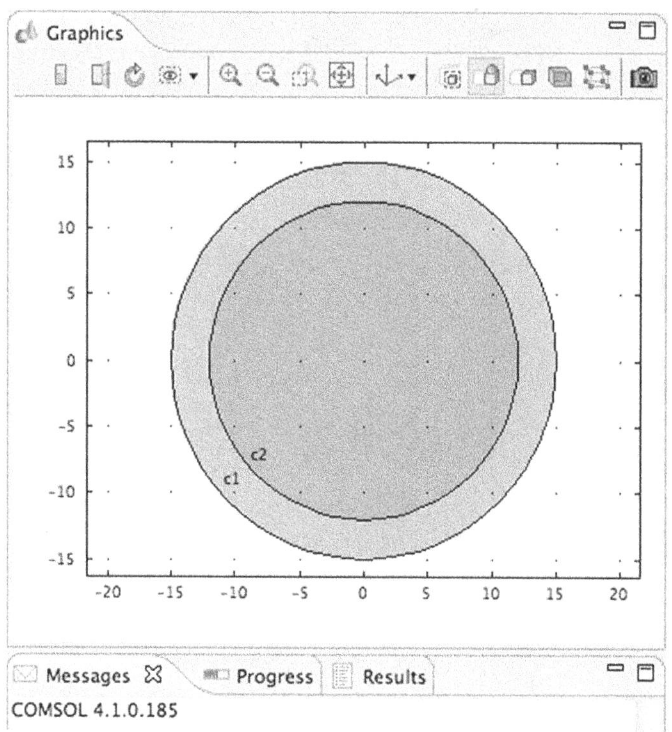

FIGURE 10.6 PML in Graphics Edit Window.

TABLE 10.4 CMM Domain

Circle	Radius	Click
3	3.0[m]	Build Selected
4	2.9[m]	Build All

Enter the coordinates shown in Table 10.4.

Click as instructed.

Repeat the sequence until completed.
See Figure 10.7.

Figure 10.7 shows the CMM Circles in the Graphics edit window.

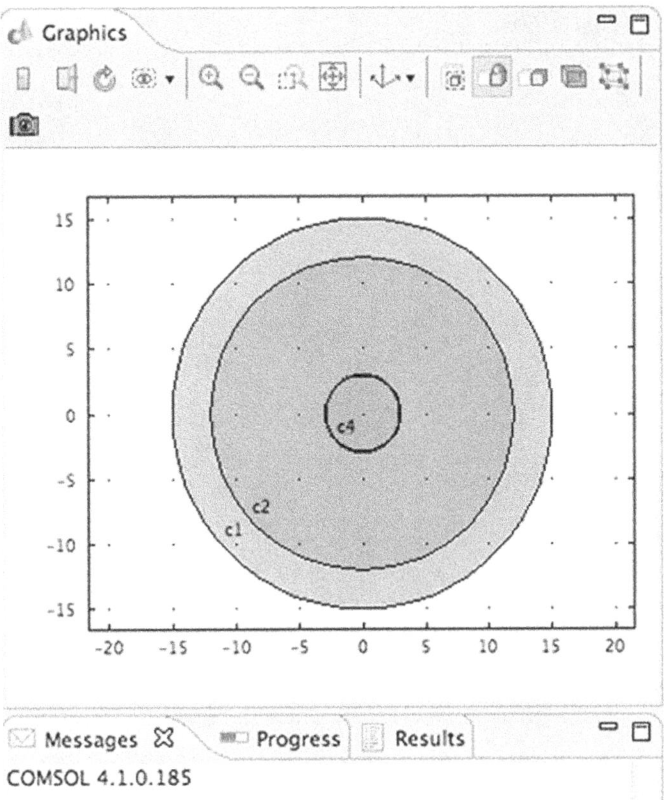

FIGURE 10.7 CMM Circles in the Graphics Edit Window.

NOTE *In order to build a metallic mirror using the CMM Circles, the modeler needs to perform two Boolean Difference operations.*

CMM Formation

Right-Click > Model Builder – Model 1 (mod1) – Geometry 1.

Select > Boolean Operations – Difference from the Pop-up menu.

Click > Zoom In.

Click > Zoom in a second time.

Click > Zoom Box.

Select > A small portion of the Edge of Circle 3 and Circle 4.

See Figure 10.8.

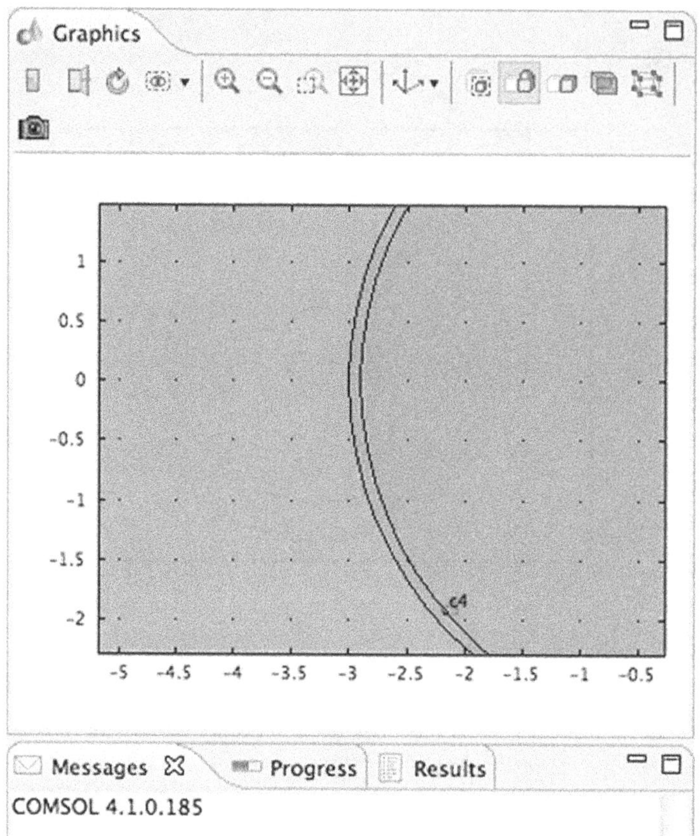

FIGURE 10.8 CMM Circles 3 & 4 in Graphics Edit Window.

Figure 10.8 shows the selected portion of the CMM Circles 3 & 4 in the Graphics edit window.

Click > Circle 3 in the Graphics window.

Click > Add to Selection in Settings – Difference – Difference – Objects to add.

Click > Activate Selection in Settings – Difference – Difference – Objects to subtract.

Click > Circle 4 in the Graphics window.

Click > Add to Selection in Settings – Difference – Difference – Objects to subtract.

See Figure 10.9.

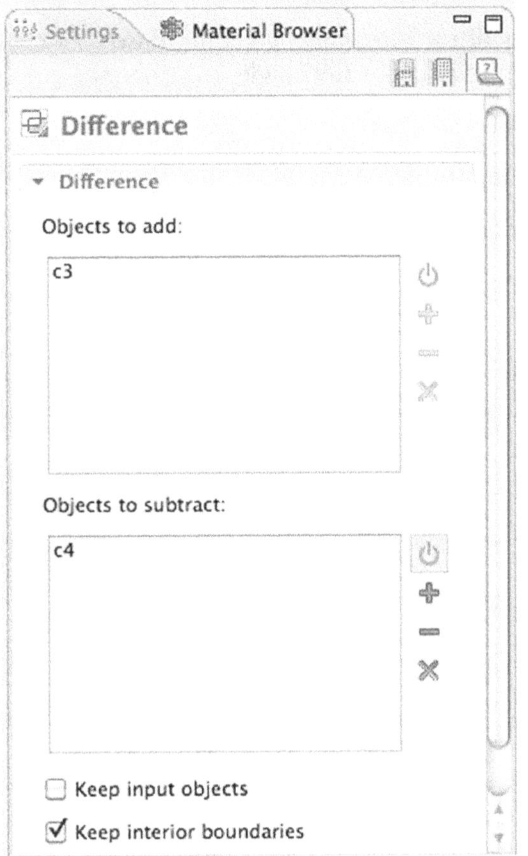

FIGURE 10.9 Settings - Difference - Difference Prepared for the Build Selected Operation.

Figure 10.9 shows Settings – Difference – Difference Prepared for the Build Selected operation.

Click > Build Selected in the Settings Toolbar.

Click > Zoom Extents.

Rectangle

NOTE *A Rectangle is needed to remove the Right Half of the difference object formed in the last step.*

Right-Click > Model Builder – Model 1 (mod1) – Geometry 1.

Select > Rectangle from the Pop-up menu.

Enter > Width = 3[m], Height = 6[m], Base = Corner, x = 0[m], and y = –3[m] in Settings – Rectangle.

Click > Build Selected.

Click > Zoom In.

Click > Zoom in a second time.

See Figure 10.10.

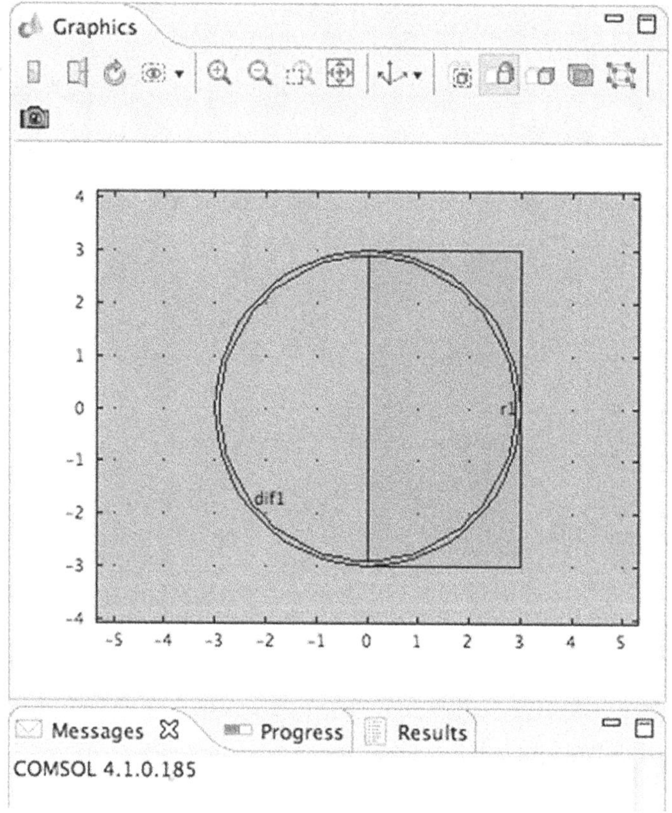

FIGURE 10.10 Graphics Components Prepared for Build Selected.

Figure 10.10 shows Graphics Components Prepared for Build Selected.

Right-Click > Model Builder – Model 1 (mod1) – Geometry 1.

Select > Boolean Operations – Difference from the Pop-up menu.

Click > dif1 in the Graphics window.

Click > Add to Selection in Settings – Difference – Difference – Objects to add.

Click > Activate Selection in Settings – Difference – Difference – Objects to subtract.

Click > r1 in the Graphics window.

Click > Add to Selection in Settings – Difference – Difference – Objects to subtract.

See Figure 10.11.

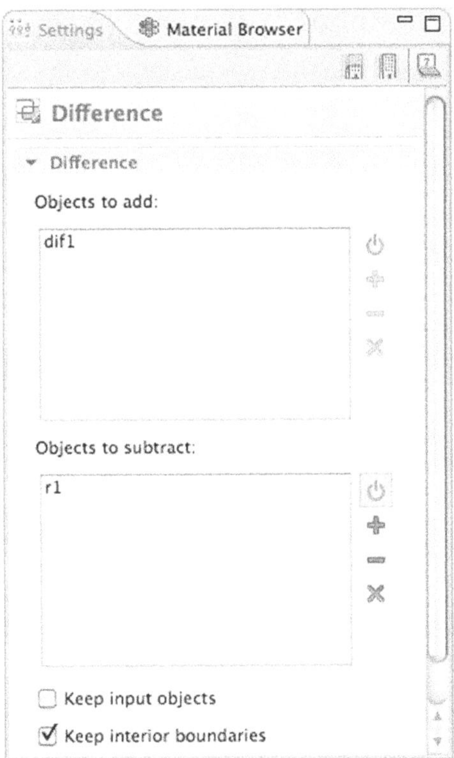

FIGURE 10.11 Settings - Difference - Difference Prepared for the Build Selected Operation.

Figure 10.11 shows Settings – Difference – Difference Prepared for the Build Selected operation.

Click > Build Selected in the Settings Toolbar.

Click > Zoom Extents.

See Figure 10.12.

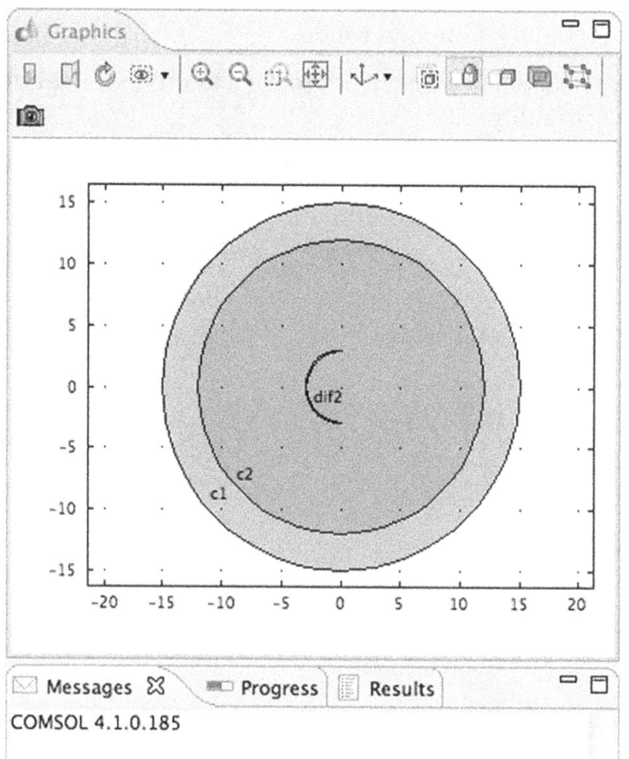

FIGURE 10.12 CMM in the Graphics Window.

Figure 10.12 shows the CMM in the Graphics window.

Selection 1

Right-Click > Model Builder – Model 1 (mod1) – Definitions.

Select > Selection from the Pop-up menu.

Click > Model Builder – Model 1 (mod1) – Definitions – Selection 1.

Right-Click > Model Builder – Model 1 (mod1) – Definitions – Selection 1.

Select > Rename from the Pop-up menu.

Enter > Mirror Boundaries in the Rename Selection edit window.

Click > OK.

Click > Domain 2, the CMM, in the Graphics window.

Click > Add to Selection in Settings – Selection – Geometric Scope – Selection input.

Click > Settings – Selection – Geometric Scope – Selection output.

Select > Adjacent boundaries from the Pop-up menu.

See Figure 10.13.

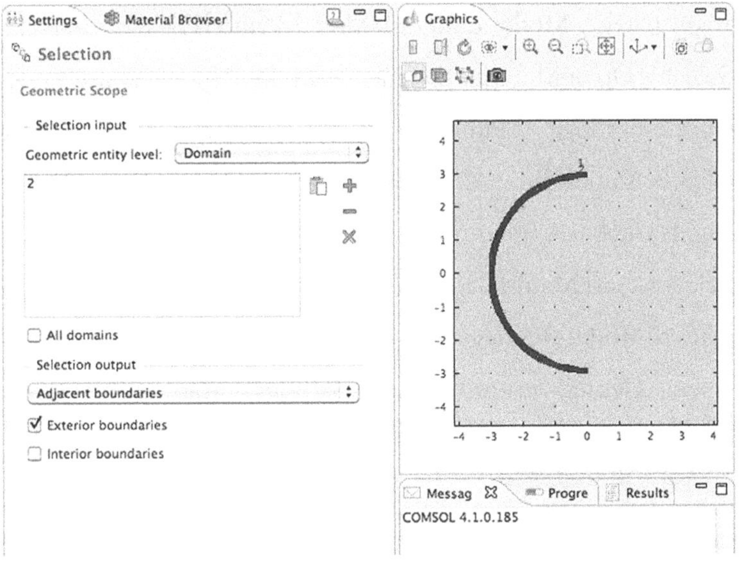

FIGURE 10.13 Mirror Boundaries Selection.

Figure 10.13 shows the Mirror Boundaries Selection.

Materials

Material 1

Right-Click > Model Builder – Model 1 (mod1) – Materials.

Select > Open Material Browser from the Pop-up menu.

Click > Settings – Material Browser – Materials – Built-in twistie.

Click > Air.

Right- Click > Air.

Select > Add Material to Model.

Select > Domain 2 in Settings – Material – Geometric Scope edit window.

Click > Remove from Selection in Settings – Material – Geometric Scope.

Material 2

Right-Click > Model Builder – Model 1 (mod1) – Materials.

Select > Open Material Browser from the Pop-up menu.

Click > Settings – Material Browser – Materials – Built-in twistie.

Click > Aluminum.

Right- Click > Aluminum.

Select > Add Material to Model.

Click > Model Builder – Model 1 (mod1) – Materials – Aluminum.

Click > Geometric entity level in Settings – Material – Geometric Scope.

Select > Boundary from the Pop-up menu.

Selection

Click > Selection in Settings – Material – Geometric Scope.

Select > Mirror Boundaries from the Pop-up menu.

Electromagnetic Waves (emw)

Domains

Click > Model Builder – Model 1 – Electromagnetic Waves (emw).

Click > 2 in Settings – Electromagnetic Waves – Domains – Selection edit window.

Click > Remove from Selection button (Minus sign) in Settings – Electromagnetic Waves – Domains.

NOTE *This model uses the boundaries of the mirror to represent the mirror. Because that is the case, the field inside the mirror is identically zero and requires no solution inside. This approach saves both time and effort.*

Settings

Click > Solve for Pull-down menu in Settings – Electromagnetic Waves – Settings – Solve for.

Select > Scattered field from the Pull-down menu.

Enter E_b in the z position of the Background electric field edit window in Settings – Electromagnetic Waves – Settings – Background electric field.

See Figure 10.14.

Figure 10.14 shows the Electromagnetic Waves Settings Selection.

Perfectly Matched Layers 1

NOTE *The Perfectly Matched Layer simulates an infinitely distant boundary layer. That eliminates possible interference of a back scattered wave with the waves in the region of interest at the center of the model.*

Right-Click > Model Builder – Model 1 – Electromagnetic Waves (emw).

Select > Perfectly Matched Layers.

Click > Perfectly Matched Layers 1.

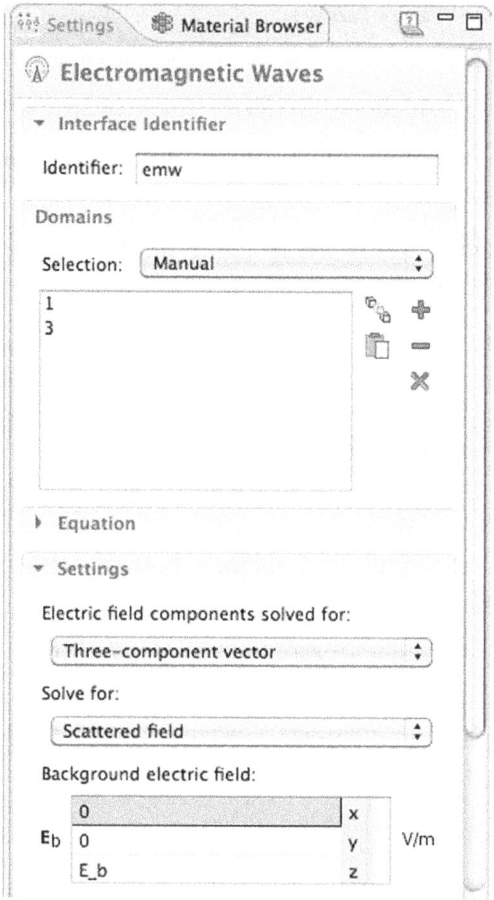

FIGURE 10.14 Electromagnetic Waves Settings Selection.

Click > Domain 3, the outer ring, in the Graphics window.

Click > Add to Selection in Settings – Perfectly Matched Layers – Domains.

Coordinates

NOTE *The coordinates used in this 2D model are Cylindrical. They are set in that mode below.*

Click > Type Pull-down menu in Settings – Perfectly Matched Layers – Geometric Settings.

Select > Cylindrical from the Type Pull-down menu.

See Figure 10.15.

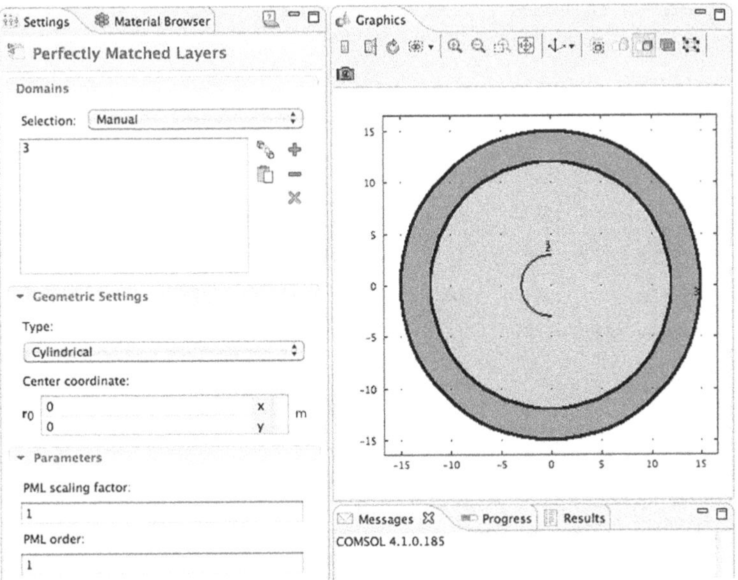

FIGURE 10.15 Perfectly Matched Layer Settings Selection.

Figure 10.15 shows the Perfectly Matched Layer Settings Selection.

Impedance Boundary Condition 1

The Impedance Boundary Condition assumes the skin depth in the material is significantly less than the material thickness. In this case it is on the order of microns.

Right-Click > Model Builder – Model 1 – Electromagnetic Waves (emw).

Select > Impedance Boundary Condition.

Click > Impedance Boundary Condition 1.

Click > Selection in Settings – Impedance Boundary Condition – Boundaries.

Select > Mirror Boundaries from the Pull-down menu.

See Figure 10.16.

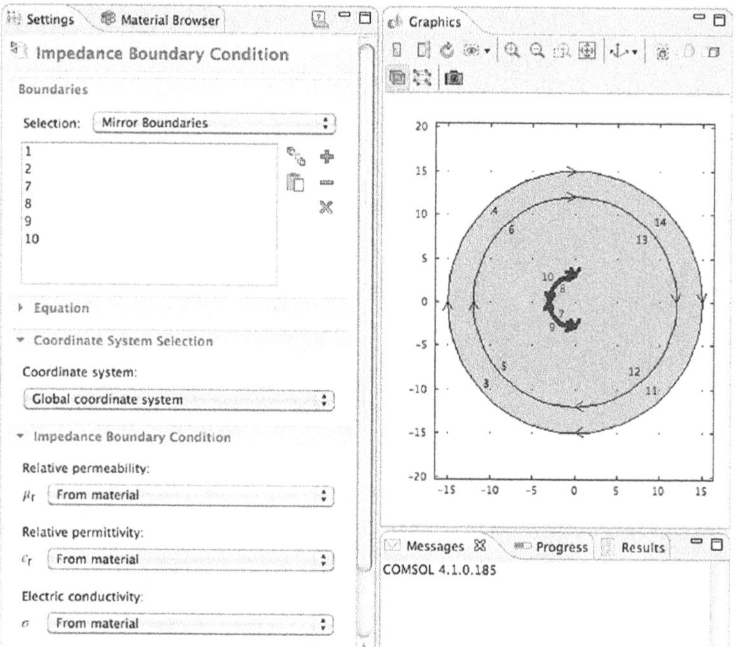

FIGURE 10.16 Impedance Boundary Condition Settings Selection.

Figure 10.16 shows the Impedance Boundary Condition Settings Selection.

Far-Field Calculation 1

NOTE *The requirement for the Far-Field Calculation is that all reflecting surfaces are surrounded, which they are.*

Right-Click > Model Builder – Model 1 – Electromagnetic Waves (emw).

Select > Far-Field Calculation.

Click > Far-Field Calculation 1.

Click > Selection in Settings – Far-Field Calculation – Boundaries.

Select > Mirror Boundaries from the Pull-down menu.

See Figure 10.17.

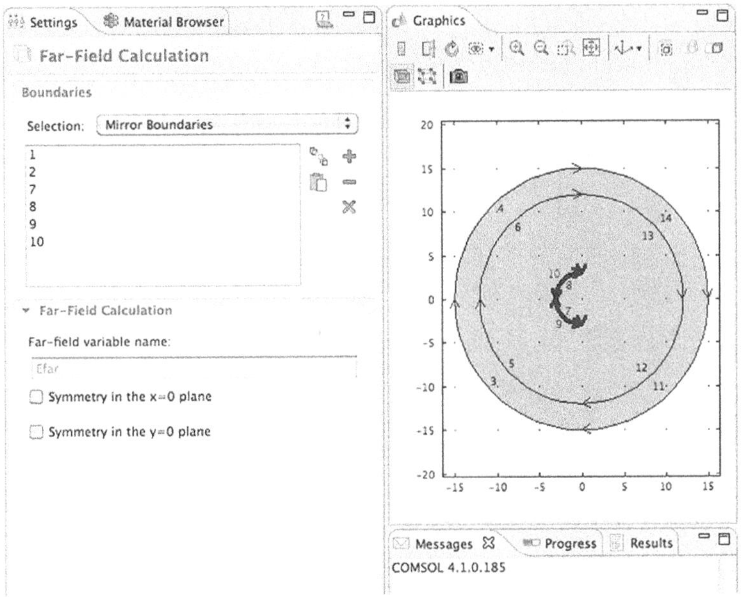

FIGURE 10.17 Far-Field Calculation Settings Selection.

Figure 10.17 shows the Far-Field Calculation Settings Selection.

Definitions

General Extrusion 1

The General Extrusion {10.13} is a coupling operator that maps an expression defined on a source domain to an expression that can be evaluated on any domain where the destination map expressions are valid.

Right-Click > Model Builder – Model 1 – Definitions.

Select > Model Couplings – General Extrusion.

Click > General Extrusion 1.

Enter > back in Settings – General Extrusion – Operator Name – Operator name.

Select > Domain 1, the inner circle, in the Graphics window.

Click > Add to Selection in Settings – General Extrusion – Source Selection – Selection.

Enter > dm_x in Settings – General Extrusion – Destination Map – x-expression edit window.

Enter > dm_y in Settings – General Extrusion – Destination Map – y-expression edit window.

See Figure 10.18.

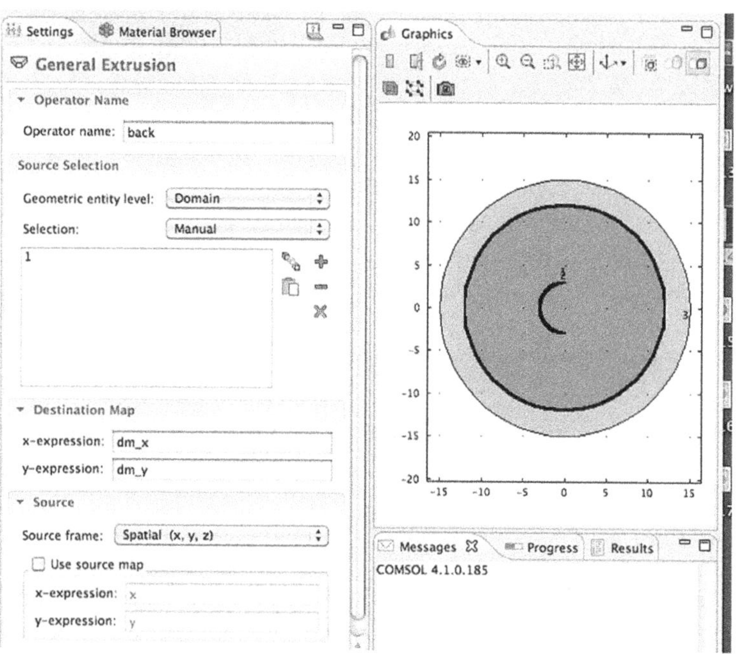

FIGURE 10.18 General Extrusion Settings Selection.

Figure 10.18 shows the General Extrusion Settings Selection.

Mesh 1

Right-Click > Model Builder – Model 1 – Mesh 1.

Select > Free Triangular from the Pop-up menu.

Click > Model Builder – Model 1 – Mesh 1 – Size.

Click > Settings – Size – Element Size Parameters twistie.

Enter > me_s in Settings – Size – Element Size Parameters – Maximum element size.

Click > Build All.

See Figure 10.19.

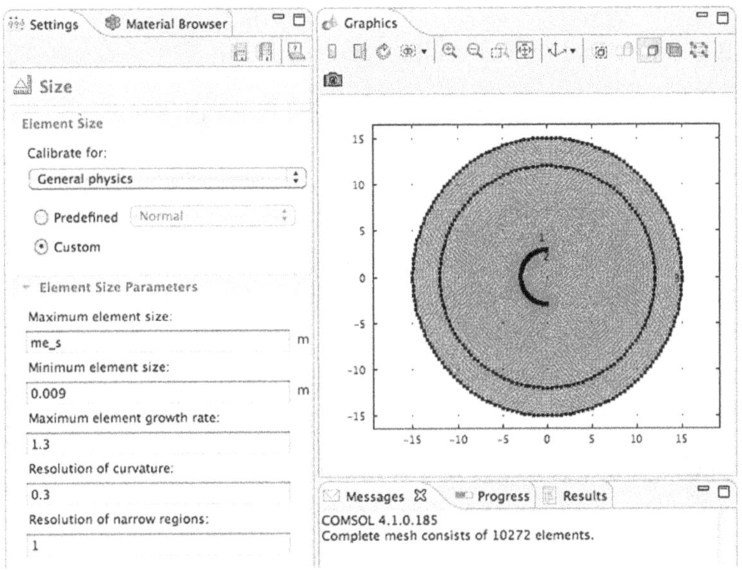

FIGURE 10.19 Graphics Window and the Size Settings after Mesh Build.

Figure 10.19 shows the Graphics Window and the Size Settings after Mesh Build.

NOTE *After the mesh is built, the model should have 10272 elements.*

Study 1

Frequencies

Click > Model Builder – Study 1 twistie.

Click > Model Builder – Study 1 – Step 1: Frequency Domain.

Enter > f_a in Settings – Frequency Domain – Study Settings – Frequencies.

Parametric Solver

NOTE *The Parametric Solver settings are being configured to solve the model at one frequency, with the impinging electromagnetic wave arriving at the CMM from each of 360 different directions. This allows the modeler to view a distributed selection of all solutions for the CMM.*

Right-Click > Model Builder – Study 1.

Select > Show Default Solver from the Pop-up menu.

Click > Model Builder – Study 1 – Solver Configurations twistie.

Click > Model Builder – Study 1 – Solver Configurations – Solver 1 twistie.

Click > Model Builder – Study 1 – Solver Configurations – Solver 1 – Stationary Solver twistie.

Click > Model Builder – Study 1 – Solver Configurations – Solver 1 – Stationary Solver – Parametric 1.

Click > Settings – Parametric – General – Defined by study step.

Select > User defined from the Pull-down menu.

Enter > phideg in the Settings – Parametric – General – Parameter names edit window.

Click > Range button in Settings – Parametric – General.

Enter > Start = 0, Stop = 360, Step = 1 in the Range Pop-up edit window.

Click > Replace button in the Range Pop-up edit window.

See Figure 10.20.
Figure 10.20 shows the Settings – Parametric Settings.

Right-Click > Model Builder – Study 1.

Select > Compute.

See Figure 10.21.
Figure 10.21 shows the CMM Computed Solution.

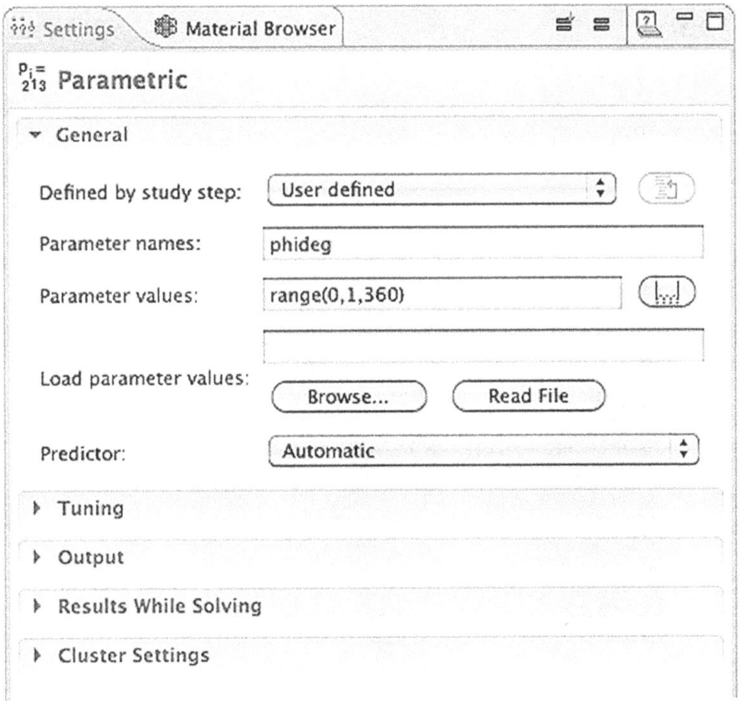

Settings | Material Browser

Parametric

General

Defined by study step: User defined

Parameter names: phideg

Parameter values: range(0,1,360)

Load parameter values: Browse... Read File

Predictor: Automatic

Tuning

Output

Results While Solving

Cluster Settings

FIGURE 10.20 Settings - Parametric Settings.

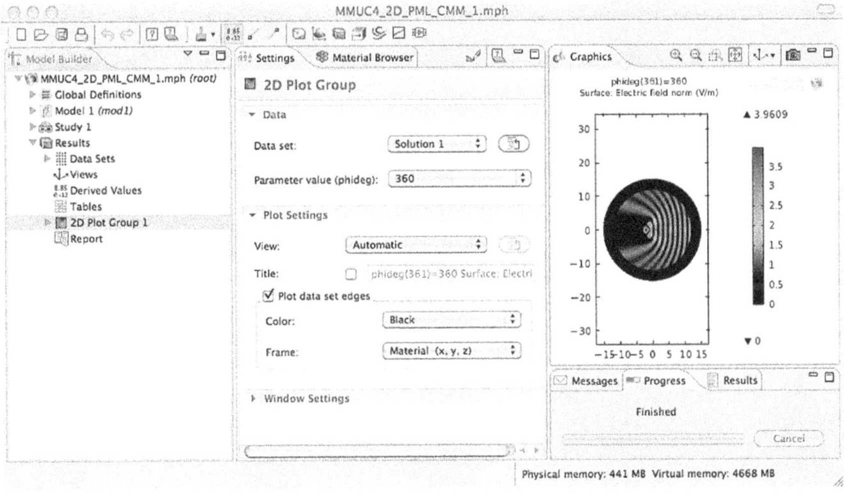

FIGURE 10.21 CMM Computed Solution.

Expanded Solution

Click > Zoom in twice.

See Figure 10.22.

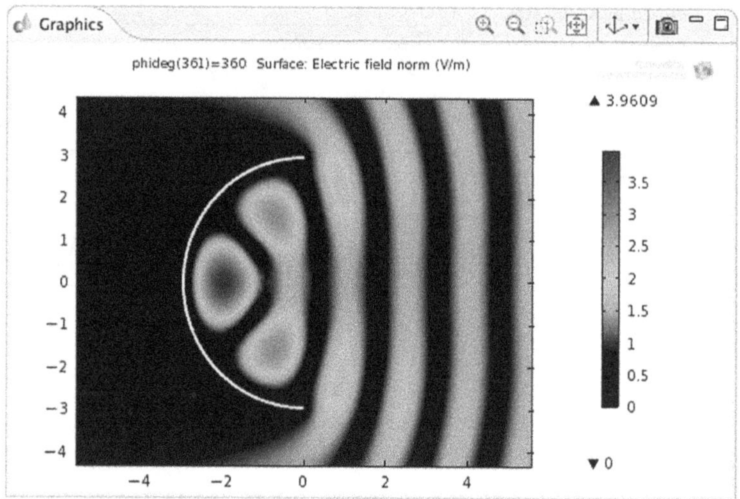

FIGURE 10.22 High Electric Field in CMM Computed Solution.

Figure 10.22 shows a High Electric Field in CMM Computed Solution.

NOTE *The modeler should note that a high electric field point forms immediately in front of the metallic mirror. The mirror contributes a gain of approximately four (4).*

2D Concave Metallic Mirror PML Model Summary and Conclusions

The 2D Concave Metallic Mirror PML Model is a powerful modeling tool that can be used to calculate the gain of various mirrors of different shapes and sizes. With this model, the modeler can easily vary all of the geometric parameters and optimize the design before the first prototype is physically built. These types of metallic mirrors are widely used in industry.

Energy Concentrator Mirror Systems

Metallic mirror energy concentration systems are widely used in renewable energy applications {10.14, 10.15, 10.16, 10.17}. This model demonstrates the building of a two trough elliptical collector, using the perfectly matched layer method.

2D Energy Concentrator PML Model

NOTE

The elliptical mirror {10.18} is a design concept widely utilized in Optical Physics. In this application, the principles of Optics are applied to lower frequency electromagnetic waves to focus an impinging wavefront into the region of a sensor. The act of focusing the wavefront effectively increases the magnitude of the impinging signal, concentrating power in the region of interest, thus making the focused signal more easily usable.

The focusing concept is demonstrated here to concentrate large-area, diffuse energy, from a renewable energy source (solar), into a smaller area, higher energy density source for more convenient application.

Startup 4.x.

Select > 2D.

Click > Next.

Click > Twistie for Radio Frequency Interface.

Click > Electromagnetic Waves (emw).

Click > Add Selected.

Click > Next.

Select > Preset Studies – Frequency Domain.

Click > Finish (Flag).

Click > Save As.

Enter MMUC4_2D_PML_EC_1.mph.

Click > Save.

See Figure 10.23.

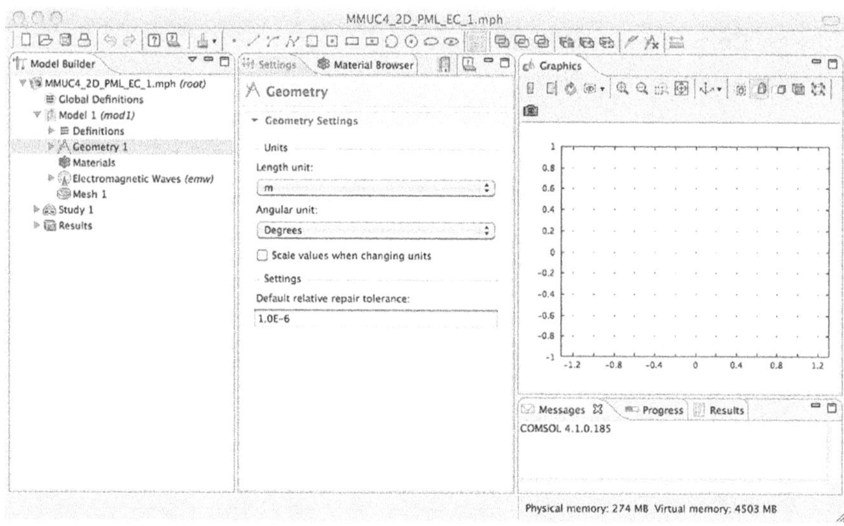

FIGURE 10.23 Desktop Display for the MMUC4_2D_PML_EC_1.mph Model.

Figure 10.23 shows the Display for the MMUC4_2D_PML_EC_1.mph model.

Global Definitions

Parameters

In Model Builder – Model 1 (mod1).

Right-Click > Model Builder – Global Definitions.

Select > Parameters from the Pop-up menu.

Enter all the Parameters, as shown, in Table 10.5.

NOTE

The speed of light in free space (vacuum) is actually 299,792,458 m/s (exactly). However, for use in this model, the approximate value of 3.0e8 m/s will be used, as a good first approximation.

The wave number is proportional to the reciprocal of wavelength.

TABLE 10.5 EC Parameters

Name	Expression	Description
f_a	100[MHz]	Frequency
c_L	3.0e8[m/s]	Speed of light in free space
k_0	2*pi*f_a/c_L	Free space wave number
E_z	1[V/m]	Electric field amplitude
me_s	c_L/f_a/6	Maximum element size

See Figure 10.24.

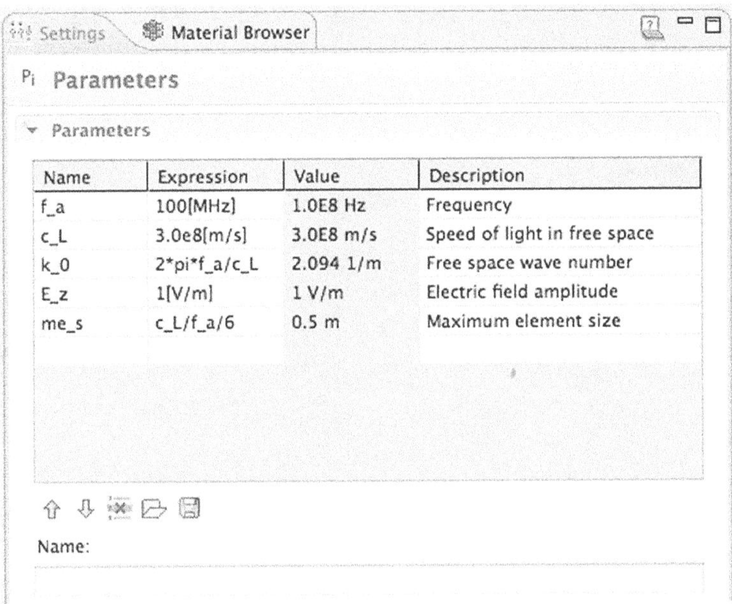

FIGURE 10.24 Settings - Parameters - Parameters Edit Window.

Figure 10.24 shows the Settings – Parameters – Parameters edit window.

Variables

In Model Builder – Model 1 (mod1),

Right-Click > Model Builder – Global Definitions.

Select > Variables from the Pop-up menu.

Enter all the Variables, as shown, in Table 10.6.

TABLE 10.6 EC variables

Name	Expression	Description
phideg	0	Initial angle
phi	phideg*pi/180	Angle of incidence, rad
E_b	exp(j*k_0*(x*cos(phi)+y*sin(phi)))*E_z	Electric field
dm_x	x*cos(phi)	Destination map x expression
dm_y	x*sin(phi)	Destination map y expression
Efar	1	Reciprocal backscatter

See Figure 10.25.

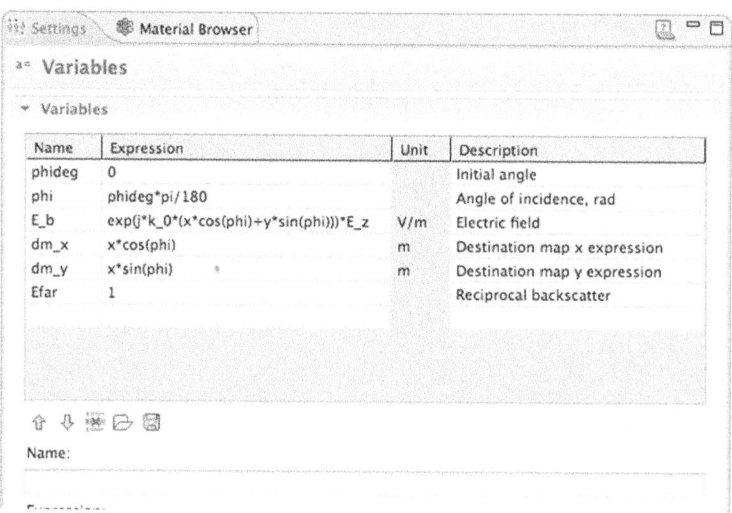

FIGURE 10.25 Settings - Variables - Variables Edit Window.

Figure 10.25 shows the Settings – Variables – Variables edit window.

Geometry

PML Domain

In Model Builder – Model 1 (mod1).

Right-Click > Model Builder – Model 1 (mod1) – Geometry 1.

Select > Circle from the Pop-up menu.

Enter the coordinates shown in Table 10.7.

Click as instructed.

Repeat the sequence until completed.

TABLE 10.7 PML Domain

Circle	Radius	Click
1	15[m]	Build Selected
2	12[m]	Build All

See Figure 10.26.

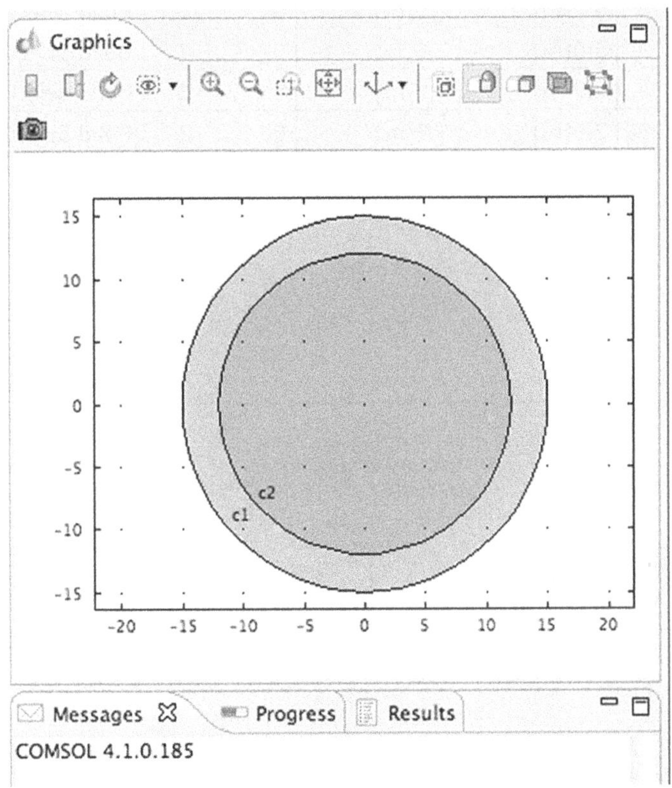

FIGURE 10.26 PML in the Graphics edit window.

Figure 10.26 shows the PML in the Graphics edit window.

EC Domain

In Model Builder – Model 1 (mod1).

Right-Click > Model Builder – Model 1 (mod1) – Geometry 1.

Select > Ellipse from the Pop-up menu.

Enter the coordinates shown in Table 10.8.

Click as instructed.

Repeat the sequence until completed.

TABLE 10.8 EC Domains

Ellipse	a-semiaxis	b-semiaxis	x	y	Click
1	3.0[m]	2.0[m]	0	-2	Build Selected
2	3.0[m]	2.0[m[	0	2	Build Selected
3	2.9[m]	1.9[m]	0	-2	Build Selected
4	2.9[m]	1.9[m]	0	2	Build Selected

See Figure 10.27.

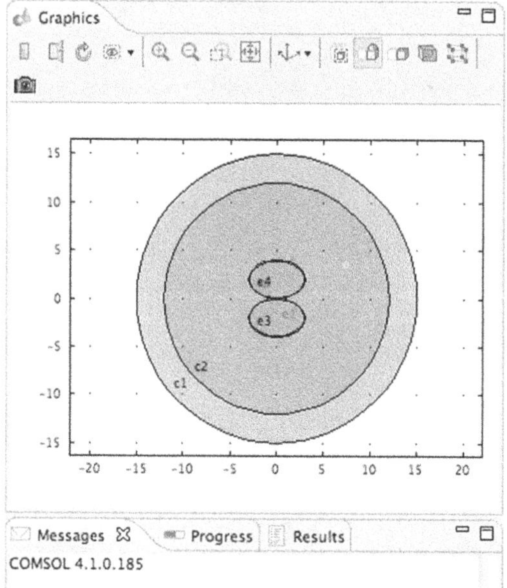

FIGURE 10.27 EC Ellipses in the Graphics Edit Window.

Figure 10.27 shows the EC Ellipses in the Graphics edit window.

In order to build the concentrators using the EC ellipses, the modeler needs to perform two Boolean Difference operations.

EC Formation

Difference 1

Right-Click > Model Builder – Model 1 (mod1) – Geometry 1.

Select > Boolean Operations – Difference from the Pop-up menu.

Click > Zoom In.

Click > Zoom in a second time.

Click > Zoom Box.

Select > A small portion of the Edge of Ellipse 1 and Ellipse 3.

See Figure 10.28.

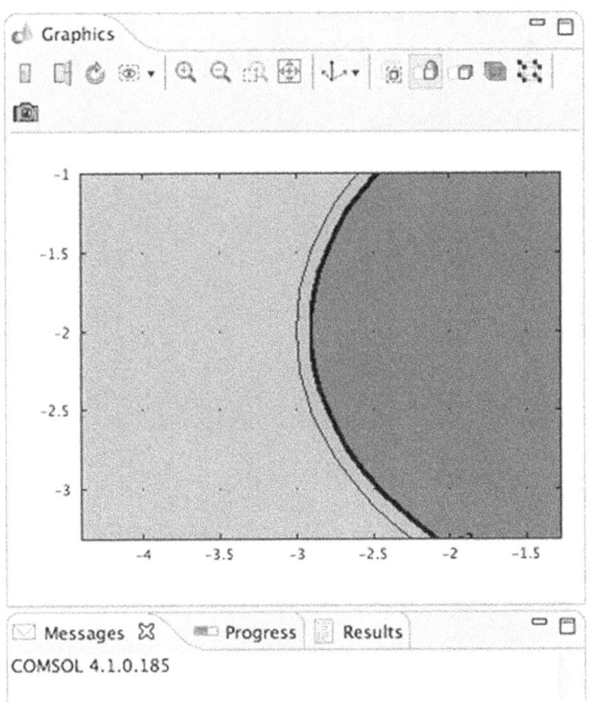

FIGURE 10.28 EC Ellipses 1 & 3 in Graphics Edit Window.

Figure 10.28 shows the EC Ellipses 1 & 3 in Graphics edit window.

Click > Ellipse 1 in the Graphics window.

Click > Add to Selection in Settings – Difference – Difference – Objects to add.

Click > Activate Selection in Settings – Difference – Difference – Objects to subtract.

Click > Ellipse 3 in the Graphics window.

Click > Add to Selection in Settings – Difference – Difference – Objects to subtract.

See Figure 10.29.

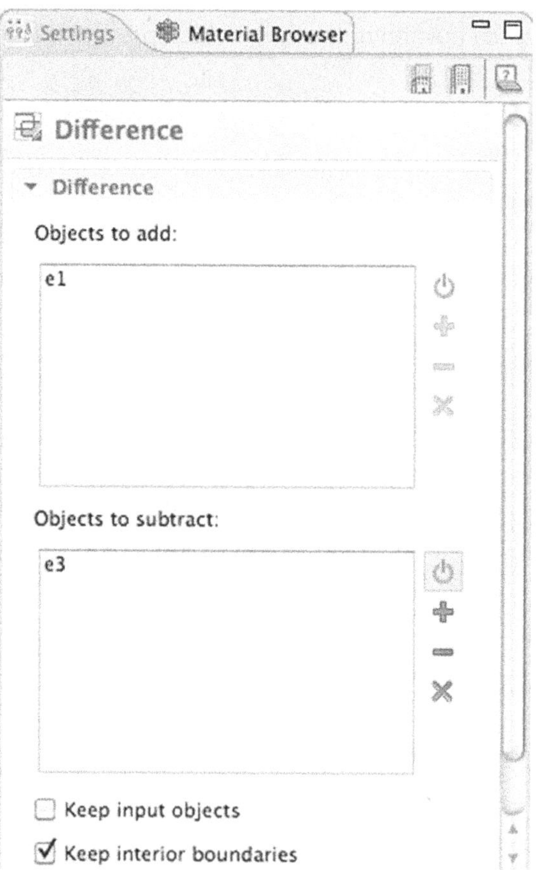

FIGURE 10.29 Settings - Difference - Difference Prepared for the Build Selected Operation.

Figure 10.29 shows Settings – Difference – Difference Prepared for the Build Selected operation.

Click > Build Selected in the Settings Toolbar.

Click > Zoom Extents.

Difference 2

Right-Click > Model Builder – Model 1 (mod1) – Geometry 1.

Select > Boolean Operations – Difference from the Pop-up menu.

Click > Zoom In.

Click > Zoom In a second time.

Click > Zoom Box.

Select > A small portion of the Edge of Ellipse 2 and Ellipse 4.

See Figure 10.30.

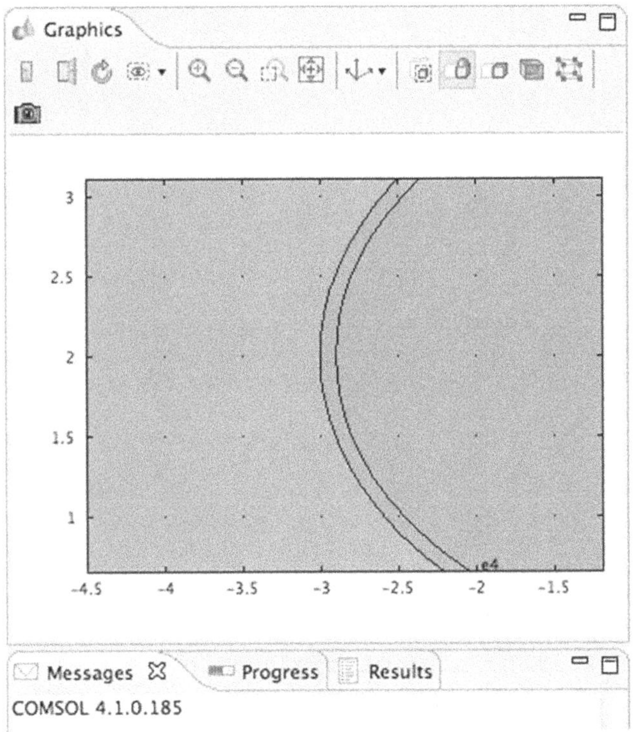

FIGURE 10.30 EC Ellipses 2 & 4 in Graphics Edit Window.

Figure 10.30 shows the EC Ellipses 2 & 4 in Graphics edit window.

Click > Ellipse 2 in the Graphics window.

Click > Add to Selection in Settings – Difference – Difference – Objects to add.

Click > Activate Selection in Settings – Difference – Difference – Objects to subtract.

Click > Ellipse 4 in the Graphics window.

Click > Add to Selection in Settings – Difference – Difference – Objects to subtract.

See Figure 10.31.

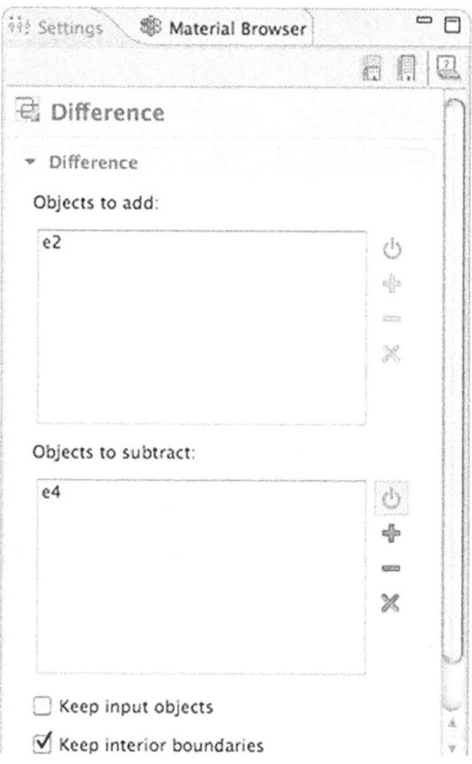

FIGURE 10.31 Settings - Difference - Difference Prepared for the Build Selected Operation.

Figure 10.31 shows Settings – Difference – Difference Prepared for the Build Selected operation.

Click > Build Selected in the Settings Toolbar.

Click > Zoom Extents.

Rectangle

NOTE *A Rectangle is needed to remove the Right Half of the difference objects formed in the last steps.*

Right-Click > Model Builder – Model 1 (mod1) – Geometry 1.

Select > Rectangle from the Pop-up menu.

Enter > Width = 3[m], Height = 8[m], Base = Corner, x = 0[m], and y = –4[m] in Settings - Rectangle.

Click > Build Selected

Click > Zoom In.

Difference 3

Right-Click > Model Builder – Model 1 (mod1) – Geometry 1.

Select > Boolean Operations – Difference from the Pop-up menu.

Shift-Click > dif1 and dif2 in the Graphics window.

Click > Add to Selection in Settings – Difference – Difference – Objects to add.

Click > Activate Selection in Settings – Difference – Difference – Objects to subtract.

Click > r1 in the Graphics window.

Click > Add to Selection in Settings – Difference – Difference – Objects to subtract.

See Figure 10.32.

Figure 10.32 shows Settings – Difference – Difference Prepared for the Build Selected operation.

Click > Build Selected in the Settings Toolbar.

Click > Zoom Extents.

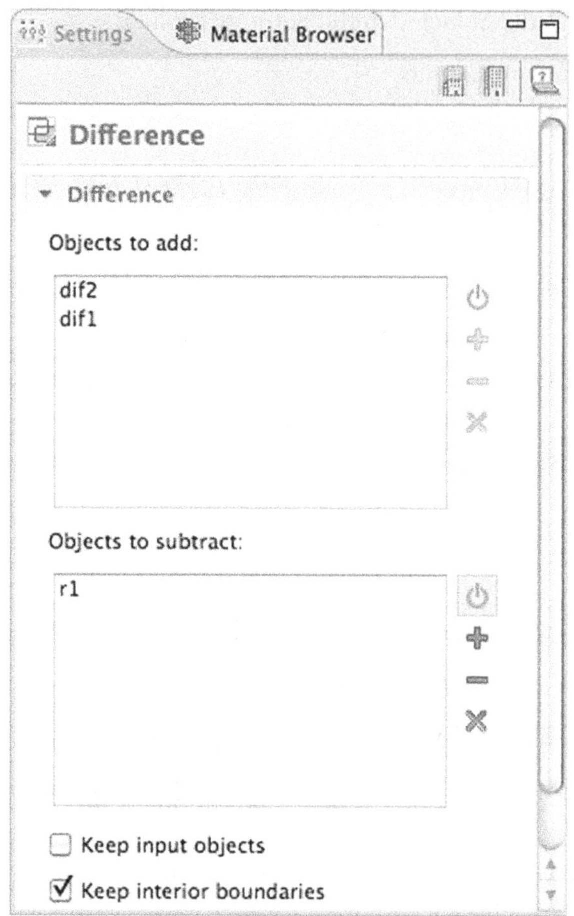

FIGURE 10.32 Settings - Difference - Difference Prepared for the Build Selected Operation.

See Figure 10.33.

Figure 10.33 shows the EC in the Graphics window.

Selection 1

Right-Click > Model Builder – Model 1 (mod1) – Definitions.

Select > Selection from the Pop-up menu.

Click > Model Builder – Model 1 (mod1) – Definitions –Selection 1.

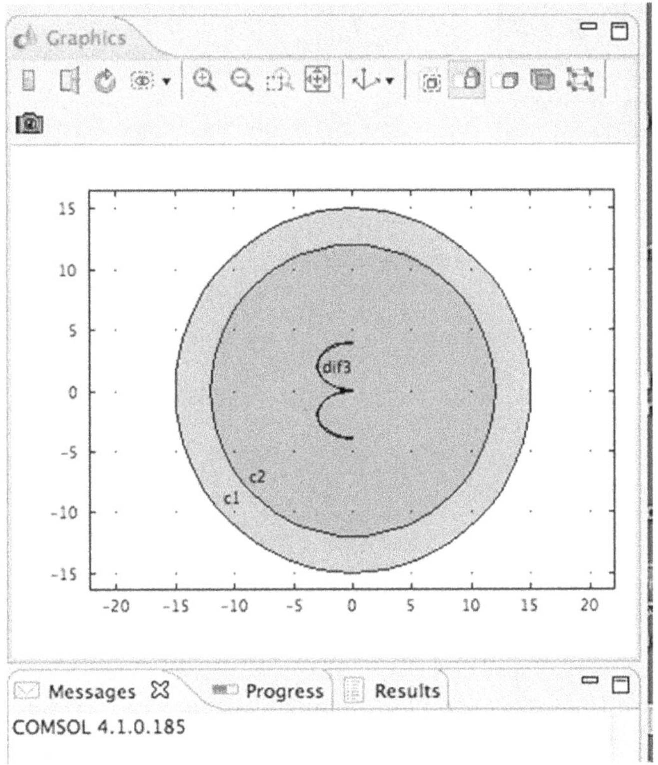

FIGURE 10.33 EC in the Graphics Window.

Right-Click > Model Builder – Model 1 (mod1) – Definitions – Selection 1.

Select > Rename from the Pop-up menu.

Enter > EC Boundaries in the Rename Selection edit window.

Click > OK.

Shift-Click > Domains 2 and 3, the two EC, in the Graphics window.

Click > Add to Selection in Settings – Selection – Geometric Scope – Selection input.

Click > Settings – Selection – Geometric Scope – Selection output.

Select > Adjacent Boundaries from the Pop-up menu.

See Figure 10.34.

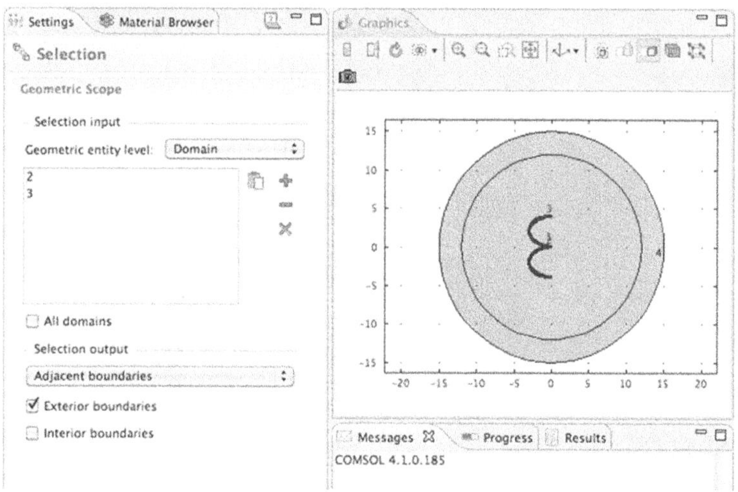

FIGURE 10.34 EC Boundaries Selection.

Figure 10.34 shows the EC Boundaries Selection.

Materials

Material 1

Right-Click > Model Builder – Model 1 (mod1) – Materials.

Select > Open Material Browser from the Pop-up menu.

Click > Settings – Material Browser – Materials – Built-in twistie.

Click > Air.

Right- Click > Air.

Select > Add Material to Model.

Select > Domain 2 and 3 in Settings – Material – Geometric Scope edit window,

Click > Remove from Selection in Settings – Material – Geometric Scope.

Material 2

Right-Click > Model Builder –Model 1 (mod1) – Materials.

Select > Open Material Browser from the Pop-up menu.

Click > Settings - Material Browser – Materials – Built-in twistie.

Click > Aluminum.

Right- Click > Aluminum,

Select > Add Material to Model.

Click > Model Builder _Model 1 (mod1) – Materials – Aluminum.

Click > Geometric entity level in Settings – Material – Geometric Scope.

Select > Boundary from the Pop-up menu.

Selection

Click > Selection in Settings – Material – Geometric Scope.

Select > EC Boundaries from the Pop-up menu.

Electromagnetic Waves (emw)

Domains

Click > Model Builder – Model 1 – Electromagnetic Waves (emw).

Shift-Click > 2 and 3 in Settings – Electromagnetic Waves – Domains – Selection edit window.

Click > Remove from Selection button (Minus sign) in Settings – Electromagnetic Waves – Domains.

NOTE *This model uses the boundaries of the EC to represent the EC. Since that is the case, the field inside the EC is identically zero and requires no solution inside. This approach saves both time and effort.*

Settings

Click > Solve for Pull-down menu in Settings – Electromagnetic Waves – Settings – Solve for.

Select > Scattered field from the Pull-down menu.

Enter E_b in the z position of the Background electric field edit window in Settings – Electromagnetic Waves – Settings – Background electric field.

See Figure 10.35.

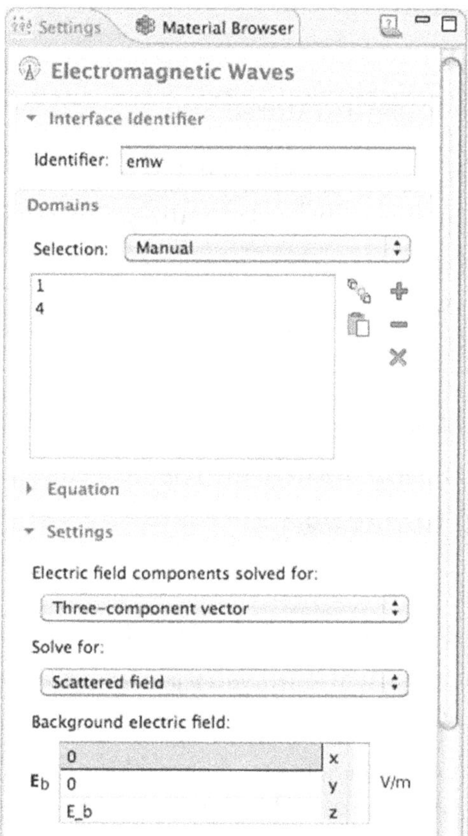

FIGURE 10.35 Electromagnetic Waves Settings Selection.

Figure 10.35 shows the Electromagnetic Waves Settings Selection.

Perfectly Matched Layers 1

NOTE *The Perfectly Matched Layer simulates an infinitely distant boundary layer. That eliminates possible interference of a back scattered wave with the waves in the region of interest at the center of the model.*

Right-Click > Model Builder – Model 1 – Electromagnetic Waves (emw).

Select > Perfectly Matched Layers.

Click > Perfectly Matched Layers 1.

Click > Domain 4, the outer ring, in the Graphics window.

Click > Add to Selection in Settings – Perfectly Matched Layers – Domains.

Coordinates

Click > Type Pull-down menu in Settings – Perfectly Matched Layers – Geometric Settings.

Select > Cylindrical from the Pull-down menu.

See Figure 10.36.

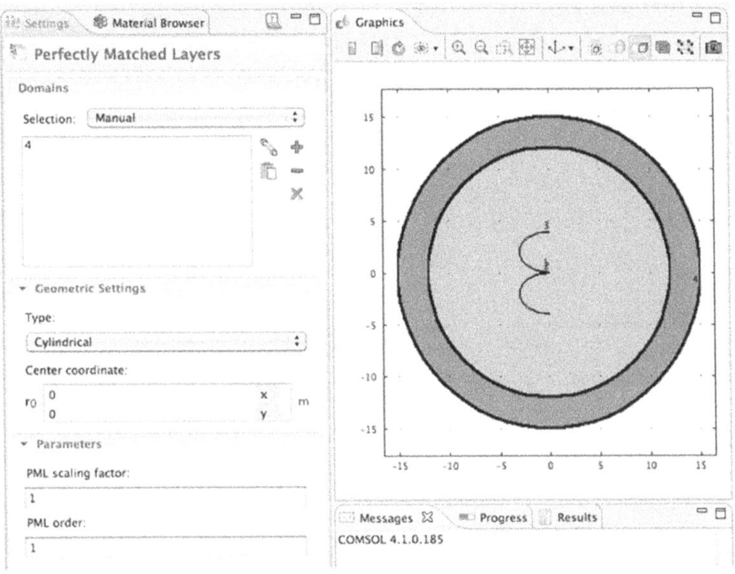

FIGURE 10.36 Perfectly Matched Layer Settings Selection.

Figure 10.36 shows the Perfectly Matched Layer Settings Selection.

Impedance Boundary Condition 1

NOTE *The Impedance Boundary Condition assumes that the skin depth in the material is significantly less than the material thickness. In this case it is on the order of microns.*

Right-Click > Model Builder – Model 1 – Electromagnetic Waves (emw).

Select > Impedance Boundary Condition.

Click > Impedance Boundary Condition 1.

Click > Selection in Settings – Impedance Boundary Condition – Boundaries.

Select > EC Boundaries from the Pull-down menu.

See Figure 10.37.

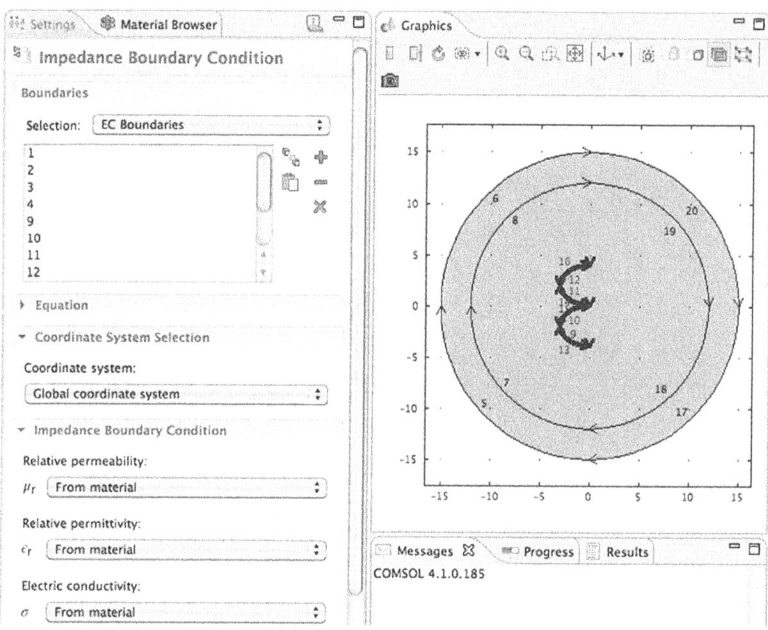

FIGURE 10.37 Impedance Boundary Condition Settings Selection.

Figure 10.37 shows the Impedance Boundary Condition Settings Selection.

Far-Field Calculation 1

NOTE *The requirement for the Far-Field Calculation is that all reflecting surfaces are surrounded, which they are.*

Right-Click > Model Builder – Model 1 – Electromagnetic Waves (emw).

Select > Far-Field Calculation.

Click > Far-Field Calculation 1.

Click > Selection in Settings Far-Field Calculation – Boundaries.

Select > EC Boundaries from the Pull-down menu.

See Figure 10.38.

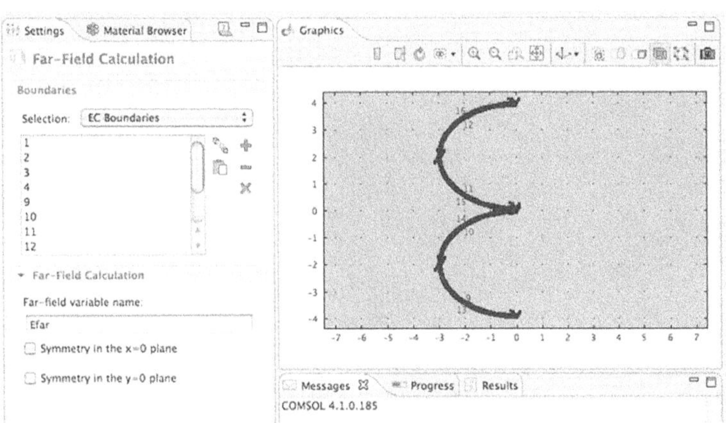

FIGURE 10.38 Far-Field Calculation Settings Selection.

Figure 10.38 shows the Far-Field Calculation Settings Selection.

Definitions

General Extrusion 1

The General Extrusion is a coupling operator that maps an expression defined on a source domain to an expression that can be evaluated on any domain where the destination map expressions are valid.

Right-Click > Model Builder – Model 1 – Definitions.

Select > Model Couplings – General Extrusion.

Click > General Extrusion 1.

Enter > back in Settings – General Extrusion – Operator Name – Operator name.

Select > Domain 1, the inner circle, in the Graphics window.

Click > Add to Selection in Settings – General Extrusion – Source Selection – Selection.

Enter > dm_x in Settings – General Extrusion – Destination Map – x-expression edit window.

Enter > dm_y in Settings – General Extrusion – Destination Map – y-expression edit window.

See Figure 10.39.

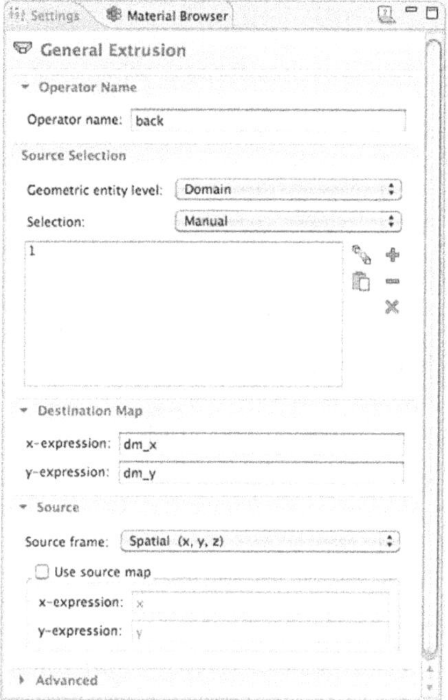

FIGURE 10.39 General Extrusion Settings Selection.

Figure 10.39 shows the General Extrusion Settings Selection.

Mesh 1

Right-Click > Model Builder – Model 1 – Mesh 1.

Select > Free Triangular from the Pop-up menu.

Click > Model Builder – Model 1 – Mesh 1 – Size.

Click > Settings – Size – Element Size Parameters Twistie.

Enter > me_s in Settings – Size – Element Size Parameters – Maximum element size.

Click > Build All.

See Figure 10.40.

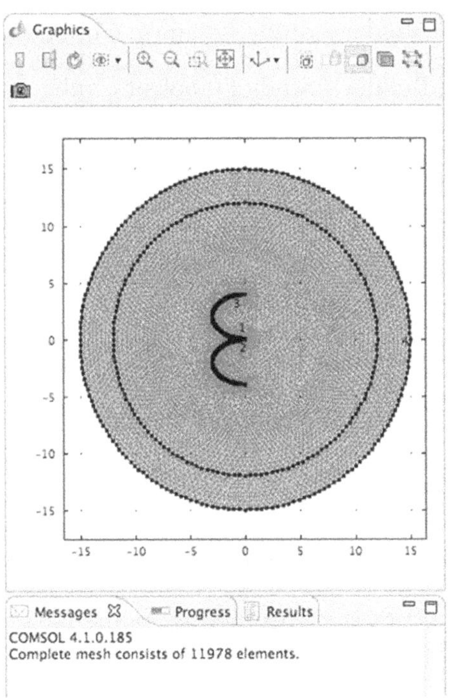

FIGURE 10.40 Graphics Window after the Mesh has been built.

Figure 10.40 shows the Graphics Window after the Mesh has been built.

NOTE *After the mesh is built, the model should have 11978 elements.*

Study 1

Frequencies

Click > Model Builder – Study 1 twistie.

Click > Model Builder – Study 1 – Step 1: Frequency Domain.

Enter > f_a in Settings – Frequency Domain – Study Settings – Frequencies.

Parametric Solver

Right-Click > Model Builder – Study 1.

Select > Show Default Solver from the Pop-up menu.

Click > Model Builder – Study 1 – Solver Configurations twistie.

Click > Model Builder – Study 1 – Solver Configurations – Solver 1 twistie.

Click > Model Builder – Study 1 – Solver Configurations – Solver 1 – Stationary Solver twistie.

Click > Model Builder – Study 1 – Solver Configurations – Solver 1 – Stationary Solver – Parametric 1.

Click > Settings – Parametric – General – Defined by study step.

Select > User defined from the Pull-down menu.

Enter > phideg in the Settings – Parametric – General – Parameter names edit window.

Click > Range button in Settings – Parametric – General.

Enter > Start = 0, Stop = 360, Step = 1 in the Range Pop-up edit window.

Click > Replace button in the Range Pop-up edit window.

See Figure 10.41.

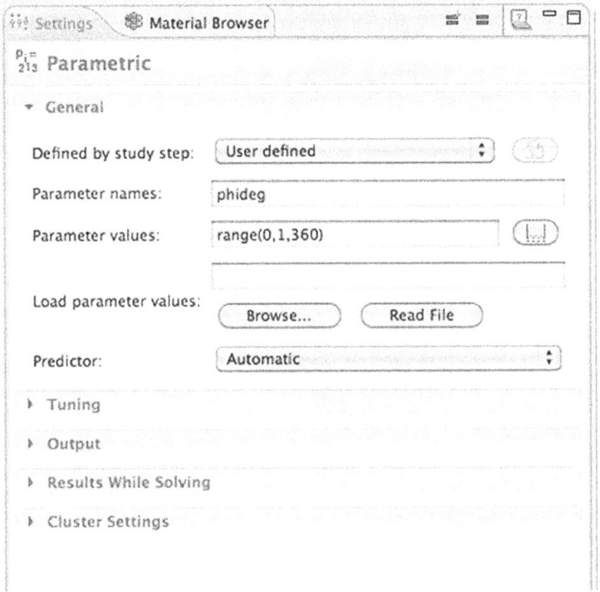

FIGURE 10.41 Settings - Parametric Settings.

Figure 10.41 shows the Settings – Parametric Settings.

Right-Click > Model Builder – Study 1.

Select > Compute.

See Figure 10.42.

Figure 10.42 shows the EC Computed Solution.

Expanded Solution

Click > Zoom in twice.

See Figure 10.43.

Figure 10.43 shows a High Electric Field in EC Computed Solution.

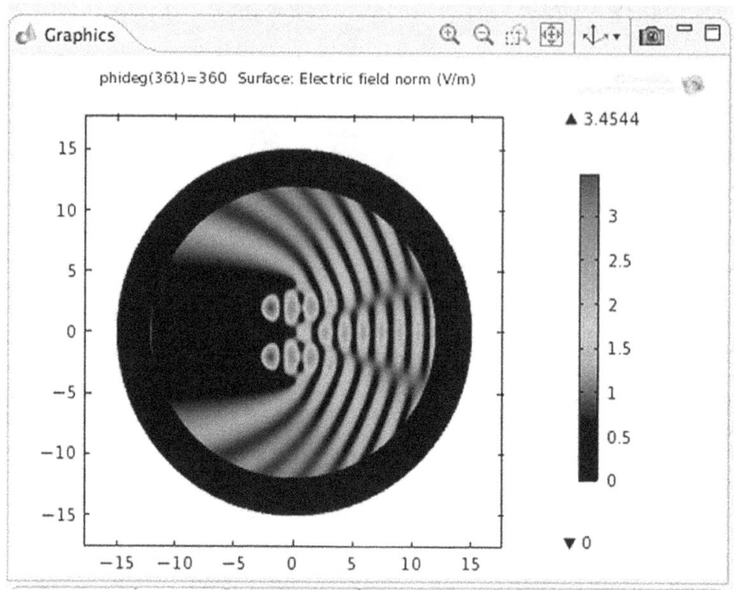

FIGURE 10.42 EC Computed Solution.

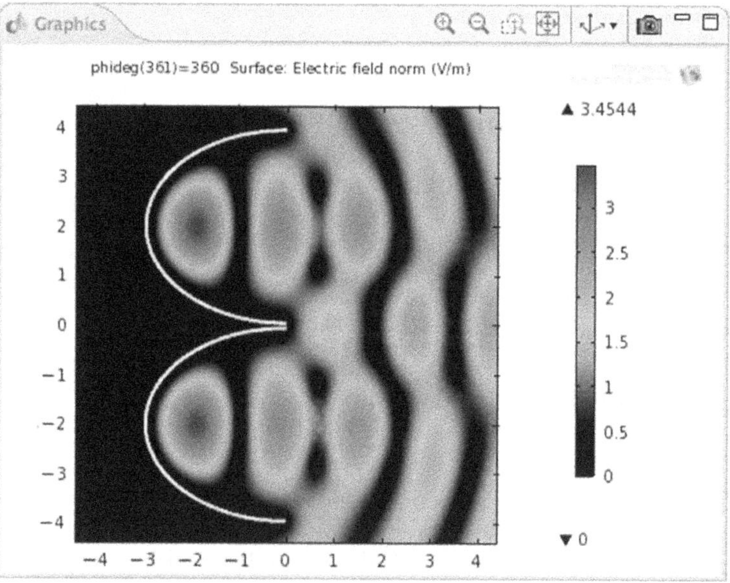

FIGURE 10.43 High Electric Field in EC Computed Solution.

The modeler should note that a high electric field point forms immediately in front of the EC. The EC contributes a gain of approximately three and one half (3.5).

2D Energy Concentrator PML Model Summary and Conclusions

The 2D Energy Concentrator PML Model is a powerful modeling tool that can be used to calculate the gain of various concentrators of different shapes and sizes. With this model, the modeler can easily vary all of the geometric parameters and optimize the design before the first prototype is physically built. These types of concentrators are widely used in industry.

FIRST PRINCIPLES AS APPLIED TO PML MODEL DEFINITION

First Principles Analysis derives from the fundamental laws of nature. In the case of models using this Classical Physics Analysis approach, the laws of conservation in physics require that what goes in (as mass, energy, charge, etc.) must come out (as mass, energy, charge, etc.) or must accumulate within the boundaries of the model.

The careful modeler must be knowledgeable of the implicit assumptions and default specifications that are normally incorporated into the COMSOL Multiphysics software model when a model is built using the default settings.

Consider, for example, the two PML models developed in this chapter. In these models, it is implicitly assumed there are no thermally related changes (mechanical, electrical, etc.), except as specified. It is also assumed the materials are homogeneous and isotropic, except as specifically indicated and there are no thin insulating contact barriers at the thermal junctions. None of these assumptions are typically true in the general case. However, by making such assumptions, it is possible to easily build a 2D First Approximation Model.

NOTE

A First Approximation Model is one that captures all the essential features of the problem that needs to be solved, without dwelling excessively on all of the small details. A good First Approximation Model will yield an answer that enables the modeler to determine if he needs to invest the time and the resources required to build a more highly detailed model.

Also, the modeler needs to remember to name model parameters carefully as pointed out in Chapter 1.

REFERENCES

10.1 COMSOL Multiphysics Users Guide, Version 4.1, pp. 288-292

10.2 http://en.wikipedia.org/wiki/Maxwell%27s_equations

10.3 http://en.wikipedia.org/wiki/Boundary_conditions

10.4 http://en.wikipedia.org/wiki/Perfectly_matched_layer

10.5 http://en.wikipedia.org/wiki/FDTD

10.6 http://en.wikipedia.org/wiki/Finite_element_method

10.7 http://math.mit.edu/~stevenj/18.369/pml.pdf

10.8 RF Module Users Guide, Version 4.1, p. 28

10.9 RF Module Users Guide, Version 4.1, p. 30

10.10 http://en.wikipedia.org/wiki/Gain

10.11 http://en.wikipedia.org/wiki/Speed_of_light

10.12 http://en.wikipedia.org/wiki/Wave_number

10.13 COMSOL Multiphysics Users Guide, Version 4.1, pp. 288-292

10.14 http://en.wikipedia.org/wiki/Solar_thermal_collector

10.15 http://en.wikipedia.org/wiki/Solar_power_plants_in_the_Mojave_Desert

10.16 http://en.wikipedia.org/wiki/Solar_furnace

10.17 http://en.wikipedia.org/wiki/Concentrated_solar_power

10.18 http://en.wikipedia.org/wiki/Anidolic_lighting

Suggested Modeling Exercises

1. Build, mesh, and solve the 2D Concave Metallic Mirror PML Model as presented earlier in this chapter.
2. Build, mesh, and solve the 2D Energy Concentrator PML Model as presented earlier in this chapter.

3. Change the values of the materials parameters and then build, mesh, and solve the 2D Concave Metallic Mirror PML Model as an example problem.
4. Change the values of the materials parameters and then build, mesh, and solve the 2D Energy Concentrator PML Model as an example problem.
5. Change the value of the frequency parameter and then build, mesh, and solve the 2D Concave Metallic Mirror PML Model as an example problem.
6. Change the shape of the geometry and then solve the 2D Energy Concentrator PML Model as an example problem.

CHAPTER 11

BIOHEAT MODELS USING COMSOL MULTIPHYSICS 4.X

In This Chapter

- Guidelines for Bioheat Modeling in 4.x
  - Bioheat Modeling Considerations
- Bioheat Transfer Models
  - 2D Axisymmetric Tumor Laser Irradiation Model
  - 2D Axisymmetric Microwave Cancer Therapy Model
- First Principles as Applied to Bioheat Model Definition
- References
- Suggested Modeling Exercises

GUIDELINES FOR BIOHEAT MODELING IN 4.X

NOTE

In this chapter, two 2D Axisymmetric Bioheat models are presented. Bioheat models, in general, have proven to be very valuable in the study and application of energy locally to terminate cancer tumors. The science, engineering, and medical communities have employed this modeling technique successfully both in the past and currently. Such models serve as first-cut evaluations of potential systemic physical behavior under the influence of complex external stimuli without hazarding the life of a patient.

Bioheat model responses and other such ancillary information can be gathered and screened early in a project for a first-cut evaluation of the physical behavior of a planned prototype.

Bioheat models are typically more conceptually and physically complex than the models that were presented in earlier chapters of this text. The calculated model (simulation) information can be used in the prototype fabrication stage as guidance in the selection of a prototype geometry and materials.

Since the models in this chapter are more conceptually complex and are potentially more difficult to solve than the models presented thus far, it is important that the modeler have available the tools necessary to most easily utilize the powerful capabilities of the 4.x software. In order to do that, if you have not done this previously, the modeler should go to the main 4.x toolbar, Click > Options – Preferences – Model builder. When the Preferences – Model builder edit window is shown, Select > Show equation view checkbox and Show more options checkbox. Click > Apply {11.1}.

Bioheat Modeling Considerations

Bioheat Equation Theory

For the new modeler or those readers unfamiliar with this topic, Bioheat Modeling is employed in the development of models to analyze heat transfer in materials (tissues, fluids, etc.) and other systems related to or derived from previously or currently living organisms. The solution of such Bioheat Equation models is most obviously important when those models are designed to explore techniques for potentially critical therapeutic applications (e.g. destroying cancer cells, killing tumors, etc.) in living entities (people, dogs, cats, cows, sheep, etc.).

Harry H. Pennes published his landmark paper "Analysis of tissue and arterial blood temperatures in the resting human forearm" {11.2} in August of 1948. He proposed in that paper that heat flow is proportional to the difference in temperature between the arterial blood and the local tissue. Pennes' work is considered foundational in this area of study and has since been cited extensively {11.3}.

In the COMSOL Multiphysics 4.x software, the Bioheat Equation (Pennes Equation) is found as a separate Interface within the Heat Transfer Interface. In the Bioheat Transfer (ht) Interface, the Bioheat Equation is formulated as follows:

$$\delta_{ts}\rho C\frac{\partial T}{\partial t}+\nabla\bullet(-\vec{k}\nabla T)=\rho_b C_b\omega_b(T_b-T)+Q_{met}+Q_{ext} \qquad (11.1)$$

Where: δ_{ts} = Time scaling coefficient (default value = 1) [dimensionless].

ρ = Tissue density [kg/m^3].

C = Tissue heat capacity [J/(kg·K)].

T = Temperature [K].

$\vec{k}$ = Tissue thermal conductivity tensor [W/(m·K)].

ρ_b = Blood density [kg/m^3].

C_b = Blood heat capacity [J/(kg·K)].

ω_b = Blood perfusion rate [m^3/(m^3·s)].

T_b = Temperature, Arterial blood [K].

Q_{met} = Metabolic heat source [W/m^3].

Q_{ext} = External environmental heat source [W/m^3].

The rate at which a fluid (e.g. blood) flows through a type of tissue (e.g. muscle, heart, liver, etc.) is the perfusion rate. It is very important, of course, to know the correct perfusion value for the tissue/fluid-type in question.

NOTE *The above equation is shown as formulated for blood flow. However, it can be also equally well employed for other fluids or fluid compositions under the appropriate circumstances (e.g. artificial blood, different animal life fluids, etc.). When the modeler employs variations of this formulation of the Bioheat Equation, he needs to carefully verify the underlying assumptions employed in his particular model.*

The Bioheat Equation is similar to the conduction heat equation. In the case of steady state heat flow, the first term on the left vanishes. That is:

$$\delta_{ts}\rho C\frac{\partial T}{\partial t} = 0 \tag{11.2}$$

In the Bioheat Equation, what would normally be written as single heat source term on the left side of the heat conduction equation (Q) is now separated into three terms.

The perfusion term:

$$\rho_b C_b \omega_b (T_b - T) \tag{11.3}$$

The metabolic term:

$$Q_{met} \tag{11.4}$$

The external source term:

$$Q_{ext} \tag{11.5}$$

The division of what would typically be written as a single heat source term in the Bioheat Equation into three terms is done to facilitate for the user the conceptual linkage and to aid in the formulation of the PDE when creating models for this type of problem (biological).

NOTE

The Pennes Bioheat Equation constitutes a good First Order Approximation to those physical processes (thermal conduction) involved in the solution of the heat transfer problems in biological specimens. The Pennes Bioheat Equation formulation is usually adequate for the modeling of most biological problems of this nature. More terms can, of course, be added, when desired, at the risk of increasing the complexity, the associated model size, and the computational time.

Since the Bioheat Equation, as configured, already delivers the needed level of accuracy for a typical decision, only slightly expanded knowledge will be gained by the addition of Second Order Effects to the equation, considering the intrinsic fundamental limits of most biological system model problems.

Tumor Laser Irradiation Theory

The optical coefficient of absorption for laser photons (irradiation) of tumors is approximately the same as the optical coefficient of absorption for the surrounding tissue. This laser irradiation technique is implemented by raising the relative absorption coefficient locally by artificial means. The change in the local absorption coefficient is accomplished by injection of a designed highly optically absorbing material {11.4} into the tumor. Implementation of this type of procedure is typically considered as a minimally invasive procedure.

The absorbed laser beam photonic energy becomes a heat source for the Bioheat Equation in the region of the tumor as follows:

$$Q_{laser} = I_0 a e^{az - \frac{r^2}{2\sigma^2}} \tag{11.6}$$

Where: I_0 = Irradiation intensity [W/m^2].

a = Absorbance [1/m].

σ = Irradiated region width parameter [m].

BIOHEAT TRANSFER MODELS

NOTE *The Pennes Bioheat Equation is a valuable approach for calculating the potential efficacy of modeled treatment techniques. The fundamental principle needs to be that tumor cells are observed to die at elevated temperatures. The literature cites temperatures that range from 42 °C (315.15 K) {11.5} to 60 °C (333.15 K) {11.6}. If the modeled method raises the local temperature of the tumor cells, without excessively raising the temperature of the normal cells, then the modeled method will most probably be successful.*

The 2D Axisymmetric Tumor Laser Irradiation Model takes advantage of the transparency of human tissue in certain infrared (IR) wavelengths {11.7}. Figure 11.1 shows the structure of the modeling domain. Since the model is created as a 2D Axisymmetric Model, only the right half of the structure will (needs to) be used in the calculations.

See Figure 11.1.

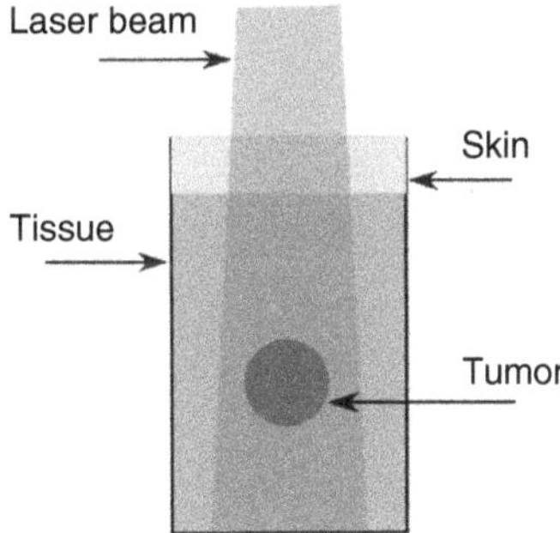

FIGURE 11.1 Tumor Laser Irradiation Physical Model.

Figure 11.1 shows the Tumor Laser Irradiation physical model.

2D Axisymmetric Tumor Laser Irradiation Model

Startup 4.x.

Select > 2D Axisymmetric.

Click > Next.

Click > Twistie for the Heat Transfer interface.

Click > Bioheat Transfer (ht).

Click > Add Selected.

Click > Next.

Select > Preset Studies – Time Dependent.

Click > Finish (Flag).

Click > Save As.

Enter MMUC4_2DAxi_TLI_1.mph.

Click > Save.

See Figure 11.2.

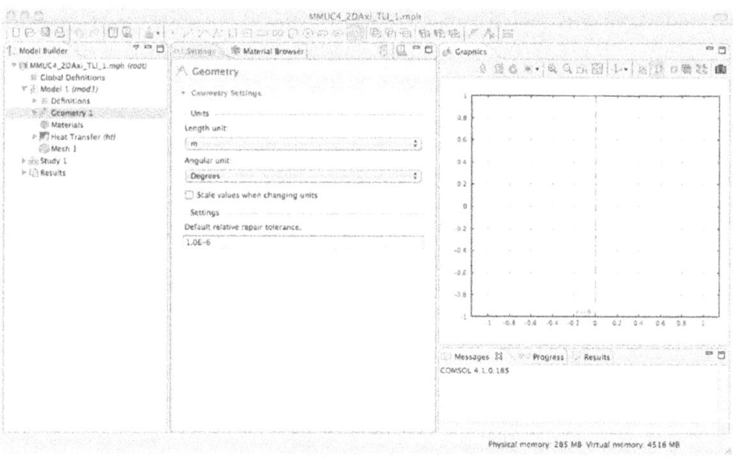

FIGURE 11.2 Desktop Display for the MMUC4_2DAxi_TLI_1.mph Model.

Figure 11.2 shows the Display for the MMUC4_2DAxi_TLI_1.mph model.

Global Definitions

Parameters

In Model Builder,

Right-Click > Model Builder – Global Definitions.

Select > Parameters from the Pop-up menu.

Enter all the Parameters, as shown, in Table 11.1.

TABLE 11.1 TLI Parameters

Name	Expression	Description
rho_blood	1000[kg/m^3]	Density blood
C_blood	4200[J/(kg*K)]	Heat capacity blood
T_blood	37[degC]	Temperature blood
k_skin	0.2[W/(m*K)]	Thermal conductivity skin
rho_skin	1200[kg/m^3]	Density skin
C_skin	3600[J/(kg*K)]	Heat capacity skin
wb_skin	3e-3[1/s]	Blood perfusion rate skin
k_tissue	0.5[W/(m*K)]	Thermal conductivity tissue
rho_tissue	1050[kg/m^3]	Density tissue
C_tissue	3600[J/(kg*K)]	Heat capacity tissue
wb_tissue	6e-3[1/s]	Blood perfusion rate tissue
k_tumor	0.5[W/(m*K)]	Thermal conductivity tumor
rho_tumor	1050[kg/m^3]	Density tumor
C_tumor	3600[J/(kg*K)]	Heat capacity tumor
wb_tumor	6e-3[1/s]	Blood perfusion rate tumor
Q_met	400[W/m^3]	Metabolic heat generation
T0	37[degC]	Temperature reference blood
h_conv	10[W/(m^2*K)]	Heat transfer coefficient skin
T_inf	10[degC]	Temperature domain boundary
I0	1.4[W/mm^2]	Laser irradiation power
sigma	5[mm]	Laser beam width coefficient

See Figures 11.3 through 11.5.

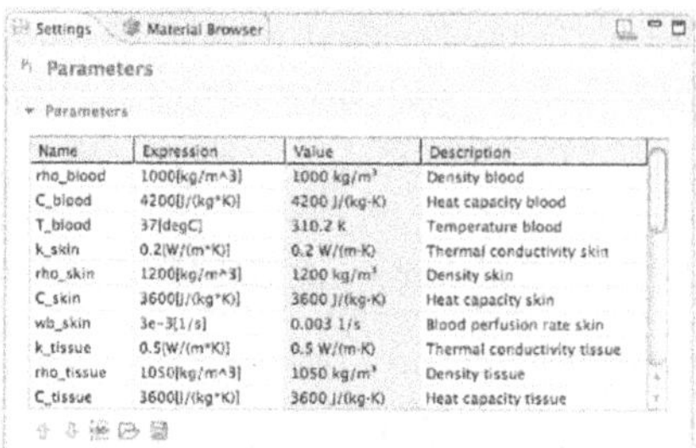

Settings | Material Browser

Parameters

Parameters

Name	Expression	Value	Description
rho_blood	1000[kg/m^3]	1000 kg/m^3	Density blood
C_blood	4200[J/(kg*K)]	4200 J/(kg·K)	Heat capacity blood
T_blood	37[degC]	310.2 K	Temperature blood
k_skin	0.2[W/(m*K)]	0.2 W/(m·K)	Thermal conductivity skin
rho_skin	1200[kg/m^3]	1200 kg/m^3	Density skin
C_skin	3600[J/(kg*K)]	3600 J/(kg·K)	Heat capacity skin
wb_skin	3e-3[1/s]	0.003 1/s	Blood perfusion rate skin
k_tissue	0.5[W/(m*K)]	0.5 W/(m·K)	Thermal conductivity tissue
rho_tissue	1050[kg/m^3]	1050 kg/m^3	Density tissue
C_tissue	3600[J/(kg*K)]	3600 J/(kg·K)	Heat capacity tissue

FIGURE 11.3 Settings - Parameters - Parameters Edit Window (Part 1).

Figure 11.3 shows the Settings – Parameters – Parameters edit window (Part 1).

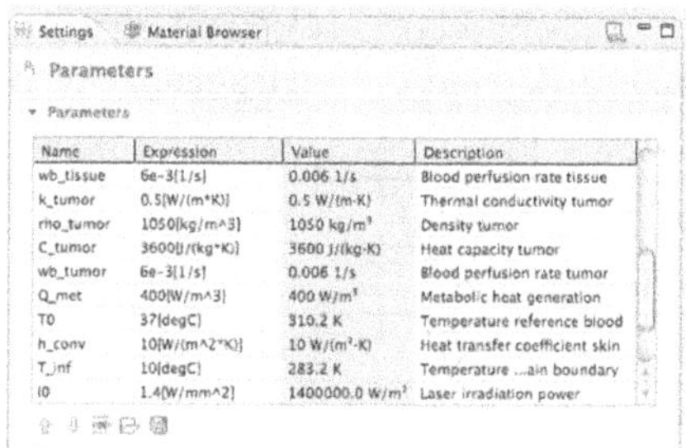

Settings | Material Browser

Parameters

Parameters

Name	Expression	Value	Description
wb_tissue	6e-3[1/s]	0.006 1/s	Blood perfusion rate tissue
k_tumor	0.5[W/(m*K)]	0.5 W/(m·K)	Thermal conductivity tumor
rho_tumor	1050[kg/m^3]	1050 kg/m^3	Density tumor
C_tumor	3600[J/(kg*K)]	3600 J/(kg·K)	Heat capacity tumor
wb_tumor	6e-3[1/s]	0.006 1/s	Blood perfusion rate tumor
Q_met	400[W/m^3]	400 W/m^3	Metabolic heat generation
T0	37[degC]	310.2 K	Temperature reference blood
h_conv	10[W/(m^2*K)]	10 W/(m^2·K)	Heat transfer coefficient skin
T_inf	10[degC]	283.2 K	Temperature ...ain boundary
I0	1.4[W/mm^2]	1400000.0 W/m^2	Laser irradiation power

FIGURE 11.4 Settings - Parameters - Parameters Edit Window (Part 2).

Figure 11.4 shows the Settings – Parameters – Parameters edit window (Part 2).

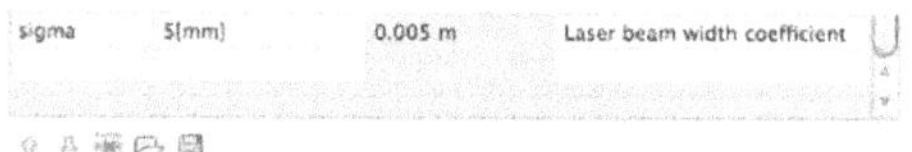

Name	Expression	Value	Description
sigma	5[mm]	0.005 m	Laser beam width coefficient

FIGURE 11.5 Settings - Parameters - Parameters Edit Window (Part 3).

Figure 11.5 shows the Settings – Parameters – Parameters edit window (Part 3).

Geometry

NOTE *The model geometry needs to be created before the local variables are entered into the model, so that the local variables can be assigned to specific domains.*

TLI Domains

In Model Builder – Model 1 (mod1),

Right-Click > Model Builder – Model 1 (mod1) – Geometry 1.

Select > Rectangle from the Pop-up menu.

Enter the coordinates shown in Table 11.2.

Click as instructed.

Repeat the sequence until completed.

TABLE 11.2 TLI Domains

Rectangles						
	Width	**Height**	**Base**	**r**	**z**	**Click**
1	0.1[m]	0.09[m]	Corner	-0.05[m]	-0.1[m]	Build Selected
2	0.1[m]	0.01[m]	Corner	-0.05[m]	-0.1[m]	Build Selected

Circle					
	Radius	**Base**	**r**	**z**	**Click**
1	0.005[m]	Center	0[m]	-0.05[m]	Build All

See Figure 11.6.

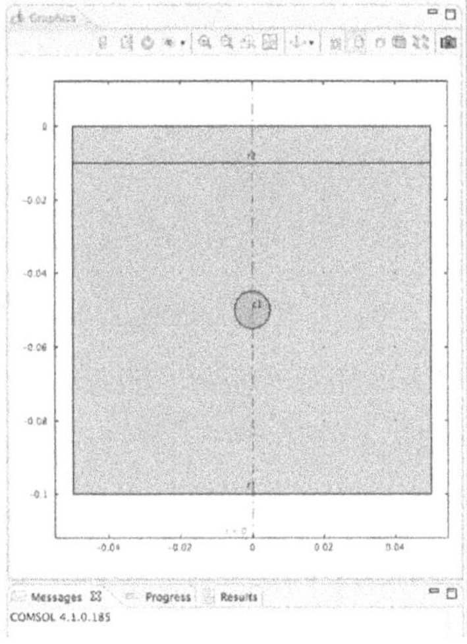

FIGURE 11.6 TLI Domains in the Graphics Edit Window.

Figure 11.6 shows the TLI Domains in the Graphics edit window.

Boolean Operations

In Model Builder – Model 1 (mod1),

Right-Click > Model Builder – Model 1 (mod1) – Geometry 1.

Select > Rectangle from the Pop-up menu.

Enter the coordinates shown in Table 11.3.

Click as instructed.

TABLE 11.3 TLI Domains

Rectangle						
	Width	**Height**	**Base**	**r**	**z**	**Click**
1	0.05[m]	0.1[m]	Corner	-0.05[m]	-0.1[m]	Build Selected

See Figure 11.7.

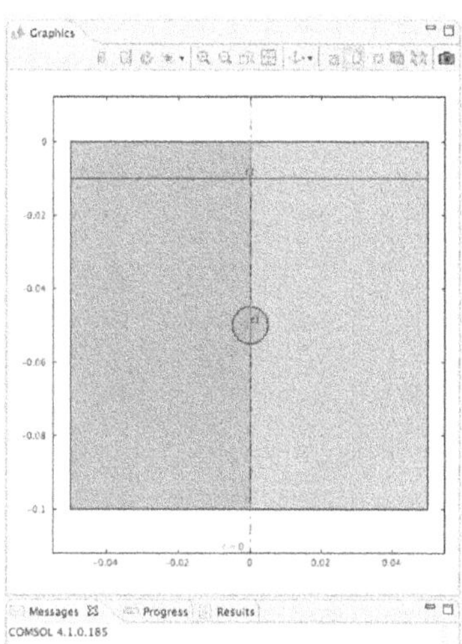

FIGURE 11.7 TLI Domains in Graphics Edit Window.

Figure 11.7 shows the TLI Domains in Graphics edit window.

NOTE *In order to build the final TLI domain using the TLI domains, the modeler needs to perform one Boolean Difference operations.*

Final TLI Domains Formation

Right-Click > Model Builder – Model 1 (mod1) – Geometry 1.

Select > Boolean Operations – Difference from the Pop-up menu.

Shift-Click > Rectangles r1 and r2 and Circle c1 in the Graphics window.

Click > Add to Selection in Settings – Difference – Difference – Objects to add.

Click > Activate Selection in Settings – Difference – Difference – Objects to subtract.

Click > Rectangle r3 in the Graphics window.

Click > Add to Selection in Settings – Difference – Difference – Objects to subtract.

See Figure 11.8.

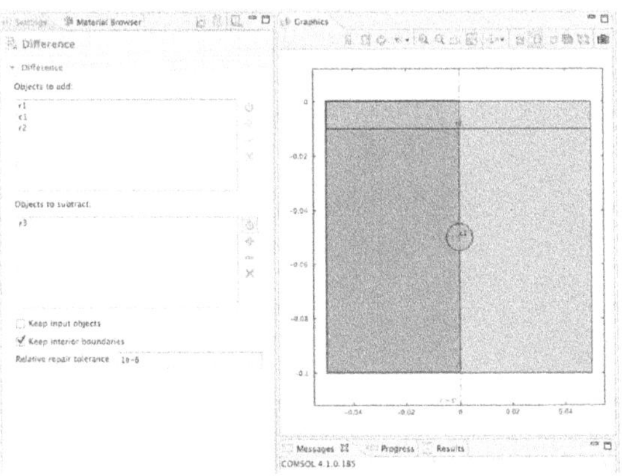

FIGURE 11.8 Settings – Difference – Difference Prepared for the Build Selected Operation.

Figure 11.8 shows Settings – Difference – Difference Prepared for the Build Selected operation.

Click > Build Selected in the Settings Toolbar.

See Figure 11.9.

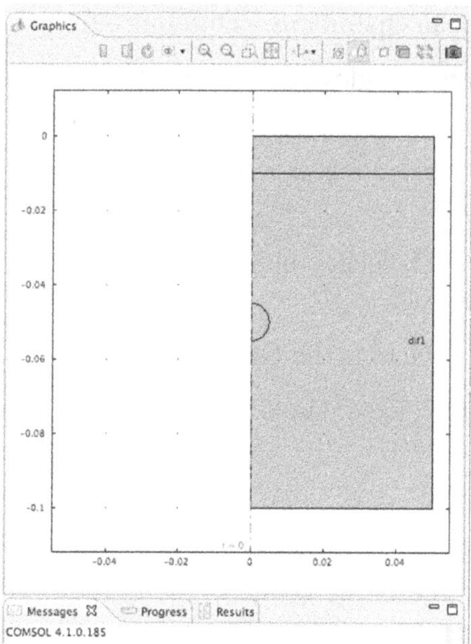

FIGURE 11.9 Graphics Window with the Final TLI Domains Built.

Figure 11.9 shows the Graphics Window with the Final TLI Domains Built.

Local Variables

Variable 1

Right-Click > Model Builder – Model 1 (mod1) – Definitions.

Select > Variables from the Pop-up menu.

Click > Variables 1.

Click > Geometric entity level in Settings – Variables – Geometric Scope.

Select > Domain.

Shift-Click > Domains 1 and 3 in the Graphics window.
See Figure 11.10.

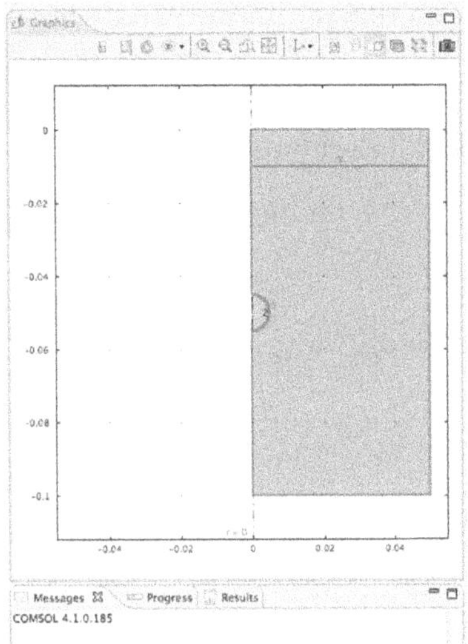

FIGURE 11.10 TLI Domains 1 and 3 Selected in the Graphics Window.

Figure 11.10 shows TLI Domains 1 and 3 selected in the Graphics Window.

Click > Add to Selection (Plus sign) in Settings – Variables – Geometric Scope.

Enter > Name = a, Expression = 0.1[1/m] Description = Absorbance in the Settings – Variables – Variables edit window.

See Figure 11.11.

FIGURE 11.11 Settings - Variables - Variables Edit Window.

Figure 11.11 shows the Settings – Variables – Variables edit window.

Variable 2

Right-Click > Model Builder – Model 1 (mod1) – Definitions.

Select > Variables from the Pop-up menu.

Click > Variables 2.

Click > Geometric entity level in Settings – Variables – Geometric Scope.

Select > Domain.

Click > Domain 2 in the Graphics window.
See Figure 11.12.

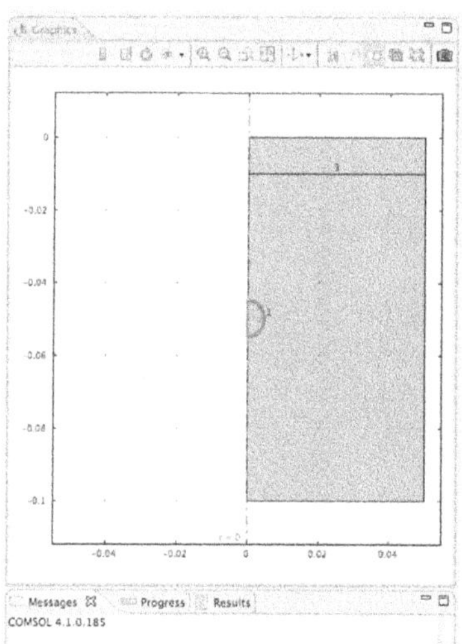

FIGURE 11.12 TLI Domain 2 Selected in Graphics Window.

Figure 11.12 shows TLI Domain 2 Selected in Graphics Window.

Click > Add to Selection (Plus sign) in Settings – Variables – Geometric Scope.

Enter > Name = a, Expression = 4[1/m] Description = Absorbance in Settings – Variables – Variables edit window.

See Figure 11.13.

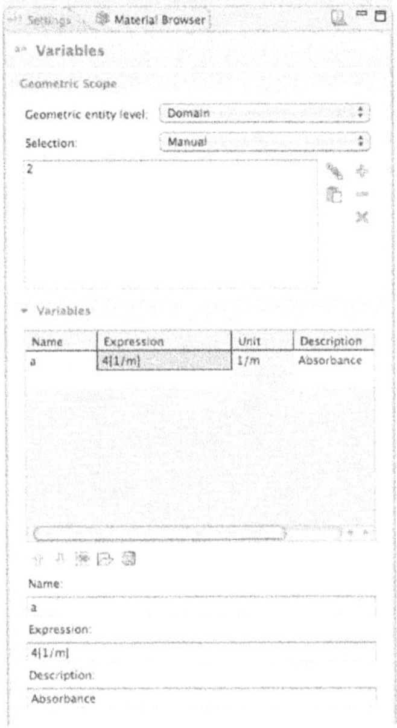

FIGURE 11.13 Settings - Variables - Variables Edit Window.

Figure 11.13 shows the Settings – Variables – Variables edit window.

Variable 3

Right-Click > Model Builder – Model 1 (mod1) – Definitions.

Select > Variables from the Pop-up menu.

Click > Variables 3.

Enter > Name = Q_laser,

Expression = I0*a*exp(a*z-r^2/(2*sigma^2)),

Description = Laser energy distribution in the Settings – Variables – Variables edit window.
See Figure 11.14.

Figure 11.14 shows the Settings – Variables – Variables edit window.

Heat Transfer (ht)

Click > Model Builder – Model 1 (mod1) – Heat Transfer (ht) twistie.

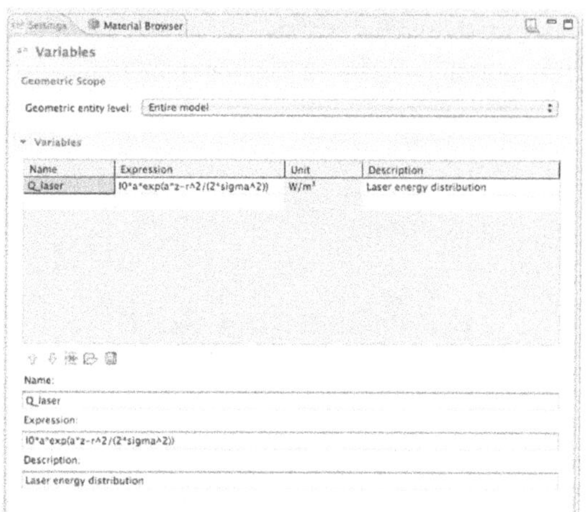

FIGURE 11.14 Settings - Variables - Variables Edit Window.

Domains

NOTE *The default condition is to apply all of these settings to the entire model. Please note that as the properties of other domains (tissues) are defined, the default settings will be overridden.*

Biological Tissue 1

Click > Biological Tissue 1.

Click > Thermal conductivity in Settings – Biological Tissue – Heat Conduction.

Select > User defined.

Enter > k_skin in Settings – Biological Tissue – Heat Conduction – Thermal conductivity (k) edit window.

Click > Density in Settings – Biological Tissue – Thermodynamics.

Select > User defined.

Enter > rho_skin in Settings – Biological Tissue – Thermodynamics – Density (ρ) edit window.

Click > Heat capacity at constant pressure in Settings – Biological Tissue – Thermodynamics.

Select > User defined.

Enter > C_skin in Settings – Biological Tissue – Thermodynamics – Heat capacity at constant pressure (C_p) edit window.

See Figure 11.15.

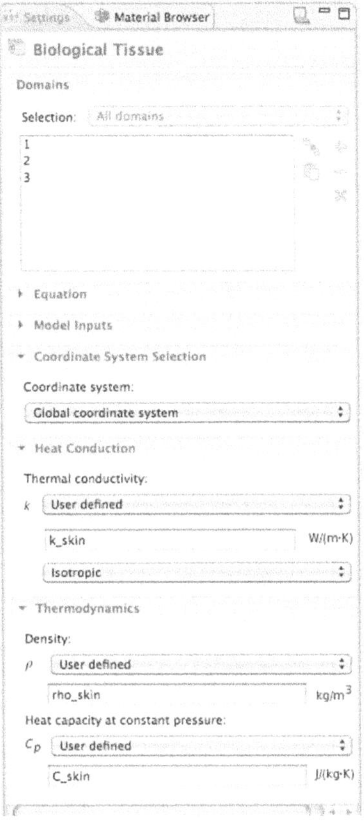

FIGURE 11.15 Settings - Biological Tissue 1 Edit Windows.

Figure 11.15 shows the Settings – Biological Tissue 1 edit windows.

Click > Model Builder – Model 1 (mod1) – Heat Transfer (ht) – Biological Tissue 1 twistie.

Click > Bioheat 1.

Enter > rho_blood in the Settings – Bioheat – Bioheat – Density, blood (ρ_b) edit window.

Enter > C_blood in the Settings – Bioheat – Bioheat – Specific heat, blood (C_b) edit window.

Enter > wb_skin in the Settings – Bioheat – Bioheat – Blood perfusion rate (ω_b) edit window.

Enter > T_blood in the Settings – Bioheat – Bioheat – Arterial blood temperature (T_b) edit window.

Enter > Q_met in the Settings – Bioheat – Bioheat – Metabolic heat source (Q_{met}) edit window.

See Figure 11.16.

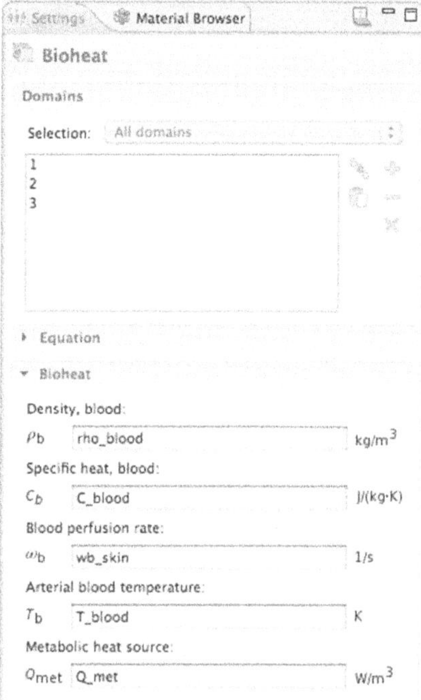

FIGURE 11.16 Settings - Bioheat 1 Edit Windows.

Figure 11.16 shows the Settings – Bioheat 1 edit windows.

Biological Tissue 2

Right-Click > Model Builder – Model 1 (mod1) – Heat Transfer (ht).

Select > Biological Tissue.

Click > Biological Tissue 2.

Click > Domain 1 in the Graphics window.

Click > Add to Selection in Settings – Biological Tissue – Domains.

Click > Thermal conductivity in Settings – Biological Tissue – Heat Conduction.

Select > User defined.

Enter > k_tissue in Settings – Biological Tissue – Heat Conduction – Thermal conductivity (k) edit window.

Click > Density in Settings – Biological Tissue – Thermodynamics.

Select > User defined.

Enter > rho_tissue in Settings – Biological Tissue – Thermodynamics – Density (ρ) edit window.

Click > Heat capacity at constant pressure in Settings – Biological Tissue – Thermodynamics.

Select > User defined.

Enter > C_tissue in Settings – Biological Tissue – Thermodynamics – Heat capacity at constant pressure (C_p) edit window.

See Figure 11.17.

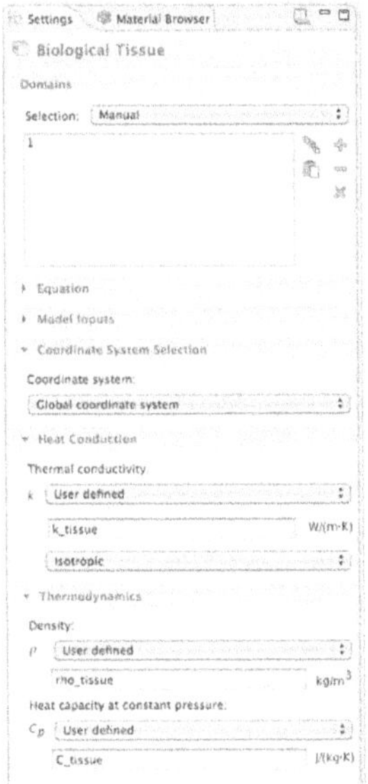

FIGURE 11.17 Settings - Biological Tissue 2 Edit Windows.

Figure 11.17 shows the Settings – Biological Tissue 2 edit windows.

Click > Model Builder – Model 1 (mod1) – Heat Transfer (ht) – Biological Tissue 2 twistie.

Click > Bioheat 1.

Enter > rho_blood in the Settings – Bioheat – Bioheat – Density, blood (ρ_b) edit window.

Enter > C_blood in the Settings – Bioheat – Bioheat – Specific heat, blood (C_b) edit window.

Enter > wb_tissue in the Settings – Bioheat – Bioheat – Blood perfusion rate (ω_b) edit window.

Enter > T_blood in the Settings – Bioheat – Bioheat – Arterial blood temperature (T_b) edit window.

Enter > Q_met in the Settings – Bioheat – Bioheat – Metabolic heat source (Q_{met}) edit window.

See Figure 11.18.

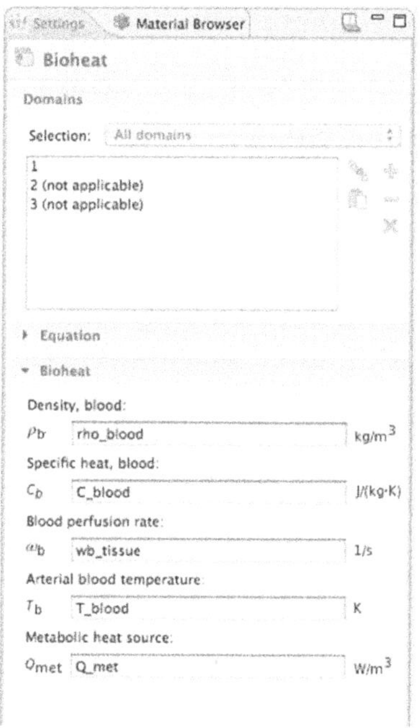

FIGURE 11.18 Settings - Bioheat 2 Edit Windows.

Figure 11.18 shows the Settings – Bioheat 2 edit windows.

Biological Tissue 3

Right-Click > Model Builder – Model 1 (mod1) – Heat Transfer (ht).

Select > Biological Tissue.

Click > Biological Tissue 3.

Click > Domain 2 in the Graphics window.

Click > Add to Selection in Settings – Biological Tissue – Domains.

Click > Thermal conductivity in Settings – Biological Tissue – Heat Conduction.

Select > User defined.

Enter > k_tumor in Settings – Biological Tissue – Heat Conduction – Thermal conductivity (k) edit window.

Click > Density in Settings – Biological Tissue – Thermodynamics.

Select > User defined.

Enter > rho_tumor in Settings – Biological Tissue – Thermodynamics – Density (ρ) edit window.

Click > Heat capacity at constant pressure in Settings – Biological Tissue – Thermodynamics.

Select > User defined.

Enter > C_tumor in Settings – Biological Tissue – Thermodynamics – Heat capacity at constant pressure (C_p) edit window.

See Figure 11.19.

Figure 11.19 shows the Settings – Biological Tissue 3 edit windows.

Click > Model Builder – Model 1 (mod1) – Heat Transfer (ht) – Biological Tissue 3 twistie.

Click > Bioheat 1.

Enter > rho_blood in the Settings – Bioheat – Bioheat – Density, blood (ρ_b) edit window.

Enter > C_blood in the Settings – Bioheat – Bioheat – Specific heat, blood (C_b) edit window.

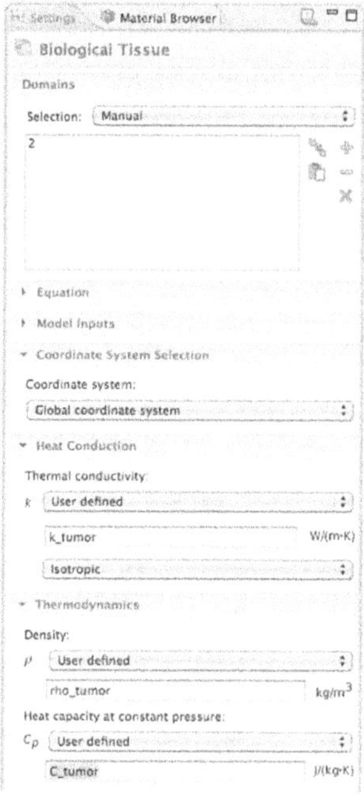

FIGURE 11.19 Settings - Biological Tissue 3 Edit Windows.

Enter > wb_tumor in the Settings – Bioheat – Bioheat – Blood perfusion rate (ω_b) edit window.

Enter > T_blood in the Settings – Bioheat – Bioheat – Arterial blood temperature (T_b) edit window.

Enter > Q_met in the Settings – Bioheat – Bioheat – Metabolic heat source (Q_{met}) edit window.

See Figure 11.20.

Figure 11.20 shows the Settings - Bioheat 3 edit windows.

Initial Values

Click > Model Builder – Model 1 (mod1) – Heat Transfer (ht) – Initial Values 1.

Enter > T0 in Settings – Initial Values – Initial Values – Temperature.

See Figure 11.21.

Figure 11.21 shows the Settings – Initial Values edit windows.

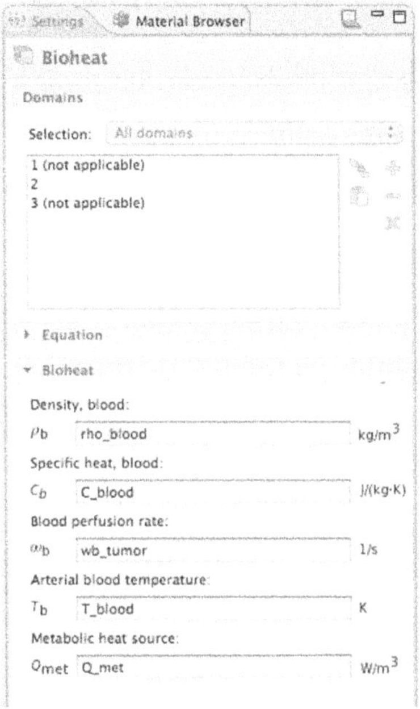

FIGURE 11.20 Settings - Bioheat 3 Edit Windows.

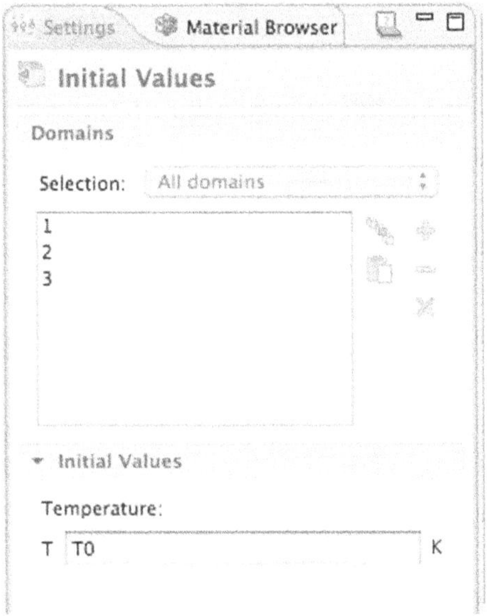

FIGURE 11.21 Settings - Initial Values Edit Windows.

Heat Source

Right-Click > Model Builder – Model 1 (mod1) – Heat Transfer (ht).

Select > Heat Source from the Pop-up menu.

Click > Heat Source 1.

Click > Settings – Heat Source – Domains – Selection.

Select > All Domains from the Pop-up menu.

Enter > Q_laser in the Settings – Heat Source – Heat Source – General source edit window.

See Figure 11.22.

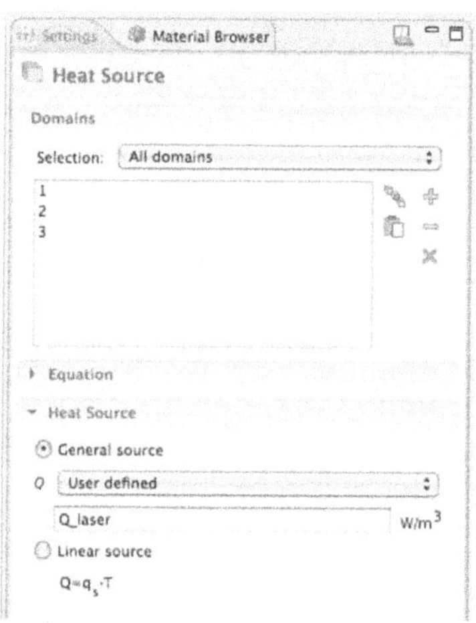

FIGURE 11.22 Settings - Heat Source Edit Windows.

Figure 11.22 shows the Settings – Heat Source edit windows.

Heat Flux

Right-Click > Model Builder – Model 1 (mod1) – Heat Transfer (ht).

Select > Heat Flux from the Pop-up menu.

Click > Heat Flux 1.

Click > Boundary 7 in the Graphics window.

See Figure 11.23.

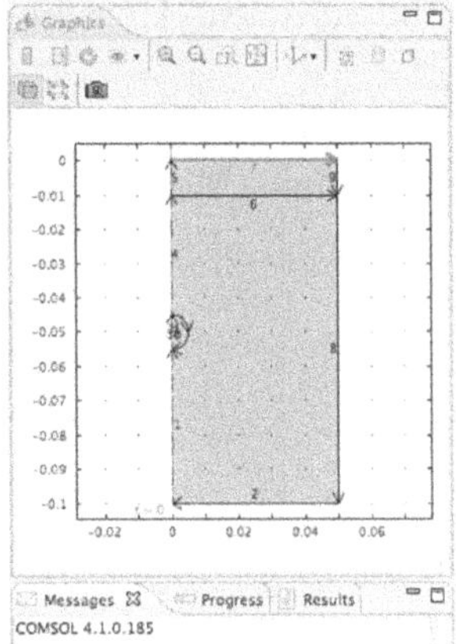

FIGURE 11.23 Boundary 7 in the Graphics Window.

Figure 11.23 shows the Boundary 7 in the Graphics window.

Click > Add to Selection (Plus sign) in Settings – Heat Flux – Boundaries.

Click > Inward heat flux button in Settings – Heat Flux – Heat Flux.

Enter > h_conv in the Settings – Heat Flux – Heat Flux – Heat transfer coefficient edit window.

Enter > T_inf in the Settings – Heat Flux – Heat Flux – External temperature (T_{ext}) edit window.

See Figure 11.24.

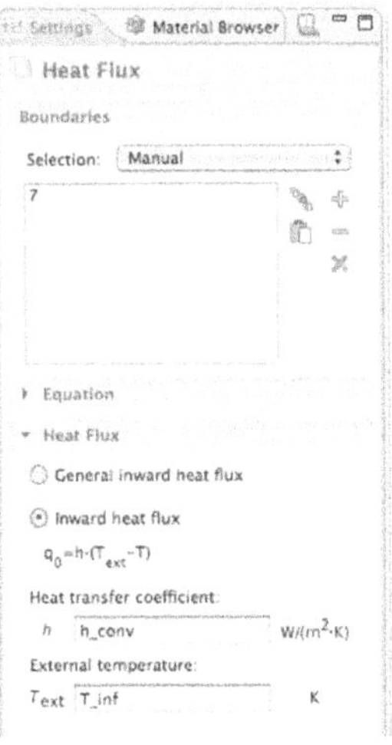

FIGURE 11.24 Settings - Heat Flux Edit Windows.

Figure 11.24 shows the Settings – Heat Flux edit windows.

Mesh 1

Right-Click > Model Builder – Model 1 – Mesh 1.

Select > Free Triangular from the Pop-up menu.

Click > Build All.

See Figure 11.25.

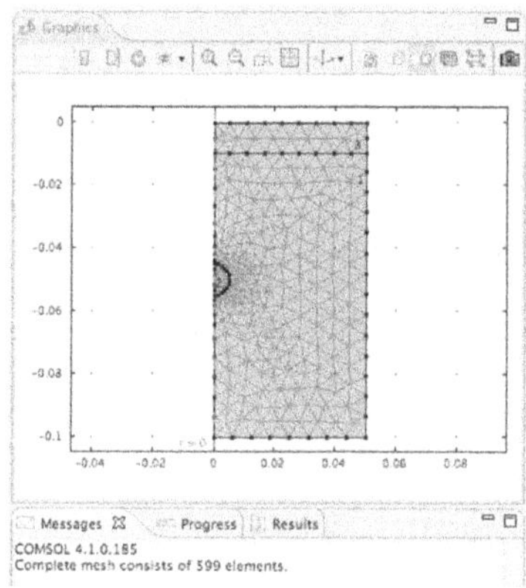

FIGURE 11.25 Graphics Window and the Size Settings after the Mesh Build.

Figure 11.25 shows the Graphics Window and the Size Settings after the Mesh Build.

NOTE *After the mesh is built, the model should have 599 elements.*

Study 1

Time Dependent Solver

Click > Model Builder – Study 1 twistie.

Click > Model Builder – Study 1 – Step 1: Time Dependent.

Click > Range button in Settings – Time Dependent – Study Settings.

Enter > Start = 0, Stop = 600, Step = 10 in the Pop-up Range edit windows.

Click > Replace button.

See Figure 11.26.

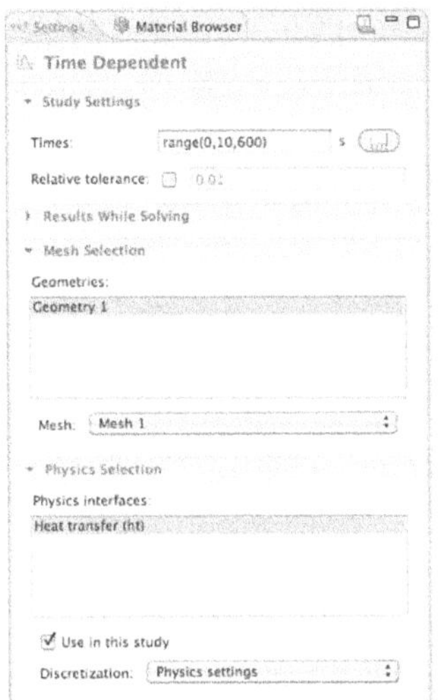

FIGURE 11.26 Settings – Time Dependent – Study Settings.

Figure 11.26 shows the Settings – Time Dependent – Study Settings.

Right-Click > Model Builder – Study 1.

Select > Compute.

TLI Solution

See Figure 11.27.

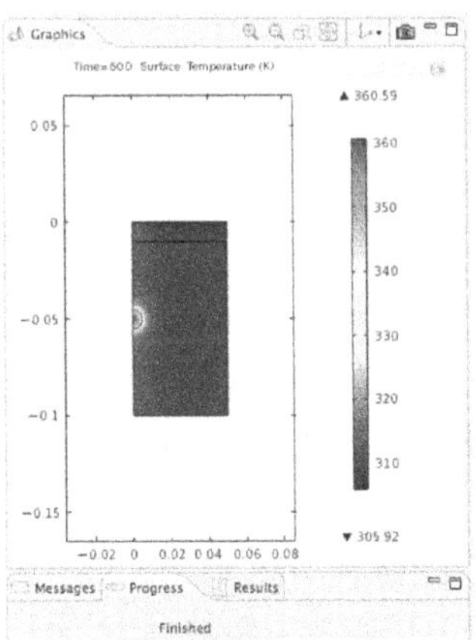

FIGURE 11.27 TLI Computed Solution, Default Parameter Plot.

Figure 11.27 shows the TLI Computed Solution, Default Parameter Plot.

Degrees C Plotted Solution

Click > Model Builder – Results twistie.

Click > Model Builder – Results – 2D Plot Group 1 twistie.

Click > Model Builder – Results – 2D Plot Group 1 – Surface 1.

Click > Settings – Surface – Expression – Unit.

Select > degC from the Settings – Surface – Expression – Unit Pop-up menu.

See Figure 11.28.

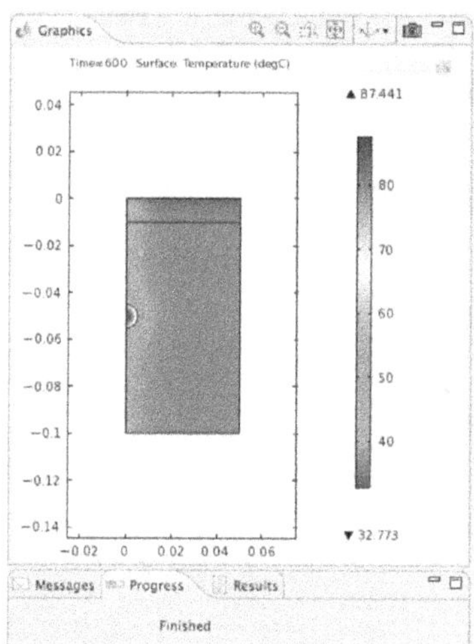

FIGURE 11.28 TLI Computed Solution Degrees C Plot.

Figure 11.28 shows the TLI Computed Solution Degrees C Plot.

NOTE *The modeler should note that the temperature of the tumor exceeds the desired 60C needed to destroy the tumor.*

2D Axisymmetric Tumor Laser Irradiation Model Summary and Conclusions

The 2D Axisymmetric Tumor Laser Irradiation Model is a powerful modeling tool that can be used to calculate the gain in temperature under laser irradiation for different injected absorbance materials. With this model, the modeler can easily vary all of the parameters and optimize the design before the first prototype is physically built or used. These types of models are widely used in industry.

Microwave Cancer Therapy Theory

Hyperthermic oncology (high temperature cancer and/or tumor treatment) {11.8} is the use of elevated temperatures to kill cancer and other tumor cells. As discussed and prototyped in the TLI model, in treatment it is necessary to locally raise the temperature of the cancer/tumor cells, while doing minimal damage to the normal (healthy) cells surrounding the tumor. In the TLI model, the energy was supplied as photothermal energy using laser irradiation. In this model, the externally applied energy is supplied through the use of a specialized microwave antenna and the application of Ohm's and Joule's Laws {11.9, 11.10}. This type of procedure is typically designated as a minimally invasive procedure {11.11}.

Figure 11.29 shows the microwave antenna in cross-section.

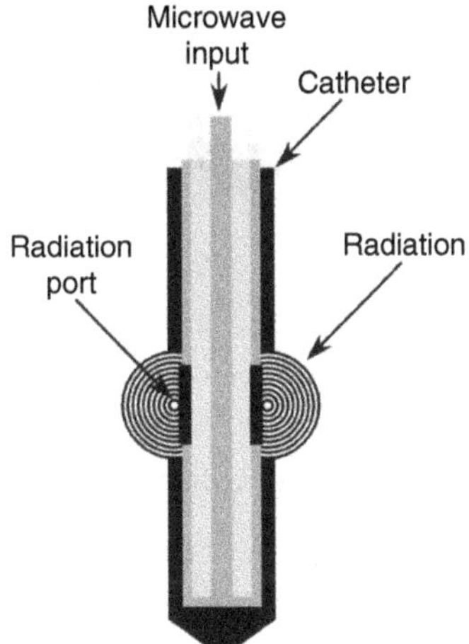

FIGURE 11.29 MCT Antenna Cross-Section.

Figure 11.29 shows the MCT Antenna Cross-section.

2D Axisymmetric Microwave Cancer Therapy Model

The following Multiphysics model solution is derived from a model that was originally developed by COMSOL as a Heat Transfer Interface Tutorial Model for the demonstration of the solution of a Bioheat Equation model. That model was developed for distribution with the Heat Transfer Module software as part of the COMSOL Heat Transfer Interface Model Library.

This model takes advantage of the conductivity of human tissue. Figure 11.30 shows the microwave antenna in cross-section radiating power, imbedded in the modeling domain (liver tissue).

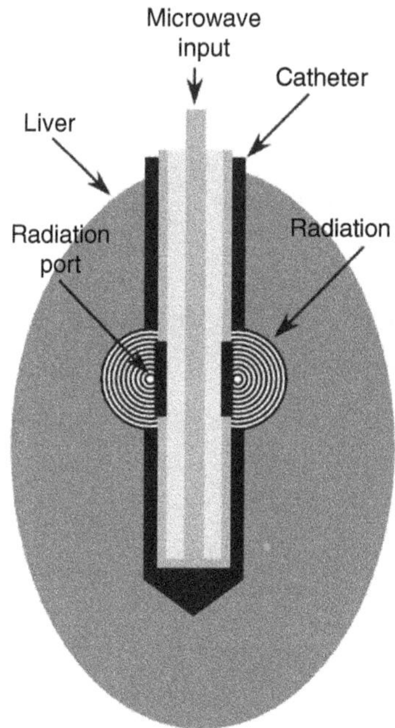

FIGURE 11.30 MCT Antenna Imbedded in Liver Tissue.

Figure 11.30 shows the MCT Antenna Imbedded in Liver Tissue.

Since the model is created as a 2D Axisymmetric Model, only the right half of the structure is used in the calculations.

Bioheat Transfer Models

NOTE

As referenced in the TLI model, the Pennes Bioheat Equation is a valuable approach for calculating the potential efficacy of modeled treatment techniques. The fundamental principle needs to be that tumor cells are observed to die at elevated temperatures. The literature cites temperatures that range from 42 °C (315.15 K) to 60 °C (333.15). If the modeled method raises the local temperature of the tumor cells, without excessively raising the temperature of the normal cells, then the modeled method will most probably be successful.

Building the 2D Axisymmetric Microwave Cancer Therapy Model

Startup 4.x.

Select > 2D Axisymmetric.

Click > Next.

Click > Twistie for the Radio Frequency interface.

Select > Electromagnetic Waves (emw).

Click > Add Selected (Plus sign) in Model Wizard – Add Physics.

Click > Twistie for the Heat Transfer interface.

Select > Bioheat Transfer (ht).

Click > Add Selected (Plus sign) in Model Wizard – Add Physics.

See Figure 11.31.

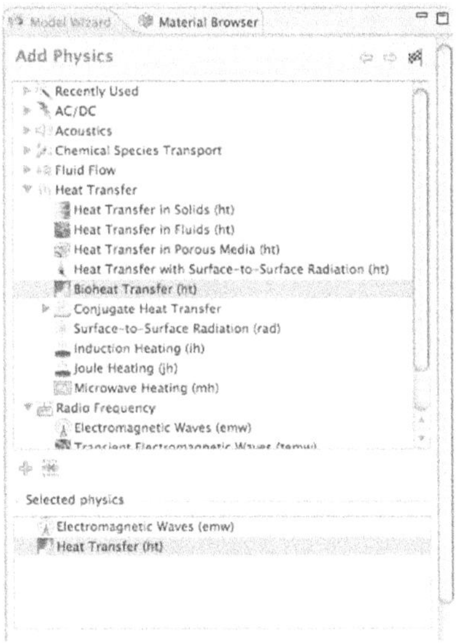

FIGURE 11.31 Model Wizard - Add Physics Selection Windows.

Figure 11.31 Model Wizard – Add Physics Selection Windows.

Click > Next.

Click > Twistie for the Custom Studies in Model Wizard.

Click > Twistie for the Preset Studies for Some Physics in Model Wizard.

Select > Frequency Domain in Model Wizard – Select Study Type – Custom Studies – Preset Studies for Some Physics.

Click > Finish (Flag).

Click > Save As.

Enter MMUC4_2DAxi_MCT_1.mph.

Click > Save.

See Figure 11.32.

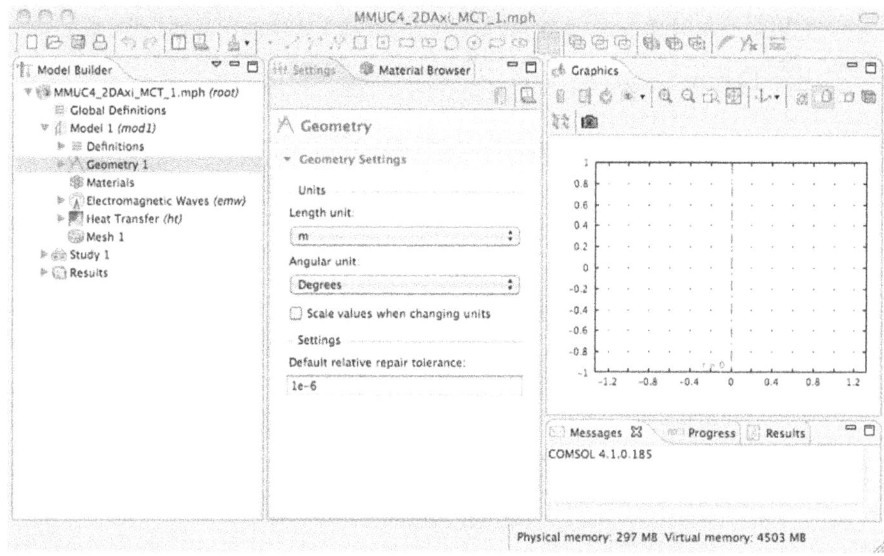

FIGURE 11.32 Desktop Display for the MMUC4_2DAxi_MCT_1.mph Model.

Figure 11.32 shows the Display for the MMUC4_2DAxi_MCT_1.mph model.

Global Definitions

Parameters

In Model Builder,

Right-Click > Model Builder – Global Definitions.

Select > Parameters from the Pop-up menu.

Enter all the Parameters, as shown, in Table 11.4.

TABLE 11.4 MCT Parameters

Name	Expression	Description
k_liver	0.56[W/(m*K)]	Thermal conductivity liver
rho_blood	1e3[kg/m^3]	Density blood
Cp_blood	3639[J/(kg*K)]	Heat capacity blood
omega_blood	3.6e-3[1/s]	Blood perfusion rate
T_blood	37[degC]	Temperature blood
P_in	10[W]	Microwave power input
f	2.45[GHz]	Microwave frequency
eps_diel	2.6[1]	Dielectric relative permittivity
eps_cat	2.6[1]	Catheter relative permittivity
eps_liver	43.03[1]	Liver relative permittivity
sigma_liver	1.69[S/m]	Conductivity liver
rho_liver	1038[kg/m^3]	Liver density
Cp_liver	4187[J/(kg*K)]	Liver heat capacity
sigma_cat	1e-6[S/m]	Catheter conductivity
mur_cat	1[1]	Catheter relative permeability
k_ptfe	0.25[W/(m*K)]	Thermal conductivity PTFE
rho_ptfe	2200[kg/m^3]	Density PTFE
Cp_ptfe	1300[J/(kg*K)]	PTFE heat capacity
sigma_diel	1e-6[S/m]	Microwave center feed
mur_diel	1[1]	Relative permeability dielectric
mur_liver	1[1]	Relative permeability liver

See Figures 11.33 through 11.35.

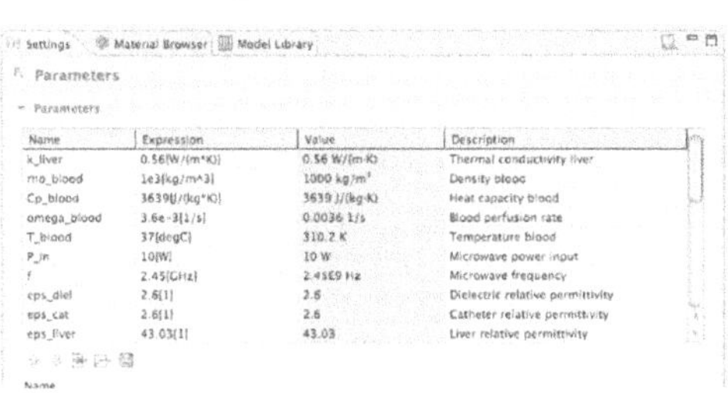

Name	Expression	Value	Description
k_liver	0.56[W/(m*K)]	0.56 W/(m·K)	Thermal conductivity liver
rho_blood	1e3[kg/m^3]	1000 kg/m³	Density blood
Cp_blood	3639[J/(kg*K)]	3639 J/(kg·K)	Heat capacity blood
omega_blood	3.6e-3[1/s]	0.0036 1/s	Blood perfusion rate
T_blood	37[degC]	310.2 K	Temperature blood
P_in	10[W]	10 W	Microwave power input
f	2.45[GHz]	2.45E9 Hz	Microwave frequency
eps_diel	2.6[1]	2.6	Dielectric relative permittivity
eps_cat	2.6[1]	2.6	Catheter relative permittivity
eps_liver	43.03[1]	43.03	Liver relative permittivity

FIGURE 11.33 Settings – Parameters – Parameters Edit Window (Part 1).

Figure 11.33 shows the Settings – Parameters – Parameters edit window (part 1).

Settings | Material Browser | Model Library

Parameters

▾ Parameters

Name	Expression	Value	Description
sigma_liver	1.69[S/m]	1.69 S/m	Conductivity liver
rho_liver	1038[kg/m^3]	1038 kg/m³	Liver density
Cp_liver	4187[J/(kg*K)]	4187 J/(kg·K)	Liver heat capacity
sigma_cat	1e-6[S/m]	1.0E-6 S/m	Catheter conductivity
mur_cat	1[1]	1	Catheter relative permeability
k_ptfe	0.25[W/(m*K)]	0.25 W/(m·K)	Thermal conductivity PTFE
rho_ptfe	2200[kg/m^3]	2200 kg/m³	Density PTFE
Cp_ptfe	1300[J/(kg*K)]	1300 J/(kg·K)	PTFE heat capacity
sigma_diel	1e-6[S/m]	1.0E-6 S/m	Microwave center feed
mur_diel	1[1]	1	Relative permeability dielectric

Name:

FIGURE 11.34 Settings - Parameters - Parameters Edit Window (Part 2).

Figure 11.34 shows the Settings – Parameters – Parameters edit window (part 2).

mur_liver	1[1]	1	Relative permeability liver

FIGURE 11.35 Settings - Parameters - Parameters Edit Window (Part 3).

Figure 11.35 shows the Settings – Parameters – Parameters edit window (part 3).

Geometry

MCT Domains

In Model Builder – Model 1 (mod1),

Right-Click > Model Builder – Model 1 (mod1) – Geometry 1.

Select > Rectangle from the Pop-up menu.

Enter the coordinates shown in Table 11.5.

Click as instructed.

Repeat the sequence until completed.

TABLE 11.5 MCT Domains

Rectangles						
	Width	**Height**	**Base**	**r**	**z**	**Click**
1	0.595e-3[m]	0.01[m]	Corner	0[m]	0[m]	Build Selected
2	29.405e-3[m]	0.08[m]	Corner	0.595e-3[m]	0[m]	Build Selected

Boolean Operations – Union

Right-Click > Model Builder – Model 1 (mod1) – Geometry 1.

Select > Boolean Operations – Union from the Pop-up menu.

Shift-Click > Rectangles r1 and r2 in the Graphics window.

Click > Add to Selection (Plus sign) in Settings – Union – Union.

Uncheck the Keep interior boundaries checkbox.

See Figure 11.36.

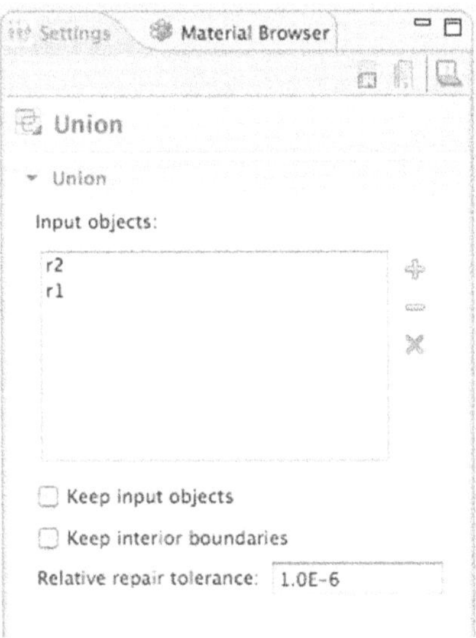

FIGURE 11.36 MCT Settings - Union - Union Edit Window.

Figure 11.36 shows the MCT Settings – Union – Union edit window.

Click > Build Selected.

See Figure 11.37.

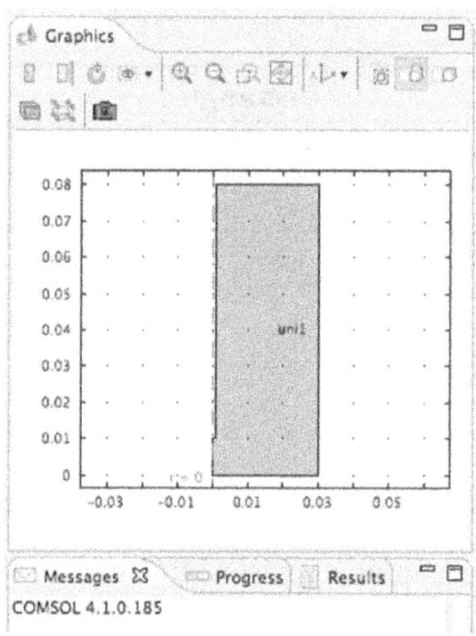

FIGURE 11.37 MCT Settings - Union - Union Edit Window.

Figure 11.37 shows the MCT Result of Union Operation on r1 and r2.

More MCT Domains

In Model Builder – Model 1 (mod1),

Right-Click > Model Builder – Model 1 (mod1) – Geometry 1.

Select > Rectangle from the Pop-up menu.

Enter the coordinates shown in Table 11.6.

Click as instructed.

Repeat the sequence until completed.

TABLE 11.6 MCT Domains

Rectangles						
	Width	Height	Base	r	z	Click
1	0.125e-3[m]	1.0e-3[m]	Corner	0.47e-3[m]	0.0155[m]	Build Selected
2	0.335e-3[m]	0.0699[m]	Corner	0.135e-3[m]	0.0101[m]	Build Selected

Bezier Polygon

Right-Click > Model Builder – Model 1 (mod1) – Geometry 1.

Select > Bezier Polygon from the Pop-up menu.

Click > Add Linear in Settings – Bezier Polygon – Polygon Segments.

Enter the Bezier Segment Coordinates and then Click as indicated in Table 11.7.

TABLE 11.7 MCT Domains

Segment and Row Number	r	z	Click
1-1	0	0	
1-2	0.03	0	Add Linear
2-2	0.03	0.08	Add Linear
3-2	0.0008	0.08	Add Linear
4-2	0.0008	0.0098	Add Linear
5-2	0	0.0092	Add Linear
6-2	0	0	Close Curve

Click > Build All in Settings Toolbar

Boolean Operations – Union

Right-Click > Model Builder – Model 1 (mod1) – Geometry 1.

Select > Boolean Operations – Union from the Pop-up menu.

The easiest method that the modeler can use to implement the next step is to use the Zoom Box to first expand the model geometry in the Graphics window about the r3 region. Then Click > uni1 at the tip of the Catheter. Next, Click b1 in the region to the Right of the Catheter boundary line (the left edge of b1).

Shift-Click > Solids uni1 and b1 in the Graphics window.

Click > Add to Selection (Plus sign) in Settings – Union – Union.

Check the Keep interior boundaries checkbox.

See Figure 11.38.

Figure 11.38 shows the MCT Settings – Union – Union edit window.

Click > Build Selected.

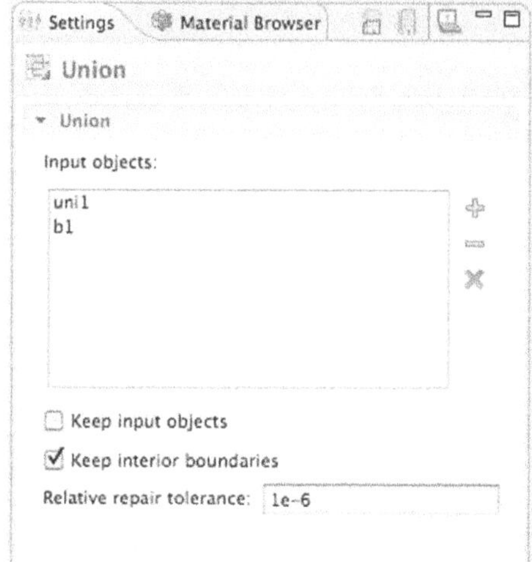

FIGURE 11.38 MCT Result of Union Operation on uni1 and b1.

See Figure 11.39.

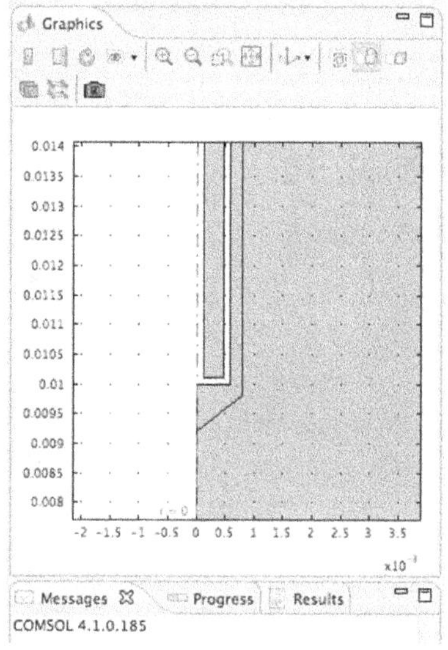

FIGURE 11.39 MCT Result of Union Operation on uni1 and b1.

Figure 11.39 shows the MCT Result of Union Operation on uni1 and b1.

Materials

Material 1

Right-Click > Model Builder – Model 1 (mod1) – Materials.

Select > Material from the Pop-up menu.

Click > Material 1.

Right-Click > Material 1.

Select > Rename from the Pop-up menu.

Enter > Liver Tissue in the Rename Material edit window.

Click > OK.

Click > Selection in Settings – Material – Geometric Scope.

Select > Manual from the Pop-up menu.

Shift-Click > All Domains in Settings – Material – Geometric Scope edit window.

Click > Clear Selection (Minus sign) in Settings – Material – Geometric Scope.

Click > Domain 1 in the Graphics window.

Click > Add to Selection (Plus sign) in Settings – Material – Geometric Scope.

See Figure 11.40.

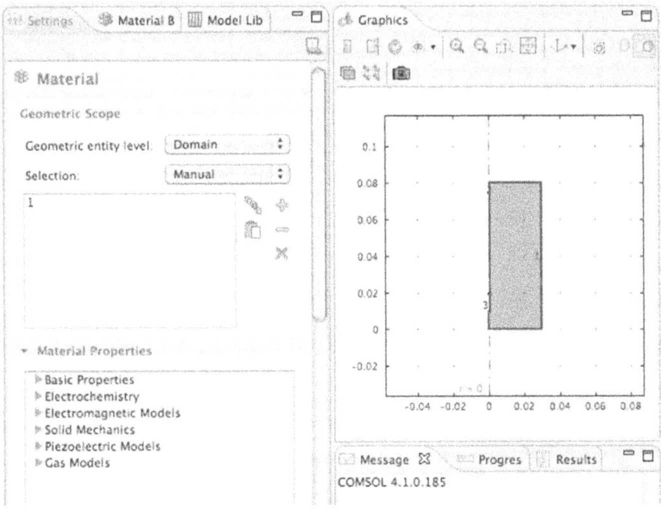

FIGURE 11.40 MCT Material 1 Domain Selection.

Figure 11.40 shows the MCT Material 1 Domain Selection.

Scroll-down in Settings – Material – Material Properties until the Material Contents edit windows are available for the entry of physical property parameters.

Enter the Liver Tissue Value Names as shown in Table 11.8.

TABLE 11.8 Liver Tissue

Property	Name	Value
Electric conductivity	sigma	sigma_liver
Relative permittivity	epsilonr	eps_liver
Relative permeability	mur	mur_liver
Thermal conductivity	k	k_liver
Density	rho	rho_liver
Heat capacity at constant pressure	Cp	Cp_liver

See Figure 11.41.

Figure 11.41 shows the MCT Material 1 Liver Tissue Properties.

Material 2

Right-Click > Model Builder – Model 1 (mod1) – Materials.

Select > Material from the Pop-up menu.

Click > Material 2.

Right-Click > Material 2.

Select > Rename from the Pop-up menu.

Enter > Catheter in the Rename Material edit window.

Click > OK.

Click > Domain 2 (the Catheter) in the Graphics window.

Click > Add to Selection (Plus sign) in Settings – Material – Geometric Scope.

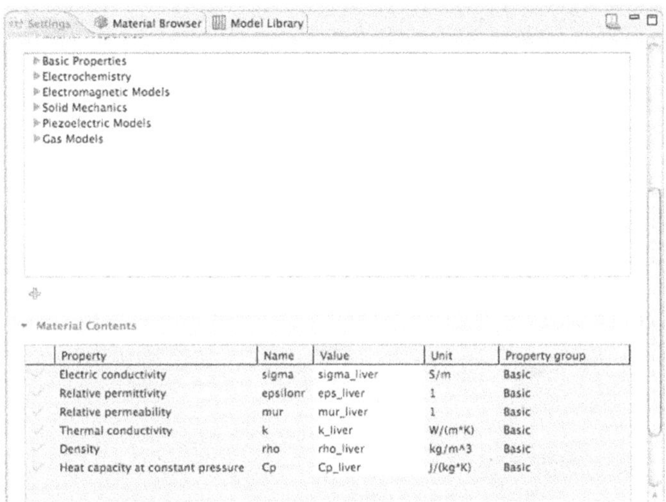

FIGURE 11.41 MCT Material 1 Liver Tissue Properties.

Scroll-down in Settings – Material – Material Properties until the Material Contents edit windows are available for the entry of physical property parameters.

Enter the Catheter Value Names as shown in Table 11.8.

TABLE 11.9 Catheter

Property	**Name**	**Value**
Electric conductivity	sigma	sigma_cat
Relative permittivity	epsilonr	eps_cat
Relative permeability	mur	mur_cat
Thermal conductivity	k	k_ptfe
Density	rho	rho_ptfe
Heat capacity at constant pressure	Cp	Cp_ptfe

See Figure 11.42.

Figure 11.42 shows the MCT Material 2 Catheter Properties.

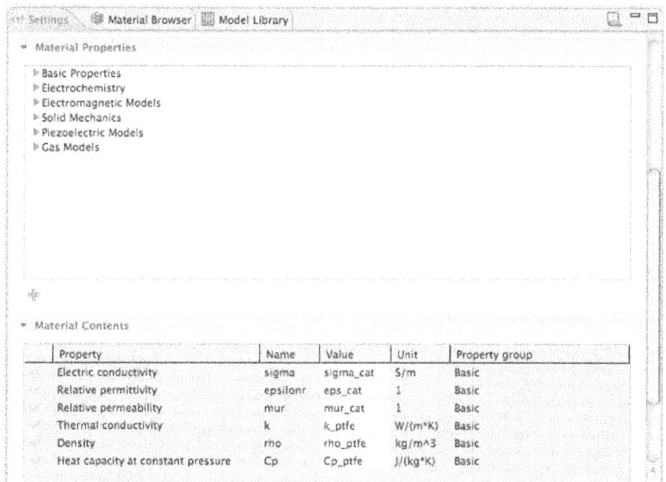

FIGURE 11.42 MCT Material 2 Catheter Properties.

Material 3

Right-Click > Model Builder – Model 1 (mod1) – Materials.

Select > Material from the Pop-up menu.

Click > Material 3.

Right-Click > Material 3.

Select > Rename from the Pop-up menu.

Enter > Dielectric in the Rename Material edit window.

Click > OK.

Click > Domain 3 in the Graphics window.

Click > Add to Selection (Plus sign) in Settings – Material – Geometric Scope.

Scroll-down in Settings – Material – Material Properties until the Material Contents edit windows are available for the entry of physical property parameters.

Enter the Dielectric Value Names as shown in Table 11.9.

TABLE 11.10 Dielectric

Property	Name	Value
Electric conductivity	sigma	sigma_diel
Relative permittivity	epsilonr	eps_diel
Relative permeability	mur	mur_diel
Thermal conductivity	k	k_ptfe
Density	rho	rho_ptfe
Heat capacity at constant pressure	Cp	Cp_ptfe

See Figure 11.43.

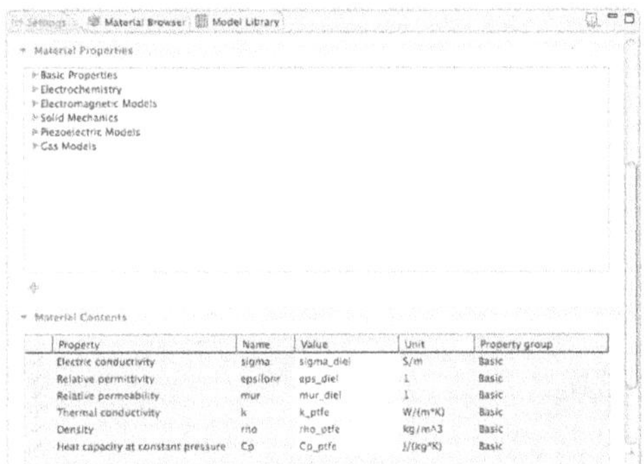

FIGURE 11.43 MCT Material 3 Dielectric Properties.

Figure 11.43 shows the MCT Material 3 Dielectric Properties.

Material 4

Right-Click > Model Builder – Model 1 (mod1) – Materials.

Select > Open Material Browser from the Pop-up menu.

Click > Settings – Material Browser – Materials – Built-in twistie.

Click > Settings – Material Browser – Materials – Built-in – Air.

Right-Click > Settings – Material Browser – Materials – Built-in – Air.

Select > Add Material to Model.

Click > Model Builder – Model 1 (mod1) – Materials – Air.

Click > Domain 4 in the Graphics window.

Click > Add to Selection (Plus sign) in Settings – Material Selection.

See Figure 11.44.

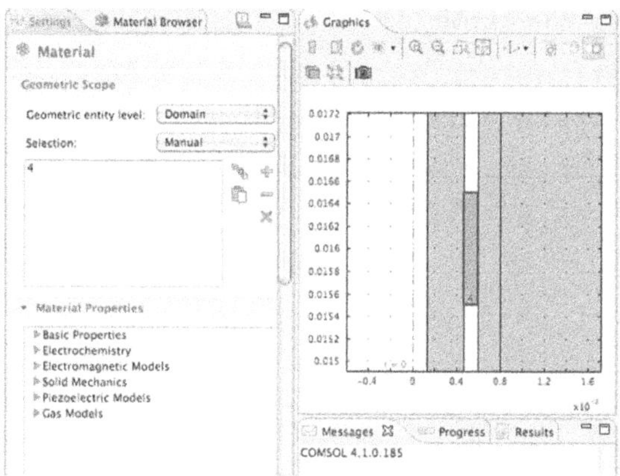

FIGURE 11.44 MCT Material 4 Air Properties.

Figure 11.44 shows the MCT Material 4 Air Properties.

Electromagnetic Waves (emw)

The electromagnetic wave step introduces the power used to heat the tumor at a single frequency.

Click > Model Builder – Model 1 (mod1) – Electromagnetic Waves (emw).

Click > Settings – Electromagnetic Waves – Equation twistie.

Click > Equation form in Settings – Electromagnetic Waves – Equation.

Select > Frequency Domain from the Pop-up menu.

Click > Frequency in Settings – Electromagnetic Waves – Equation.

Select > User defined from the Pop-up menu.

Enter > f in the Settings – Electromagnetic Waves – Equation – Frequency edit window.

See Figure 11.45.

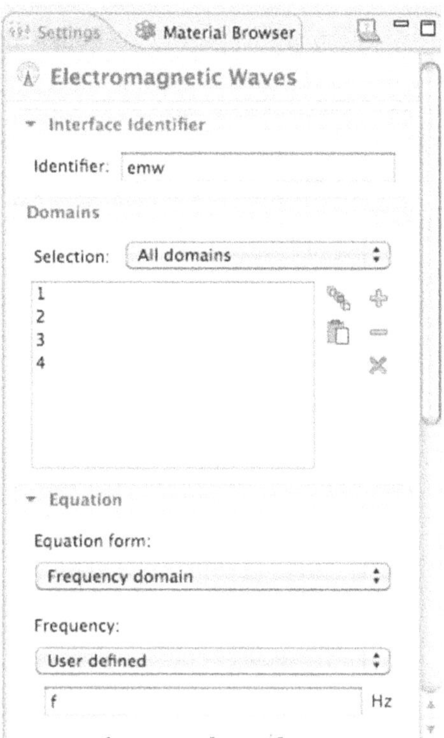

FIGURE 11.45 MCT Electromagnetic Waves Settings.

Figure 11.45 shows the MCT Electromagnetic Waves Settings.

Port 1

Right-Click > Model Builder – Model 1 (mod1) – Electromagnetic Waves (emw).

Select > Port from the Pop-up menu.

Click > Zoom Box in the Graphics Toolbar.

Expand the top left corner of the Graphic twice.

Select > Boundary 8 in the Graphics window.

Click > Add to Selection in Settings – Port – Boundaries.

Click > Type of Port in Settings – Port – Port Properties.

Select > Coaxial.

Click > Wave excitation at this port in Settings – Port – Port Properties.

Select > On.

Enter > P_in in the Settings – Port – Port Properties – Port input power (P_in) edit window.

See Figure 11.46.

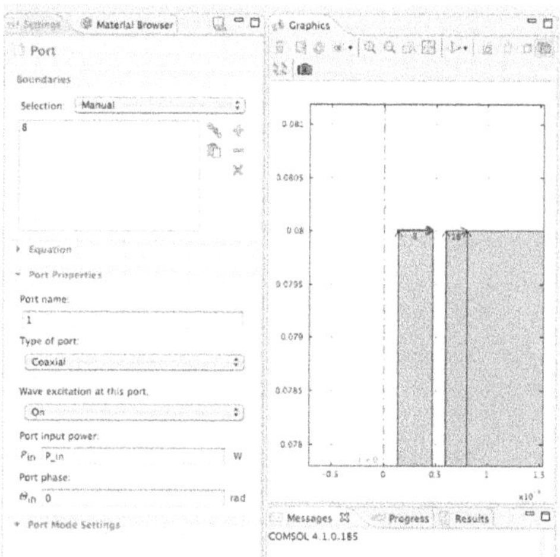

FIGURE 11.46 MCT Electromagnetic Waves Port Settings.

Figure 11.46 shows the MCT Electromagnetic Waves Port Settings.

Scattering Boundary Condition 1

Right-Click > Model Builder – Model 1 (mod1) – Electromagnetic Waves (emw).

Select > Scattering Boundary Condition from the Pop-up menu.

NOTE *The easiest method that the modeler can use to implement the next step is to use the Zoom Box to first expand the model geometry in the Graphics window as needed. Then Shift-Click the desired boundary.*

Shift-Click > Boundaries 14, 18, 20, 21 in the Graphics window.

Click > Add to Selection (Plus sign) in Settings – Scattering Boundary Condition – Boundaries – Selection.

See Figure 11.47.

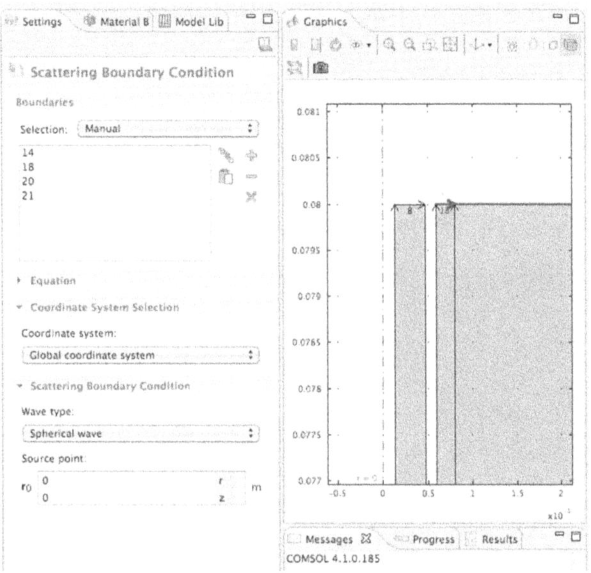

FIGURE 11.47 MCT Electromagnetic Waves Scattering Boundary Conditions Settings.

Figure 11.47 shows the MCT Electromagnetic Waves Scattering Boundary Conditions Settings.

Heat Transfer (ht)

Click > Model Builder – Model 1 (mod1) – Heat Transfer (ht) twistie.

Domains

The default condition is to apply all of these settings to the entire model. Please note that as the properties of other domains (tissues) are defined, the default settings will be overridden.

Click > Model Builder – Model 1 (mod1) – Heat Transfer (ht).

Click > Clear Selection (X sign) in Settings – Heat Transfer – Domains.

Click > Domain 1 in the Graphics window.

Click > Add to Selection in Settings – Heat Transfer – Domains.

Heat Source 1

Right-Click > Model Builder – Model 1 (mod1) – Heat Transfer (ht).

Select > Heat Source from the Pop-up menu.

Click > Heat Source 1.

Click > Selection in Settings – Heat Source – Domains.

Select > All domains from the Pop-up menu.

Click > Q list in Settings – Heat Source – Heat Source.

Select > Total power dissipation density (emw/wee1) from the Pop-up menu.

Bioheat 1

Click > Model Builder – Model 1 (mod1) – Heat Transfer (ht) – Biological Tissue 1 twistie.

Click > Model Builder – Model 1 (mod1) – Heat Transfer (ht) – Biological Tissue 1 – Bioheat 1.

Enter > ρ_b = rho_blood, C_b =Cp_blood and ω_b = omega_blood in the Settings – Bioheat – Bioheat edit windows.
See Figure 11.48.

Figure 11.48 shows the Settings – Bioheat – Bioheat edit windows.

Mesh 1

Right-Click > Model Builder – Model 1 – Mesh 1.

Select > Free Triangular from the Pop-up menu.

Click > Model Builder – Model 1 – Mesh 1 – Size.

Click > Settings – Size – Element Size Parameters twistie.

Enter > 3[mm] in Settings – Size – Element Size Parameters – Maximum element size.

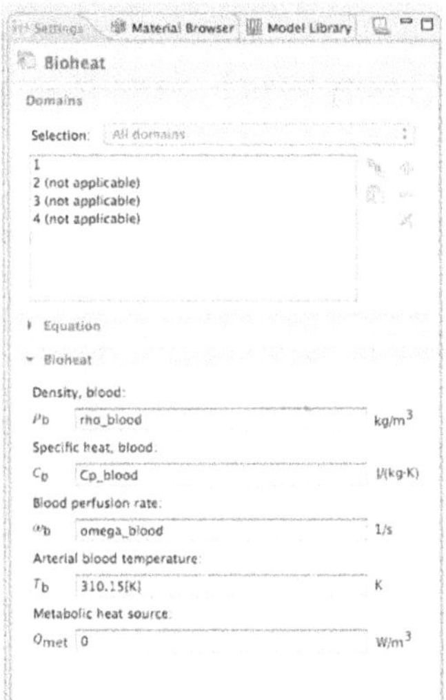

FIGURE 11.48 Settings - Bioheat - Bioheat Edit Windows.

See Figure 11.49.

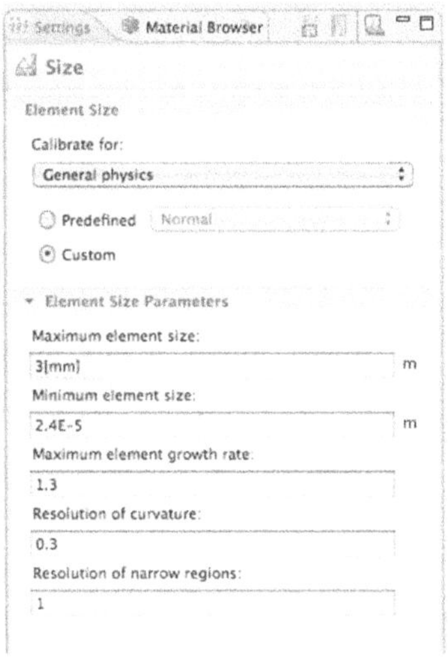

FIGURE 11.49 Size Settings.

Figure 11.49 shows the Size Settings.

Right-Click > Model Builder – Model 1 – Mesh 1 – Free Triangular 1.

Select > Size from the Pop-up menu.

Click > Size 1.

Click > Geometric entity level in Settings – Size – Geometric Scope.

Select > Domain from the Pop-up menu.

Expand the figure in the Graphics window twice with the Zoom Box tool.

Click > Domain 3 in the Graphics window.

Click > Add to Selection (Plus sign) in Settings – Size – Geometric Scope.

Click > Custom in Settings – Size – Element Size.

Click > Maximum element size check box in Settings – Size – Element Size Parameters.

Enter > 0.15[mm].

See Figure 11.50.

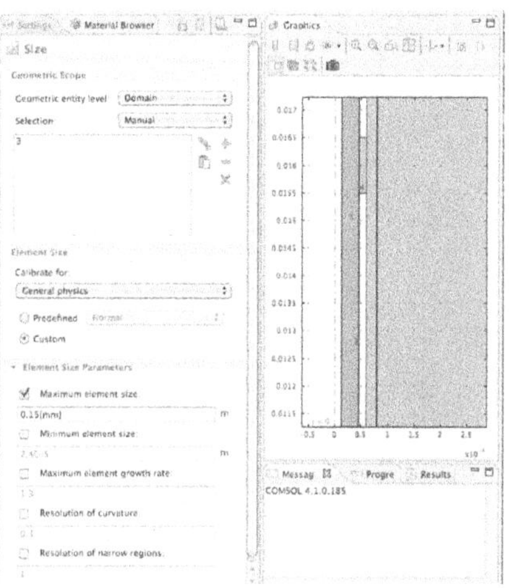

FIGURE 11.50 Size 1 Settings and Domain 3.

Figure 11.50 shows the Size 1 Settings and Domain 3.

Click > Build All.

Click > Zoom Extents.

See Figure 11.51.

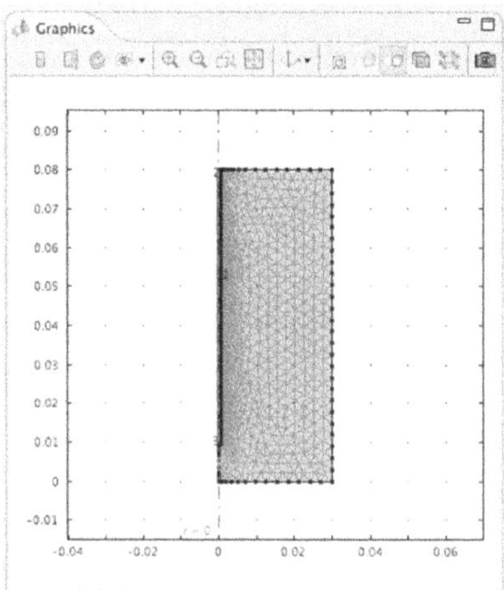

FIGURE 11.51 MCT Meshed Model.

Figure 11.51 shows the MCT Meshed Model.

NOTE *After the mesh is built, the model should have 7113 elements.*

Study 1

Frequency Domain Solver

Click > Model Builder – Study 1 twistie.

Click > Model Builder – Study 1 – Step 1: Frequency Domain.

Enter > f in Settings – Frequency Domain – Study Settings – Frequencies edit window.

Heat Transfer stationary Solver

Right-Click > Model Builder – Study 1.

Select > Study Steps – Stationary from the Pop-up menu.

Right-Click > Model Builder – Study 1.

Select > Compute.

Results

2D Plot Group 1

See Figure 11.52.

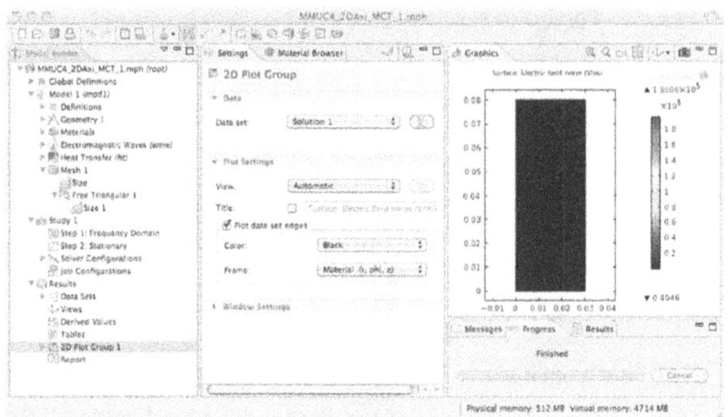

FIGURE 11.52 MCT Model Solution using Default Plot Parameters.

Figure 11.52 shows the MCT Model Solution using Default Plot Parameters.

NOTE *The solution shown in Figure 11.52 is the sequential solution, first for the microwave power dissipation and then for the resultant heating. This solution assumes that the material properties are not a function of temperature.*

Click > Model Builder – Results – 2d Plot Group 1 twistie.

Click > Model Builder – Results – 2d Plot Group 1 – Surface 1.

Enter > log10(mod1.emw.normE) in Settings – Surface – Expression – Expression.

Click > Plot in the Settings Toolbar.

Click > Zoom Extents.

See Figure 11.53

Figure 11.53 shows the MCT Model Solution using log10(normal electric field).

Surface Temperature Plot

Click > Replace Expression in Settings – Surface – Expression.

Select > Heat Transfer – Temperature.

Click > Unit in Settings – Surface – Expression.

Select > degC.

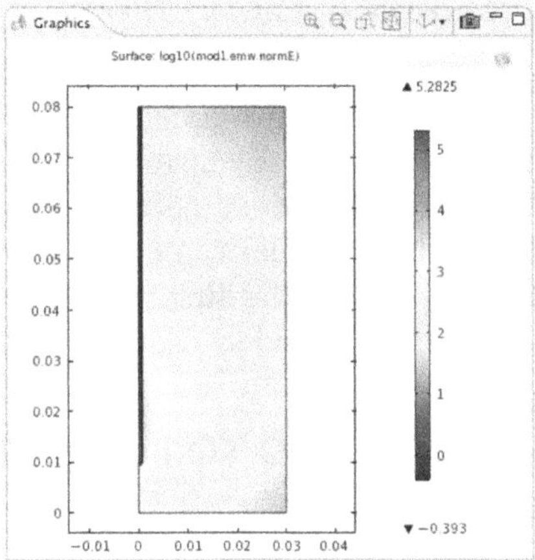

FIGURE 11.53 MCT Model Solution using log10(normal electric field).

Click > Plot.

See Figure 11.54

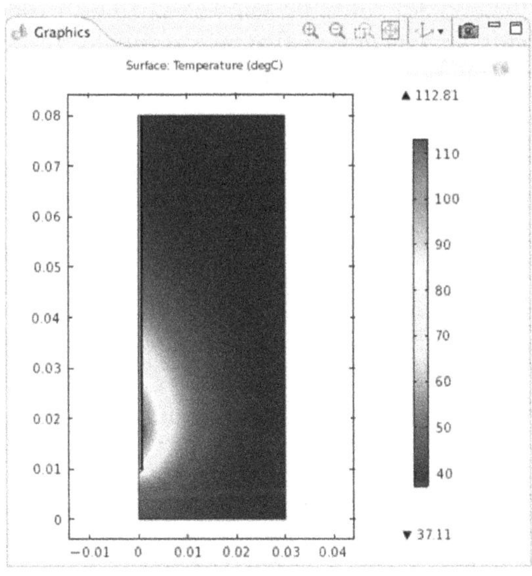

FIGURE 11.54 MCT Model Solution Local Temperatures.

Figure 11.54 shows the MCT Model Solution Local Temperatures.

The modeler should note that the temperature of the tumor exceeds the desired 60C needed to destroy the tumor.

2D Axisymmetric Microwave Cancer Therapy Model Summary and Conclusions

The 2D Axisymmetric Microwave Cancer Therapy Model is a powerful modeling tool that can be used to calculate the gain in temperature under microwave dissipation for different geometry designs, materials, and power levels. With this model, the modeler can easily vary all of the parameters and optimize the design before the first prototype is physically built or used. These types of models are widely used in industry.

FIRST PRINCIPLES AS APPLIED TO BIOHEAT MODEL DEFINITION

First Principles Analysis derives from the fundamental laws of nature. In the case of models using this Classical Physics Analysis approach, the laws of conservation in physics require that what goes in (as mass, energy, charge, etc.) must come out (as mass, energy, charge, etc.) or must accumulate within the boundaries of the model.

The careful modeler must be knowledgeable of the implicit assumptions and default specifications that are normally incorporated into the COMSOL Multiphysics software model when a model is built using the default settings.

Consider, for example, the two Bioheat models developed in this chapter. In these models, it is implicitly assumed that there are no thermally related changes (mechanical, electrical, etc.), except as specified. It is also assumed the materials are homogeneous and isotropic, except as specifically indicated and there are no thin insulating contact barriers at the thermal junctions. None of these assumptions are typically true in the general case. However, by making such assumptions, it is possible to easily build a 2D Axisymmetric First Approximation Model.

NOTE

A First Approximation Model is one that captures all the essential features of the problem that needs to be solved, without dwelling excessively on all of the small details. A good First Approximation Model will yield an answer that enables the modeler to determine if he needs to invest the time and the resources required to build a more highly detailed model.

Also, the modeler needs to remember to name model parameters carefully as pointed out in Chapter 1.

REFERENCES

11.1 COMSOL Multiphysics Users Guide, Version 4.1, pp. 288-292

11.2 H.H. Pennes, J. Appl. Physiology, V. 1, No.2, pp. 93-122

11.3 http://en.wikipedia.org/wiki/Bioheat_transfer

11.4 L.R. Hirsch, et al., Engineering in Medicine and Biology, 2002. 24th Annual Conference, pp. 530-531

11.5 L.R. Hirsch, et al., Proceedings of the 25' Annual International Conference of the IEEE EMBS Cancun, Mexico. September 17-21,2003

11.6 Saito, et al., Antennas, Propagation and EM Theory, 2000. Proceedings. ISAPE 2000. 5th International Symposium

11.7 D.P. O'Neal et al., j.canlet.2004.02.004

11.8 http://www.cancer.gov/cancertopics/factsheet/Therapy/hyperthermia

11.9 http://en.wikipedia.org/wiki/Ohm%27s_Law

11.10 http://en.wikipedia.org/wiki/Joule%27s_laws

11.11 http://en.wikipedia.org/wiki/Minimally_invasive

Suggested Modeling Exercises

1. Build, mesh, and solve the 2D Axisymmetric Tumor Laser Irradiation Model as presented earlier in this chapter.
2. Build, mesh, and solve the 2D Axisymmetric Microwave Cancer Therapy Model as presented earlier in this chapter.
3. Change the values of the materials parameters and then build, mesh, and solve the 2D Axisymmetric Tumor Laser Irradiation Model as an example problem.
4. Change the values of the materials parameters and then build, mesh, and solve the 2D Axisymmetric Microwave Cancer Therapy Model as an example problem.
5. Change the value of the absorbance parameter and then build, mesh, and solve the 2D Axisymmetric Tumor Laser Irradiation Model as an example problem.
6. Change the design of the geometry and then solve the 2D Axisymmetric Microwave Cancer Therapy Model as an example problem.

www.ingramcontent.com/pod-product-compliance
Lightning Source LLC
Chambersburg PA
CBHW080238230326
41458CB00096B/2625

* 9 7 8 1 9 3 6 4 2 0 0 9 4 *